Gene-Edited Plants for Crop Improvement

Sana Basharat
https://orcid.org/0000-0002-6737-5611
Hainan University, China

Muhammad Waseem
Hainan University, China

IGI Global
Scientific Publishing
Publishing Tomorrow's Research Today

Vice President of Editorial Melissa Wagner
Director of Acquisitions Mikaela Felty
Director of Book Development Jocelynn Hessler
Production Manager Mike Brehm
Cover Design Jose Rosado

Published in the United States of America by
IGI Global Scientific Publishing
701 East Chocolate Avenue
Hershey, PA, 17033, USA
Tel: 717-533-8845 | Fax: 717-533-7115
Website: https://www.igi-global.com E-mail: cust@igi-global.com

Library of Congress Cataloging-in-Publication Data

Names: Basharat, Sana, 1992- editor | Waseem, Muhammad, 1988- editor
Title: Gene-edited plants for crop improvement / edited by Sana Basharat, Muhammad Waseem.
Description: Hershey, PA : IGI Global Scientific Publishing, 2027. | Includes bibliographical references and index. | Summary: "This book aims to consolidate the rapidly expanding knowledge on gene editing in plants"-- Provided by publisher.
Identifiers: LCCN 2026007422 (print) | LCCN 2026007423 (ebook) | ISBN 9798337395340 hardcover | ISBN 9798337395357 paperback | ISBN 9798337395364 ebook
Subjects: LCSH: Crops--Genetic engineering | Crop improvement | CRISPR (Genetics) | Agricultural biotechnology | Gene editing
Classification: LCC SB123.57 .G4265 2026 (print) | LCC SB123.57 (ebook)
LC record available at https://lccn.loc.gov/2026007422
LC ebook record available at https://lccn.loc.gov/2026007423

British Cataloguing in Publication Data
A Cataloguing in Publication record for this book is available from the British Library.

Editorial Advisory Board

Table of Contents

Detailed Table of Contents

Sustainable agriculture is essential for meeting the growing global demand for food while minimizing environmental impacts and preserving natural resources. Gene editing technologies, particularly CRISPR-Cas systems, have emerged as transformative tools for improving crop productivity, resilience, and nutritional quality. Unlike conventional breeding methods, gene editing enables precise modifications of specific genetic sequences, accelerating the development of desirable traits in plants and livestock. This technology offers significant opportunities to enhance resistance to pests and diseases, improve tolerance to drought, salinity, and temperature extremes, and reduce reliance on chemical fertilizers and pesticides. Consequently, gene editing can contribute to increased agricultural efficiency, reduced environmental degradation, and strengthened food security under changing climatic conditions. In addition, gene-edited crops have the potential to improve nutritional content and post-harvest quality, supporting healthier and more sustainable food systems.

Gene editing has emerged as a new tool in crop biotechnology, allowing for precise, efficient, and targeted change of plant genomes. Advanced techniques like CRISPR-Cas9, CRISPR-Cas12a, TALENs, and Zinc Finger Nucleases (ZFNs) have allowed for quick advances in major crops such as rice, wheat, maize, soybean, potato, and tomato. Applications include improved yield, disease resistance, abiotic stress tolerance, nutritional quality, and post-harvest properties. Gene editing speeds up breeding programs by directly targeting genes responsible for important agronomic features, eliminating reliance on traditional, labor-intensive breeding cycles. Despite its benefits, problems exist in regulatory frameworks, ethical issues, and public acceptance. Ongoing research and technical improvements are projected to increase efficiency and expand applications. Gene editing has great promise for generating climate-resilient crops, promoting sustainable agriculture, and enhancing global food security in the face of population expansion and environmental change.

Gene editing can substantially enhance crop yield, strengthening global food security where traditional breeding is often too slow to match climate change, pests, and diseases. Technologies such as CRISPR-Cas9, TALENs, and ZFNs enable precise, targeted changes that improve drought and pest resistance, nutrient-use efficiency, and growth rate. By identifying and modifying genes controlling key traits, researchers have developed higher-performing crops, including rice, wheat, maize, and tomato. Major applications include boosting stress tolerance, improving pest and disease resistance, increasing nutrient uptake, and shortening growth cycles. Key constraints remain—editing efficiency, off-target effects, cost, and regulatory barriers—alongside ethical concerns about ecological impacts and public perceptions. Future progress will likely depend on multi-gene editing, climate-adaptive trait stacking, and integration with precision agriculture and AI-driven decision tools, supporting higher yields alongside sustainability and resilience.

strand breaks at sites adjacent to a Protospacer Adjacent Motif (PAM). These breaks initiate cellular healing processes. Non-Homologous End Joining (NHEJ) results in insertion and deletion gene mutations. The targeted insertion or replacements accomplish the same goal as HDR, or homology-directed repair. This allows several modifications to be made without the need to add extra DNA. Compared to earlier techniques like TALENs and Zinc Finger Nucleases (ZFNs), CRISPR-Cas9 is more user-friendly, flexible, and specific. Because of improved gRNAs and Cas variants, it has fewer off-target effects.

Gene editing is rapidly transforming plant biology and crop improvement by enabling precise, efficient genome modifications to address food security, climate change, and sustainability. CRISPR/Cas systems, alongside TALENs and base editors, target genes controlling yield, nutritional quality, and resistance to drought, salinity, pests, and diseases—delivering traits that often exceed traditional breeding speed and scope. Despite strong lab and greenhouse progress, field-scale deployment remains limited, with few varieties commercialized and performance still being tested across diverse environments. Key barriers include off-target effects, regulatory uncertainty, and public acceptance. Achieving durable gains will require integrating gene editing with sustainable agronomy, genomic selection, high-throughput phenotyping, and data-driven analytics, supported by responsible governance and societal engagement.

The problem of disease outbreaks and the dearth of crop types resistant to disease have made modern agriculture less reliable. A disease outbreak that damages commercially significant crops have a substantial effect on the economy. However, it is predicted that agriculture will be seriously threatened by the advent of new pathovars or the alteration in pathogen host specificity brought on by climate change. In agricultural crops, genome editing techniques include mutagenesis, knockout, knockdown, modification, repression, and activation of the target gene. As a result, genome editing techniques are being used more frequently. Because of its minimal off-target risk and simplicity of use, the CRISPR/Cas9 system is more efficient for gene editing than other genome editing techniques like TALENs and ZFNs. This review aims to illustrate how gene editing technologies can produce climate-resilient crops.

CRISPR/Cas9 gene editing has rapidly advanced agriculture by enabling precise, efficient, and stable crop genome changes without introducing foreign DNA. This chapter reviews key developments and applications for improving yield, nutritional quality, disease resistance, abiotic stress tolerance, and post-harvest traits. Reported

Preface

Agriculture faces unprecedented challenges as global food demand continues to rise while climate change, environmental degradation, emerging pathogens, and diminishing natural resources threaten agricultural productivity. Conventional breeding has contributed significantly to crop improvement over the past century; however, the pace of developing resilient, high-yielding, and nutritionally enhanced crop varieties often falls short of the rapidly evolving demands of modern agriculture. Recent advances in genome editing technologies, particularly CRISPR/Cas systems, have transformed plant science by enabling precise, efficient, and targeted modifications of plant genomes. These innovations offer unprecedented opportunities to accelerate crop improvement while supporting sustainable agricultural development.

Gene-Edited Plants for Crop Improvement presents a comprehensive overview of the practical applications of genome editing technologies in modern crop improvement. The volume highlights how precise genome engineering is being used to enhance crop productivity, improve nutritional quality, strengthen resistance against diseases and pests, and develop tolerance to abiotic stresses such as drought, salinity, heat, and temperature extremes. The book also presents successful case studies that demonstrate the translation of laboratory discoveries into commercially relevant crop varieties, illustrating the growing impact of gene editing on global agriculture.

Rather than focusing solely on technological advances, this volume emphasizes the integration of genome editing into sustainable agricultural systems. It explores how gene-edited crops can contribute to food security, climate resilience, resource-use efficiency, and environmentally responsible farming practices. By combining recent scientific developments with practical applications, the book illustrates the transformative potential of genome editing in addressing contemporary agricultural challenges.

This volume brings together contributions from researchers and experts working across plant biotechnology, molecular genetics, crop science, plant breeding, plant pathology, and agricultural biotechnology. Each chapter has been carefully selected to provide current knowledge while identifying future opportunities for research and innovation.

We hope this book will serve as a valuable resource for researchers, graduate students, educators, plant breeders, policymakers, biotechnology professionals, and others interested in the application of genome editing for sustainable crop improvement. We also anticipate that it will stimulate further interdisciplinary research and encourage the responsible development and deployment of genome editing technologies to support global food security and sustainable agriculture.

CHAPTER OVERVIEW

Chapter 1: Gene Editing in Agriculture - A Sustainable Approach to Improving Crop Productivity and Resilience

Sustainable agriculture is essential for meeting the growing global demand for food while minimizing environmental impacts and preserving natural resources. Gene editing technologies, particularly CRISPR-Cas systems, have emerged as transformative tools for improving crop productivity, resilience, and nutritional quality. Unlike conventional breeding methods, gene editing enables precise modifications of specific genetic sequences, accelerating the development of desirable traits in plants and livestock. This technology offers significant opportunities to enhance resistance to pests and diseases, improve tolerance to drought, salinity, and temperature extremes, and reduce reliance on chemical fertilizers and pesticides. Consequently, gene editing can contribute to increased agricultural efficiency, reduced environmental degradation, and strengthened food security under changing climatic conditions. In addition, gene-edited crops have the potential to improve nutritional content and post-harvest quality, supporting healthier and more sustainable food systems.

Chapter 2: Applications of Gene Editing in Major Crops

Gene editing has emerged as a new tool in crop biotechnology, allowing for precise, efficient, and targeted change of plant genomes. Advanced techniques like CRISPR-Cas9, CRISPR-Cas12a, TALENs, and Zinc Finger Nucleases (ZFNs) have allowed for quick advances in major crops such as rice, wheat, maize, soybean, potato, and tomato. Applications include improved yield, disease resistance, abiotic stress tolerance, nutritional quality, and post-harvest properties. Gene editing speeds up breeding programs by directly targeting genes responsible for important agronomic features, eliminating reliance on traditional, labor-intensive breeding cycles. Despite its benefits, problems exist in regulatory frameworks, ethical issues, and public acceptance. Ongoing research and technical improvements are projected to increase efficicncy and expand applications. Gene editing has great promise for generating climate-resilient crops, promoting sustainable agriculture, and enhancing global food security in the face of population expansion and environmental change.

Chapter 3: Enhancing Crop Yield through Gene Editing: Enhancing Crop Yield

Gene editing can substantially enhance crop yield, strengthening global food security where traditional breeding is often too slow to match climate change, pests, and diseases. Technologies such as CRISPR-Cas9, TALENs, and ZFNs enable precise, targeted changes that improve drought and pest resistance, nutrient-use efficiency, and growth rate. By identifying and modifying genes controlling key traits, researchers have developed higher-performing crops, including rice, wheat, maize, and tomato. Major applications include boosting stress tolerance, improving pest and disease resistance, increasing nutrient uptake, and shortening growth cycles. Key constraints remain—editing efficiency, off-target effects, cost, and regulatory barriers—alongside ethical concerns about ecological impacts and public perceptions. Future progress will likely depend on multi-gene editing, climate-adaptive trait stacking, and integration with precision agriculture and AI-driven decision tools, supporting higher yields alongside sustainability and resilience.

Chapter 4: Improving Nutritional Quality of Crops with Gene Editing

Crop nutritional quality is shaped by environment, variety, management, and post-harvest handling, with genetic variation driving differences in nutrient composition and adaptability. Gene editing is a targeted biotechnology that inserts, deletes, or replaces DNA sequences to redesign crop traits, advancing breeding toward genome-informed design. Compared with conventional gene transfer, gene editing is generally more precise and efficient, enabling site-specific changes without broadly altering gene copy number or relying on random DNA integration. As a leading genome-targeted tool, it is reshaping crop breeding and improving selection efficiency, with major implications for agricultural development and food security. This chapter reviews key genome-editing platforms used in crop improvement and outlines three strategies to enhance nutritional quality: redirecting metabolic pathways, activating stress-response genes that influence nutrient profiles, and systematically optimizing multi-gene networks.

Chapter 5: Gene Editing and Crop Improvement Using CRISPR-CAS9 Technology

The CRISPR-Cas9 (Clustered Regularly Interspaced Short Palindromic Repeats paired with Cas9), genome editing techniques have significantly advanced. These technologies have changed functional genomics and agricultural breeding. This RNA-guided method makes use of a synthetic guide RNA (gRNA) that corresponds to the target DNA sequences. It instructs the Cas9 endonuclease to make double strand breaks at sites adjacent to a Protospacer Adjacent Motif (PAM). These breaks initiate cellular healing processes. Non-Homologous End Joining (NHEJ) results in insertion and deletion gene mutations. The targeted insertion or replacements accomplish the same goal as HDR, or homology-directed repair. This allows several modifications to be made without the need to add extra DNA. Compared to earlier techniques like TALENs and Zinc Finger Nucleases (ZFNs), CRISPR-Cas9 is more user-friendly, flexible, and specific. Because of improved gRNAs and Cas variants, it has fewer off-target effects.

Chapter 6: Gene Editing for Abiotic Stress Tolerance

Gene editing is rapidly transforming plant biology and crop improvement by enabling precise, efficient genome modifications to address food security, climate change, and sustainability. CRISPR/Cas systems, alongside TALENs and base editors, target genes controlling yield, nutritional quality, and resistance to drought, salinity, pests, and diseases—delivering traits that often exceed traditional breeding speed and scope. Despite strong lab and greenhouse progress, field-scale deployment remains limited, with few varieties commercialized and performance still being tested across diverse environments. Key barriers include off-target effects, regulatory uncertainty, and public acceptance. Achieving durable gains will require integrating gene editing with sustainable agronomy, genomic selection, high-throughput phenotyping, and data-driven analytics, supported by responsible governance and societal engagement.

Chapter 7: Disease Resistance In Gene – Edited Crops

The problem of disease outbreaks and the dearth of crop types resistant to disease have made modern agriculture less reliable. A disease outbreak that damages commercially significant crops have a substantial effect on the economy. However, it is predicted that agriculture will be seriously threatened by the advent of new pathovars or the alteration in pathogen host specificity brought on by climate change. In agricultural crops, genome editing techniques include mutagenesis, knockout, knockdown, modification, repression, and activation of the target gene.

As a result, genome editing techniques are being used more frequently. Because of its minimal off-target risk and simplicity of use, the CRISPR/Cas9 system is more efficient for gene editing than other genome editing techniques like TALENs and ZFNs. This review aims to illustrate how gene editing technologies can produce climate-resilient crops.

Chapter 8: Case studies - Successful Gene-Edited Crops: Case studies

CRISPR/Cas9 gene editing has rapidly advanced agriculture by enabling precise, efficient, and stable crop genome changes without introducing foreign DNA. This chapter reviews key developments and applications for improving yield, nutritional quality, disease resistance, abiotic stress tolerance, and post-harvest traits. Reported examples include drought-tolerant maize, iron/vitamin-enriched rice, low-gluten wheat, and disease-resistant potato and grape. Emerging platforms such as base and prime editing expand options for safer, more accurate edits. Regulatory and public acceptance is growing, with approvals and commercialization increasing in countries such as the United States, Canada, Brazil, and Japan. Beyond biosafety, gene editing can reduce chemical inputs and support environmental and economic gains. The chapter concludes by emphasizing the need for responsible governance, ethics, and equitable access—especially in developing regions—to realize sustainable agriculture and food security.

Muhammad Waseem
Hainan University, China

Sana Basharat
Hainan University, China

Chapter 1
Gene Editing in Agriculture:
A Sustainable Approach to Improving Crop Productivity and Resilience

Anum Mehmood
http://orcid.org/0009-0004-8105-1690
College of Food Science and Engineering, Hainan University, Haikou, China

Uzair Bhatti
http://orcid.org/0000-0002-8743-2783
School of Information and Communication Engineering, Hainan University, China

Mukhayya Ruzieva
Department of Sport and Psychology, Mamun University, Khiva, Uzbekistan

Nargiza Abdusamiyeva
Department of Medicine, Termez University of Economics and Service, Termez, Uzbekistan

Yonis Gulzar
http://orcid.org/0000-0002-6515-1569
Department of Management Information Systems, King Faisal University, Al-Ahsa, Saudi Arabia

Yazeed Yasin Ghadi
http://orcid.org/0000-0002-7121-495X
Department of Computer Science and Software Engineering, Al Ain University, Al Ain, UAE

ABSTRACT

Sustainable agriculture is essential for meeting the growing global demand for food while minimizing environmental impacts and preserving natural resources. Gene editing technologies, particularly CRISPR-Cas systems, have emerged as transformative tools for improving crop productivity, resilience, and nutritional quality. Unlike conventional breeding methods, gene editing enables precise modifications of specific genetic sequences, accelerating the development of desirable traits in plants and

DOI: 10.4018/979-8-3373-9534-0.ch001

livestock. This technology offers significant opportunities to enhance resistance to pests and diseases, improve tolerance to drought, salinity, and temperature extremes, and reduce reliance on chemical fertilizers and pesticides. Consequently, gene editing can contribute to increased agricultural efficiency, reduced environmental degradation, and strengthened food security under changing climatic conditions. In addition, gene-edited crops have the potential to improve nutritional content and post-harvest quality, supporting healthier and more sustainable food systems.

1. INTRODUCTION

Although it has been anticipated by all farmers and also highly thought of in the agricultural sector, the global food system is presently one of the most significant ones in human history as more food comes from the sector with less natural resources being used and causing less damage to the environment. Sustainable food production is under serious pressure due to large-scale population growth, climate change, soil degradation, water scarcity, loss of biodiversity and pressure of pests and diseases. Conventional farming practices have played a vital role in the achievement of food security, but they have also resulted in a higher consumption of chemical fertiliser, pesticides, irrigation water, fossil fuel-based inputs etc. Consequently, today's agriculture—not only has to improve productivity, but also ensure a more effective use of resources, resilience to climate change, environmental safety, and nutritional value. In this context, gene editing has proved to be a highly effective and targeted biotechnology which may play a role in sustainable agriculture by enhancing specific properties of crops that correspond directly to yield, stress resistance, nutritional value, resistance to diseases and resource utilization efficiencies (Tripathi et al., 2022).

Gene editing is the collective name given to various molecular methods which enable scientists to introduce targeted and precise changes within the genetic material of an organism. Gene editing is an alternative to conventional breeding, which involves offspring generations and many crossing and selection generations, which can use specific genes linked to desirable traits of an organism to make changes in a relatively shorter period. With the advent of zinc-finger nucleases, transfection activator-like effector nucleases, and particularly clustered regularly interspaced short palindromic repeats/CRISPR-associated protein systems, plant biotechnology has been revolutionized. In particular, CRISPR/Cas systems have been attracting a lot of attention describing features that they are relatively simple, efficient, versatility and cost of implementation. By using the CRISPR technique, scientists can cut out any unwanted genes, enhance useful properties, manipulate gene expression, or introduce specific minimal changes without the need for an additional piece of non-native material. This precision is extremely relevant for creating new crop varieties with better traits that would adapt to farmers' and consumers' demands, through genetic manipulation (Rajput et al., 2021).

Of course, one of the most critical roles gene editing will play in sustainable agriculture is that of boosting crop yield. There has been the constant aim of increasing yield in plant breeding but this has become more of a need, as land for agriculture is scarce and often not conducive to the environment. Plant architecture, flowering time, grain size, photosynthetic efficiency, nutrient transport and reproductive development are areas in which genes can be targeted for editing. Engineered changes in these traits will allow creation of crops that produce increased biomass, improve harvest index, yield more seeds, and produce higher quality grain. Yield related gene editing has, for instance, been investigated in some of the leading crop plants like rice, wheat, maize, soybean, tomato, and potato. Such improvements are particularly warmly appreciated in areas with high food insecurity and ones where traditional breeding is hindered by resources or the environment being too demanding for traditional cross-breeding, such as because of long crop cycles (Anyshchenko, 2022).

Gene editing has become relevant in agricultural research in several other instances such as climate change. Many of the world already feel the impact of the impact of rising temperatures and irregular rainfall, drought, salinity, floods and extreme weather events already seriously affecting crop yields. The potential for gene editing is that it can be used to create climate-smart crops through gene modification to enhance water-use efficiency, heat tolerance, tolerance to salt, root development, regulation of stomata, and pathways for stress responses. Water availability may be deeper in the soil for the deeper rooted crop, for instance, and water loss during drought conditions can be reduced by the stomatal control of the plants. Likewise, genetic modifications of heat shock proteins, osmotic adjustment or adjustments of ions can make crops tolerant to high temperatures and salinity. Adaptable varieties can help ensure food security, reduce crop losses and aid farmers in at-risk agricultural areas (Sarfraz et al., 2023).

Pathogens and diseases are also the important challenge for agriculture worldwide and loss in production are significant annually. Chemical pesticides and fungicides are commonly used to address these threats, but high levels of chemicals in the environment can result in undesirable effects, such as detrimental impacts on beneficial organisms, soil and water contamination, and pest resistance. With gene editing, a more specific method of crop protection can be developed by altering susceptibility genes, boosting the plant's immune system or enhancing defence against insects, fungi, bacteria and viruses. For example, disease resistance may occur after pathways in plants' genes are disabled that are exploited by the pathogens. For viral diseases, likewise, resistance can be increased by targeting host factors that are necessary for the viral replication. This can help minimize reliance on chemical pesticides and improve sustainable agricultural practices (Dima & Inzé, 2021).

Additionally, the targeting technology of gene editing could now provide significant opportunities for the enhancement of the nutritional properties of crops and food, the central tenet of sustainable agriculture. Food security is more than an abundance of calories, it's also the availability of access to food for nutrition and health. There remains a significant micronutrient deficiency burden in many populations, including vitamin A deficiency (VAD), iron deficiency (ID), zinc deficiency, and folate deficiency. Gene editing can be used to enhance the nutritional profile of crops and add vitamin, minerals, essential amino acids, healthy oils, antioxidants, and bioactive compounds. Also, it can have a negative effect on unwanted molecules like allergens, toxins, anti-nutrients or molecules that makes food unsafe. As such, gene-edited crops have the potential to enhance agricultural output and human health, especially in low-income nations where staple crops make a predominant part of their diet (Das et al., 2023).

Use of resources is also critical. Crops should be more productive in sustainable agriculture with less production input. The potential uses include enhancing NUE, PUE, WUE and photosynthetic performance using gene editing. More effectively utilizing the nutrients can minimize the use of artificial fertilizers, thereby lowering production expenses and pollution levels in the environment. Likewise, water efficient and drought-resistant cultivars require less irrigation, particularly in water-limited areas. Gene editing can be used to enhance economic and environmentally sustainable agriculture, by increasing efficiencies in using resources (Wan et al., 2022).

While it has potential benefits, there are also ethical, regulatory, biosafety, and socioeconomic issues associated with gene editing. There is concern about public acceptance, particularly as most confuse gene editing with the use of older genetic modified organism technologies. Countries apply different regulatory approaches to geneedited crops, some of which consider them to be the same as conventionally bred crops if no foreign DNA is added.Different countries have different approaches to regulating geneedited crops, some of which may classify them as like those traditionally bred crops if no foreign DNA is introduced. Possible off-target effects, gene flow to wild relatives, effects on biodiversity and long-term ecological consequences are all issues of biosafety. Some of the socioeconomic issues include access to technology, intellectual property rights, a smallholder's right to seed ownership and benefit from it, and the power of smallholder farmers to benefit from the innovations. Hence, enabling governance, non-biased risk evaluations, participative processes for developmental policies, and public involvement are key issues in the success of the gene editing in agricultural application (Rehman et al., 2022).

2. OVERVIEW OF GENE EDITING TECHNOLOGIES

Gene editing technologies are sophisticated molecular tools for specific and targeted changes in the sequence of living organisms. In agriculture, these technologies have led to the creation of new technologies to enhance agricultural crops on a faster and more accurate level than traditional plant breeding. Conventional traits are bred into plants over many generations, whereas gene editing enables researchers to target specific plant genes linked to valuable traits, like yield, drought resistance, disease resistance, nutrition and resource use efficiency. Precision is particularly relevant to sustainable agriculture, where a goal is to boost productivity without augmenting environmental stresses (Kitenge et al., 2025).

The development of gene editing has passed through several important stages. Other genome editing technologies available prior to CRISPR-Cas systems were designed as zinc finger nucleases (ZFN) and transcription activator-like effector nucleases (TALEN). They were significant in showing that a targeted DNA modification is possible in these systems. However, they were very complicated, costly, and hard to design. The field was later transformed by the discovery and development of CRISPR/Cas systems, which make use of simpler, more flexible and more efficient CRISPR-based editing. CRISPR/Cas, base editing, and prime editing have emerged from recent biotechnology studies in agriculture as key tools that have broadened the possibilities for modifying the crop genome (Amoah et al., 2024). Figure 1 shows the gene editing applications.

One of the first programmable genome editing tools to be developed is zinc-finger nucleases (ZFNs). They do this by binding to a sequence of DNA with zinc-finger proteins and then making a cut in the DNA with a nuclease enzyme at the target site. After cutting, the natural repair mechanism of the plant cells repairs the break in the DNA. While undergoing this repair, these mutations, or targeted changes, can be made. Despite the great success of the zinc-finger technology, the applications in crop improvement were restricted due to the fact that it is difficult and time consuming to design zinc-finger proteins that recognize each target gene. Because of this, ZFNs are less utilized than their successor, newer and simpler systems, are, and are instead historically significant (Panigrahi et al., 2023).

After ZFNs, transcription activator-like effector nucleases (TALENs) were developed, and were more flexible in targeting. TALENs rely on a suite of DNA-binding proteins also found in plant pathogenic bacteria that encode the ability to bind to specific DNA sequences. These are associated to a nuclease enzyme that cleaves the DNA, as the ZFNs. TALENs work more reliably than ZFNs, and have been applied in a few selected crops. They have assisted with the modification of disease resistance, plant quality and stress tolerance traits. But, the process of constructing TALEN is still time consuming in comparison to CRISPR/Cas systems.

This is why TALENs are still valuable tools, but are no longer as preferred crop research gene editing platform (Waghmode et al., 2024).

At present, the people working on CRISPR/Cas technology in the agriculture sector are the most widely used. CRISPR, which is clustered regularly interspaced short palindromic repeats, is followed by Cas (CRISPR-associated proteins) and Cas9 and Cas12 are examples of such proteins. This system uses a guide RNA to direct the Cas enzyme to a particular piece of DNA to which the enzyme cuts. This then heals the cut forming the desired mutation or change to the cell. One big benefit of CRISPR/Cas is that it could be rapidly and simply designed through changing the guide RNA sequence. This results in superior speed, cost, and usage in comparison to ZFNs and TALENs. CRISPR/Cas technologies have been extensively used in rice, wheat, maize, tomato, potato, soybeans, and other plant species to boost yield, stress resistance, resistance to diseases, and nutritional value. Gene editing has also received recognition from FAO as a cutting-edge technology which has great potential to speed up breeding and contribute to food security, when applied responsibly (Nor A'azizam et al., 2024).

CRISPR/Cas9 is the most popular CRISPR system, but other Cas proteins are also significant. Thus, for some genome editing applications, Cas12 may have distinct cutting capabilities. Unlike DNA, Cas13 targets RNA, creating new opportunities for temporary changes in gene regulation rather than permanent genetic changes. Researchers have more options depending on the crop species, target gene type and desired outcomes with these different CRISPR systems. Several new types of Cas are being identified, which increase the accuracy, flexibility and range of applications with plant biotechnology. Commonly called transformative tools, CRISPR/Cas systems have been recently reviewed as important means to facilitate fundamental plant investigations and applied crop improvement (Ali et al., 2022).

A newer type of gene editing that allows for alteration of one base in the DNA, but no double-strand DNA breaks. This is crucial because a lot of beneficial features are governed by minute genetic changes. Typically, base editors consist of a modified Cas and a DNA-modifying enzyme. One of their functions is to allow them to change one DNA base to another at an exact point in the targeted site. Cytosine base editors and adenine base editors, for instance, can make accurate point mutations. Base editing is beneficial in agriculture to enhance herbicide resistance, disease resistance, grain quality and stress response. Base editing does not cause double-strand breaks, which could minimize unwanted mutations and increase the accuracy of editing in comparison to traditional CRISPR-C Sherlock cutting approaches (Brower-Toland et al., 2024).

Another exciting and sophisticated technology aimed at gene editing is known as prime editing. It employs a modified Cas enzyme, reverse transcriptase, and a unique prime editing guide-RNA for the introduction of targeted insertions and deletions and substitutions. By not necessitating a donor DNA template or a double strand break as usually found in CRISPR/cas9 editing, prime editing avoids the complexity of inserting foreign DNA.Unlike standard CRISPR/cas9 editing, prime editing does not require a donor DNA template or double strand breaks. Therefore, it seems to be a good resource for identifying various situations where precise improvement of crops could be a useful strategy. Prime editing could have particular significance for the correction of specific genetic variants as well as the improvement of quality traits, and the generation of resistant crops to abiotic or biotic stresses. Currently, prime editing is under development in plant systems; however, it is also considered as one of the most promising technologies in the future for precision breeding in plants (Clapp & Ruder, 2020).

Beyond DNA editing, there are other technologies of gene regulation. CRISPR interference and CRISPR activation have the capacity to suppress or enhance the expression of a given gene, but not alter the DNA sequence. These are valuable for genetic function studies and for genetic determination of complex traits. Another promising technique that is becoming available is epigenome editing - changing the activity of a gene by altering an epigenetic attribute rather than altering its sequence. Examples of such technologies include situations where long-term genome changes are not needed. They also give opportunities for the understanding of how plants react to environmental stresses, nutrient availabilities and developmental signals. One of the key technical challenges is the transfer of parts of the gene editing system into the cells of plants. These include Agrobacterium transformation, particle bombardment, protoplast transfection and viral vector systems (Karavolias et al., 2021).

3. GENE EDITING AND CROP YIELD IMPROVEMENT

In view of the ever growing demands of food, enhancing crop yields is one of the most crucial goals of agricultural research, under immense pressure of limited cultivable land, diminishing water availability and decreasing soil fertility. Yield is an extremely complex trait involving various genetic, physiological and environmental factors. It relies on plant architecture, flowering time, grain number, grain size, seed weight, photosynthetic efficiency, nutrient's acquisition, stress tolerance and reproductive success. While many of the genetic traits and characteristics desirable for increased crop productivity have been achieved through traditional breeding, it can take a large number of generations to develop an improved line. Also, traits may be genetically linked to those less desirable, breeding cycles may be lengthy, and environmental variation can disadvantage traditional breeding. Gene editing is a more precise, and faster technique, which directly targets genes which regulate yield related traits. Hence, it has emerged as an essential method for improving varieties of crops with greater yield for sustainable farming (Pattnaik et al., 2023).

A significant characteristic of crop yield that can be enhanced with the help of gene editing is related to grain size and grain weight. Grain size is an important factor in the ultimate yield in the cereal crops rice, wheat and maize. There is great potential for gains in overall productivity through modest modification of genes associated with length, width and filling of grains. CRISPR/Cas has been used to modify genes related to the division of cells, growth of cells, and starch deposition in developing grains. Researchers can use these genes to boost seed mass and yield that can be harvested. This is particularly true in staple food crops where even marginal yield gains can yield significant gains at the national/ global level due to a size advantage for the obtainable grain (Mestanza et al., 2025).

Architecture of plants is also a significant aspect that can be used to improve the yield, which is another aspect of gene editing. Temperament plays important role in crop productivity, including plant height, number of tillers, branching structure, leaf angle and canopy structure. Good plant architecture facilitates light interception, photosynthesis and minimizes lodging and space utilization. In cereals, for instance, a significant factor in the success of the Green Revolution was the use of semi-dwarfing genes, as the plants were shorter and did not fall over and were able to produce higher yields. Transcription factors and genes that contribute to plant height, tillering, stem strength and leaf direction can be modified to further tailor architecture of plants for gene-editing. These improvements can help increase crop intensity, or the production on the same parcel of land (Fernandes et al., 2022).

Gene editing can also enhance their flowering time and maturity length, which are also yield-related traits. Flowering should occur at an appropriate time in order to avoid environmental stresses like drought, heat, frost or higher levels of precipitation. Earlier-maturing cultivars might be better for avoiding terminal drought, and later-flowering cultivars might help maximize biomass production in some climates. Using 'gene editing' techniques, floral genes could be modified very precisely, thereby enabling adaptation of crop plants for specific areas, seasons and cropping systems. This can be especially beneficial when considering climate change and the more unpredictable growing seasons. Gene-edited crops can control flowering time, which can help reduce crop yield fluctuations and maintain stable production under variations of environmental conditions (Singh et al., 2021).

Another use of gene editing is to increase photosynthetic efficiency, which can improve yield. Plants can produce biomass through photosynthesis (directly influences biomass production). But, usually natural photosynthesis is not efficient. In certain crops, the efficiency of the carbon fixation or due to poor light distribution in the canopy or due to a poor ability to reduce carbon dioxide to carbohydrates. Target genes for chlorophyll synthesis, carbon assimilation, stomatal control, and leaf development can all be used for gene editing. Greater photosynthetic potential can result in greater biomass production and in the end, higher grain or fruit production. Photosynthesis is an intricate function; and while it's a long shot, with gene editing it opens a very promising avenue for boosting this basic function in crop plants (Qaim, 2020).

Gene editing also has a significant impact on enhancing the reproductive efficiencies. Pollen viability and viability of seeds, or sensitivity of reproductive organs in response to environmental stress are the causes of yield loss in several sets of crops. Fertility and yield can be decreased due to various high temperature-humidity stress factors such as drought, salinity. The gene editing technologies can be used to alter genes that play a role in pollen development, fertilization, embryo formation, and seed development. Gene edited crops can yield greater grain, fruit, or seed yields even under stress due to reproductive stability. It is especially important in crops that are grown in suboptimal environmental conditions where stresses at reproductive stage are a major potential factor in yield loss (Roy et al., 2026).

Gene editing can also enhance stress tolerance, which would be a yield benefit, but not a direct yield trait. Drought, heat, salinity, flooding, nutrient deficiency, pest and disease are some of the limiting factors that prevent crops from achieving their genetic yield potential. Through modification of these stresses, gene editing can create a more tolerant plant, thereby enabling the plant to grow and produce under less than optimal conditions. As an example, modification of genes involved with root growth can promote better uptake of water and nutrients; alteration of genes involved in stress response can make plants more resilient against water stress or extreme heat. Likewise, disease resistant cultivars will insure yield through pathogen inhibition. Gene editing can contribute to increased yields and yield stability this way (Mehta et al., 2020).

Adverse effect of low efficiency use of nutrients on crop productivity is also associated with nutrient use efficiency. Nitrogen, phosphorus, potassium and micronutrients are essential for plant growth and development, but excessive use of fertilizer can lead to production cost and high levels of environmental pollution. Crop improvement to enhance nutrients uptake, transport and utilization capacity is possible by gene editing. More efficient N or P use efficiency yields higher crop production for less fertilizer. It is very relevant in the context of Sustainable Agriculture since it allows to reduce the usage of chemical fertilisers and loses of nutrients in soil and water systems. Thus, editing of genes can add to both productivity and environmental protection (Gordon et al., 2021).

Important crops like rice, wheat, maize, soybeans, tomato, and potato are already key crops in gene editing for yield. Genes involved in stress tolerance, panicle structure, tiller number and grain size have been modified in rice to enhance its productivity. The generation of wheat with high-quality grain, resistance to diseases, and yield-related characteristics is underway using gene editing techniques despite the complexity of the highly large wheat genome. In maize, editing has been tried for drought and kernel development and plant architecture. The gene editing has helped enhance tomato fruit size, tomato plant shape, potato maturity, and potato quality in tomato and potato. The examples demonstrate how gene editing can be used in both crops—cereal and horticulture. Yield is a polygengetic and environment dependent trait, which makes the improvement of yield through the use of gene editing difficult, despite its great potential (Camerlengo et al., 2022).

4. GENE EDITING FOR CLIMATE-RESILIENT AGRICULTURE

Climate change poses as one of the largest threats to modern agriculture as it directly influences the growth, productivity, quality and food security of crops. In many parts of the world, a warmer climate, irregular rainfall, longer periods of drought, soil salinity, floods, heat waves and the introduction of new pests and diseases are limiting crop yields and performances. The impacts of these stresses are particularly devastating to smallholder farmers and to developing countries, in which agriculture is highly reliant on natural rainfall and on a limited supply of resources. Climate-resilient agriculture is designed to foster resilient agricultural systems and susceptible crop varieties in sub-optimal and changing conditions. In this context, gene editing has proven to be a very powerful tool for creating a crop that is resilient to abiotic stress and thus produces stable yield under climatic pressure (Lalotra et al., 2024).

A major use of gene editing in climate resilient agriculture is drought-tolerant crop development. The effect of drought stress is due to a lack of water availability that impacts the growth rate, photosynthesis and transport of nutrients and is known to reduce flowering, grain filling and yield as well. With the regular method of breeding, there has been some drought resistant varieties developed and drought tolerance is a multifactorial disease trait that is influenced by environment factors. The scientists can target specific genes associated with root growth, stomatal regulation, water transport, osmotic adjustment and signalling of stresses using the process of gene editing. Ultimately, edited crops can be created to have deeper roots, better water absorption, less water loss and enhanced protection of cells in dry conditions, among others. These qualities may enable crops to withstand dry periods and retain productivity without increasing the amount of irrigation (Zheng et al., 2025).

There is a big section devoted to climate-resistant food - another key goal of gene editing is to design crops resistant to temperature and high humidity. High temperature affects seed production by causing any of the following: damage to plant proteins, damage to pollen viability, disturbance to flowering, reduction in seed filling time and a reduction in seed yield. Heat stress at certain developmental stages of many crops can result in significant loss in yield production. Genes involved in various areas of heat-shock protein function, antioxidant enzyme activity, flowering control, membrane stability, and reproduction can be edited by gene editing. Thus, such improvement of their innate defense against high temperatures might enable crops to continue to grow and reproduce at high temperatures. The ability of gene edited crops to withstand high temperatures is crucial given rising heat wave frequency and severity in global warming (Thankappan et al., 2024).

Salinity is present in many ag fields, and these are becoming worse due to poor irrigation practices, rising temperatures, seawater intrusion, and land degradation. Extremely high salt concentration in the soil decrease its ability to absorb water, lead to ion toxicity, disrupt the plant's' nutrient balance, and restrict plant growth. Potential targets for the gene editing approach to increase salt tolerance include genes relating to sodium transport, potassium balance, ion compartmentalization within plants, root architecture and pathways for plant stress response. Sodium exclusion from shoots and storage in vacuoles are, for instance, two processes that can be modified by change of genes to prevent salt damage. The value of salt-tolerant crops is immense to be assessed in the coastal areas, arid regions and irrigated lands of increasing salinity (Penna et al., 2023).

Further stresses of climate change include flooding and waterlogged crop areas, which decreases crop productivity. There must be sufficient oxygen in the soil for plants to absorb nutrients, if the water level is too high, it will not be able to provide adequate oxygen.If water level is too high it will not be capable of supplying adequate oxygen for absorption, which may lead to plant death. Many crop varieties, such as maize, wheat, soybeans and vegetables are waterlogging-sensitive, whereas rice is somewhat tolerant. The contributions of genes to tolerance to waterlogging, including in aspects of root aeration, anaerobic metabolism, ethylene signalling and survival in low oxygen, can be targeted for improving tolerance. In areas where intense rainfall is common, monsoon variability, flash flood and sub-optimal drainage conditions, flood-tolerant varieties can be particularly beneficial (Tripathi et al., 2022).

Crop traits to resist simultaneous multiple stress are also important parts of climate-resilient agriculture. Crop growing has become more complex than in these controlled studies because, in field conditions, there is no exposure to a single stress. Drought can also be accompanied by heat stress, salinity with water stress, and disease pressure fluctuates as temperatures and humidities change. Thus, editing one stress related gene may not be enough, in all cases. Scientists can edit multiple genes at once with modern gene editing techniques, particularly with multiplex CRISPR editing. Combined resistance to drought, heat, salinity, and disease can be achieved in this way. The multistress resilience of a crop is relevant to the productivity of agriculture in the face of climate uncertainty (Karavolias et al., 2021).

Another area of potential climate resilience through gene editing is through root system architecture improvements. Roots are at the heart of the acquisition of water and nutrients, anchorage, soil interactions and stress tolerance. Water and nutrients become available under the drought or nutrient (P or K) deficient conditions within the deeper soil profile in deeper, stronger and more branched rooting systems of crops, resulting in improved drought and nutrient stress resistances. Roots can be bred to develop shortened, angled, and to cut down on lateral root growth and development of root hairs. The better the post traits, the better water use efficiency and less need for frequent irrigation and application of fertilizer. This not only helps in adapting to climate change but also in sustainable management of the resources.

In addition, the antioxidant defense system of plants could be enhanced. Various abiotic stresses like drought, heat, salinity etc. induce oxidative stress in plants through generation of ROS. If not controlled, these molecules destroy cell membranes and DNA, protein, and lipids. The antioxidant activity can be improved by gene editing the antioxidant enzyme (superoxide dismutase, catalase, peroxidase, ascorbate peroxidase and others) genes. Increases in antioxidant activity can improve antioxidant protection to enhance cellular functions and physiology in plants under environmental stress.

Climate resilience could also be enhanced via genes editing as a mechanism to control plant hormones. Plant responses to stress are mediated by hormones like abscisic acid, ethylene, auxin, cytokinin, gibberellin and jasmonic acid. For instance, abscisic acid regulates the stomatal closure during drought and ethylene regulates flooding and stress responses. Mutating genes for hormones influencing biosynthesis, signaling or sensitivity, will enhance plant response to unfavorable environments. Hormone pathways, however, are complex – changes must be carefully carried out and controlled as it can cause growth, flowering and yield changes.

5. GENE EDITING FOR PEST AND DISEASE RESISTANCE

Pests and diseases are one of the largest challenges damaging agricultural production and achieving food security. Giant losses are sustained annually in crops due to insects, fungi, bacteria, viruses, nematodes and other plant diseases. Such issues are typically managed using chemical pesticides, fungicides and insecticides, which, however, can raise a number of environmental and health issues when used in excess, in farmers' fields. They include pesticide resistance, damage to good insects, contaminating beneficial insects, disrupting ecological balance, and increasing production expenses. It is therefore strongly needed to substitute or improve existing crop protection measures for sustainable agriculture by other and more effective means. Recent advances in gene editing have attracted a high amount of interest as a means to engineer crop tolerance for higher levels of pest and disease and less reliance on chemical control.

A critical approach to editing genes to increase resistance to a disease is to just modify the susceptibility or S- genes. These are genes of plants that pathogens exploit to penetrate, infect and replicate within the plant. If such genes are manipulated or removed, then the infectious pathogen may not be able to successfully infect the plant. This is extremely beneficial as it targets the elimination of the plants' susceptibility versus the introduction of foreign resistance genes. The susceptibility genes can be precisely targeted using CRISPR/Cas system and create resistant crop varieties. One such way is by mutating genes in the host or the disease to make them resistant; for example, the genes involved in how a pathogen enters the host, how it obtains nutrients or how the host and pathogen interact. This is one of the most promising options for long lasting resistance to crop diseases.

Plants can also have their natural immune cellular system fortified by gene editing. Plants possess sophisticated defense mechanism which detects the pathogen invaders and triggers defense responses. They entail pattern-triggered immunity, effector triggered immunity, production of antimicrobial compounds, reinforcement of cell walls and activation of defense-related genes. By editing genes involved in these immune pathways, researchers can make plants even more responsive and react more rapidly and strongly when attacked by a pathogen. This can lead to increased tolerance to fungal, bacterial and virus resistance and leave chemical spraying to a minimum. Intensifying plant immunity is particularly relevant for plants that are cultivated in conditions of high disease pressure in humid tropical and subtropical environments.

Fungal diseases are significant pathogens of grain, fruits, vegetables and legumes and are a big target of gene editing. Numerous economically important crops are host to disease organisms like powdery mildew, blast, rust, blight, and wilt. A strategy for increasing resistance is to target genes that fungi use in fighting plants, or to boost plant defense responses, using gene editing technology. For instance, wheat, tomato and grape are plants in which susceptibility genes have been targeted for editing to achieve mildew resistance. Again, genes that are thought to be responsible for pathogen recognition and defense signaling can be manipulated to provide resistance against rice blast and other fungal diseases. The improvements can result in decreased fungicide use and contribute to environmentally friendly crop production.

Bacterial diseases decrease productivity of crops and are more difficult to control as there are fewer chemical control products. Often diseases like bacterial blight, bacterial wilt, citrus canker and bacterial spot result in serious economic losses. By altering the genes of the host plant, which bacteria exploit when they invade, gene editing can help produce bacterial disease resistant crops. For some pathogens, the efficacy in promoting disease is achieved by the introduction of effector proteins into plant cells to direct plant gene expression. Plants can escape from these bacterial effectors by editing the target sequences they are able to recognize without being affected by growth. The technique is well suited to transfer resistance to rice, and even to tomato, citrus and other significant crops.

Another important constraint in agriculture is viral diseases; once the plant is an infected product, the disease spreads quickly and it is very difficult to control the disease. Viruses can spread from plant to plant by insect vectors, which include aphids, whiteflies and leafhoppers. There are two ways that gene editing can help enhance viral resistance. The first is that it can attack the genes of a plant that are necessary for viral replication or mobility. Second, it can base on the CRISPR system directly attack a sequence in viral DNA/RNA. For instance, CRISPR / Cas9 can be employed to target DNA viruses and CRISPR / Cas13 could now be applied to RNA viruses. These provide some new tools for the management of viral diseases like tomato yellows, cassava mosaic, rice black, and cucurbit yellow.

Gene editing can also be used to create resistance against insect pests to prevent their attack on the plants. Insects cause injury to plants through feeding on leaves, stems, flowers, fruits, and grains. They are also pests of many plants diseases. The traditional pest control also depends largely on insecticides, which over time can become antibiotic against insects. Crop resistance to insect attack through gene editing can be achieved by disrupting genes related to the production of plant defense compounds, surface structure, volatile signals or insect feeding preference. These can also be enhanced such as to improve the cuticles, cell walls, or to produce natural anti-insect compounds. These changes can make it less easy for pests to feed and reproduce, and result in less pest damage to the crop.

In yet another likely strategy, genes associated with plant secondary metabolites are altered. Plants do produce a number of chemical compounds as an innate defence against pests and disease. Among these are alkaloids, terpenoids, phenolics, glucosinolates and other defense-related compounds. Gene editing can be employed to boost levels of positive compounds and/or control their creation more efficiently. Some compounds might be negativ and it needs to be evaluated carefully for their effects on taste, nutrition and safety, however. Hence, pest resistance and novel food properties and consumer acceptance should be taken into consideration when developing the gene editing variants.

Multiplex gene editing is especially valuable in crop pest and disease resistance since pest and disease resistance in crops is often determined by multiple genes. A pathogen or pest may create resistance in a single gene but the crop may not be resistant to multiple threats. When multiple genes are modified at once, it's called multiplex editing, and can be done via CRISPR. This may lead to exposure to different strains or species of pathogens and develop a broad-spectrum resistance. It can also limit the danger of pests/pathogens being able to quickly kill resistance.

Because of the need to apply pesticides multiple times in sustainable agriculture, broad-spectrum and durable resistance are particularly advantageous.

6. APPLICATIONS OF GENE EDITING IN MAJOR CROPS

In key crops, gene editing is a technology that has been extensively used for yield and quality, resistance, resource-use efficiency and climate adaptation. Even modest gains in the productivity and resilience of staple crops will have an impact on food security as they contribute to most of the world's food supply. CRISPR/Cas, base editing, and prime editing have become increasingly popular for crop research due to their ability to precisely edit genes that provide important agronomic traits. The gene editing of major crops like rice, wheat, maize, soybean, potato, tomato, cotton, and oilseed crops has emerged as a clear focus for gene editing, because of their economic significance and nutritional importance.

Rice is one of the major cereal crops and it is a primary food of more than 50% of population internationally. Rice has been widely subjected to gene editing, due to its remarkable connectivity of gene information and efficient transformation. CRISPR/Cas has been used to modify grain size, grain number, increased plant height, flowering time, disease resistance, drought tolerance, salinity tolerance, and nutritional quality. Genes involved, for instance, in panicle architecture, tillering and grain weight can be modified for enhanced yield potential. To find susceptibility genes targeted by disease-causing organisms, particularly in development of disease on rice in bacterial blight and blast, disease resistance has also been improved. Moreover, gene editing can be useful to enhance drought, salinity and flood tolerance as well as the development of more tolerant rice plants, which will be especially relevant in the context of climate-resilient agriculture.

Wheat is another important staple crop that presents more complexity in gene editing because the genome of wheat is a large and polyploid one. CRISPR/Cas technology has the potential to create new opportunities to enhance wheat productivity and quality, despite its complexity. Genes that control grain size, grain protein content, gluten quality, disease resistance and stress tolerance have all been targeted with gene editing. Genetic modification for resistance to fungal diseases (e.g., powdery mildew, rust, Fusarium head blight) can help prevent loss of yield and the need to use pesticides on wheat. Genetic editing is also a possibility to enhance drought resistance or heat tolerance, which are crucial traits, given that wheat may be cultivated in areas with projected increases in temperatures and drought.

Maize is the crop of major consideration for food, feed and industrial use. Improvements in maize architecture, kernel development, drought and disease resistance and nutritional properties have applied gene editing. Maize yield is density sensitive, plant architecture (root system) and reproductive development have a significant effect on yield per area and editing of genes related to them can contribute to increased yield per area. Being able to withstand drought is particularly crucial for maize, because water stress starts during flowering can have a severe impact on grain formation. Gene Editing has the potential to reduce water-use efficiency and stress-response pathways and enhance root growth. Maize can also be improved for use in animal feed and human consumption due to its merits of starch composition, oil content and protein quality.

Soybean is a large-scale legume crop which is harvested for protein, oil, animal food and industrial materials. The soybean has been edited to enhance the plant's ability to mature (timing), adapt to the environment, resist disease, and make oil and protein more nutritious as well as enhancing plant architecture. A significant one is to introduce superior fatty acid composition and stability for the soybean oil produced. Another way that gene editing can be used is in order to be more resistant to plants that cause diseases, like soybean cyst nematode, bacterial pustule and fungal diseases. Additionally, soybean has been a part of biological nitrogen fixation; hence, enhancements of its root characteristics and symbiotic efficiency can also help support sustainable agriculture. Stress tolerant edited soybeans can support continued productivity during drought, heat and soil salinity challenges.

Potato is one of the most important tuber crops and its importance for carbohydrate for the global population. Because of the vegetative propagation and genetic complexity of potato, there are several traits that are difficult or impossible to improve using conventional breeding methods, and gene editing holds significant promise to enhance potato improvement. CRISPR/Cas has been successfully applied to increase resistance against late blight, to decrease in enzymatic browning to enhance tuber quality, and to modify the starch composition. Late blight is a potato disease that is one of the most devastating and does need frequent fungicide application. Chemical input can be reduced and susceptibility towards this disease can be reduced with gene-editing. Moreover, potatoes that are likely to be produced in an edited format, with a reduced level of browning and better storage characteristics, can help reduce post harvest losses, as well as food waste.

Tomato is currently one of the most thoroughly research horticultural crops for gene editing. It is a good model and valued as an economic vegetable. Tomato has been used to enhance shelf life, resistance to disease, improve fruit size and number, plant architecture, flowering time, and nutritional quality through the use of gene editing. Consumers' assessment of the quality and market value of the fruit and vegetable can be enhanced by improving certain characteristics, including lycopene level, sugar-acid balance, firmness and ripening phenomena. There are numerous bacterial, fungal, and viral diseases in tomato, and the genes for susceptibility and immunity are candidates for editing to achieve resistance. Tomato is one of the most successful crops for potential practical applications of gene editing, as it has a relatively short cultivation period, and proven transformation procedures.

Cotton is one of the most economically important fibre crops. Genetic modification has been employed in cotton to increase fiber length, fibers strength, fiber fineness, plant structure, disease resistance and stress tolerance. Cotton may be under attack from insect pests, short of rain, salt, and degradation of the soils. Genes related to the development of the fibers can be edited to yield better quality fiber, and genes related to stress-response can be edited to make it more drought and salt tolerant. Another way that gene editing will benefit cotton is by enhancing natural resistance to diseases and pests or allowing plants to be more resistant to some of the greatest ones.

Other crops of interest for gene editing are oilseed crops like canola, mustard, sunflower and groundnut. These crops have spilled over to become important sources of edible oil, protein and raw material for industry. Oil quality can be enhanced with gene editing by altering the pathways associated with fatty acid biosynthesis. For instance, increasing oleic acid and reducing undesirable fatty acid increases the stability and nutrition value of oil. Another area of desirable gene editing is the resistance in oilseed crops against salinity, drought and disease. Oilseed crops are significant for improvement, as there is an increasing demand for plant based oils globally and a decline in both yield and quality of these crops under stressful climate conditions.

7. CHALLENGES AND LIMITATIONS

While gene editing holds significant promise for sustainable agriculture, there are still some challenges and limitations that need to be addressed. The technical, biological, regulatory, ethical, environmental and socioeconomic challenges are mentioned. Through the use of gene editing techniques including CRISPR/Cas, base editing and prime editing, crops have become easier and more precise to improve than through conventional breeding, but with a few pitfalls. However, the success of a gene-edited crop hinges significantly on the efficiency of the laboratorial manipulation of crops and on their regeneration in the field, as well as in terms of biosafety testing, acceptance by the public, and the availability of the crops to farmers. So, when it comes to adopting gene editing as a secure tool for sustainable agriculture, its restrictions should be given the appropriate thought.

One of the major technical challenges is off-target editing. The gene editing system can result in off-target mutations when it mutates and alters a sequence of DNA in a region near the one involved in the desired alteration. While the CRISPR/Cas system is very accurate, complete accuracy is not guaranteed. If the mutations occur in important genes, then they can be problematic by having an effect on plant growth, yield, quality or safety. Depending on guide RNA design, editing enzyme used, crop genome size and similarity of DNA sequences in the genome, there is a risk of off-target effects. It is particularly significant in crop species like wheat, cotton and potato that are polyploid and have multiple, similar-copying genes. To minimize this risk the new guide RNA design, high fidelity Cas enzymes, and careful screening are needed to ensure whole genome sequencing.

Introducing the gene editing materials into plant cells is a challenge, too. To introduce the gene editing, CRISPR/Cas proteins, guide RNAs or gene editing constructs have to be introduced into target cells. Common techniques consist of Agrobacterium-mediated transformation, particle bombardment, protoplast transfection and viral delivery systems. But these techniques do not equally well adapt to all crops. There are some important crops that are not easy to be transformed and some tissues do not regenerate. That's due in large part to the fact that creating a full-edited plant, even if the target gene is known and understood, can still be challenging. Problems with efficient delivery and regeneration still represent a significant challenge for many crops.

Another big limitation is plant tissue culture regeneration. Once tissue is edited, plant cells need to be regenerated to whole plants. This may take time, be genotype dependent, and be technically challenging. The crop species vary in how easily they can regenerate, from high to low. Unwanted genetic and/or epigenetic changes or variations, referred to as somaclonal variations, may occur in the tissue culture. Such alterations may impact performance of the plants and make assessment of edited lines more difficult. Hence, the need for better tissue-culture technology or finding more tissue-culture free alternatives to editing.

Another challenge is related to the complexity of agricultural traits. There are several major characters that are polyreнic, including yield, environmental tolerance to water, heat, nutrients and disease resistance, which are strongly environment-dependent. A single gene may not yield a desirable improvement or expression. Sometimes the improvement of one characteristic could have a negative impact on another characteristic. For instance, higher stress tolerance can correlate with lower growth rates and larger grains can cause a reduction in grain numbers. Thus, to carry out gene editing, there is a need for comprehensive understanding of gene functions, interactions between genes and plant physiology. Multi-gene editing can facilitate, but also raises the design, validation and regulatory challenge.

Another critical drawback is that performance in real field is also problematic. But, many of the gene editing experiments conducted in the lab, greenhouse and in controlled conditions are not consistently field successful. In field situations, the plants are subjected to variable soil conditions, climatic conditions, pests, diseases and competition. In one location, a well-behaved genetic modification of a crop can be effective, but not necessarily at the same efficiency in another location. Therefore, multi-location and multi-season field trials are required before introduction of gene edited varieties is recommended for farmers. To what extent can gene editing be of practical benefit if there is no field validation?

One major hurdle to the commercialization of gene-edited crops is regulatory uncertainty. How gene-edited crops are regulated varies across countries. Whether or not foreign DNA is included, some countries regulate some gene-edited crops differently from GMOs, and, in other countries, GM and GEEEP regulations are similar. This lack of global regulation harmony is causing challenges for research, trade in seed, global collaborations and commercialisation. Creating an app can be costly and the approval process can also be time-consuming, particularly in countries with stringent app store regulations. Greater clarity, science and transparency is required in setting regulations to promote responsible innovation with safeguards for human health and the environment.

Caution also needs to be taken with biosafety and environmental, issues and concerns, too. If not suitably evaluated, gene-based crops might behave unpredictably in the environment. Potential issues depending on the nature of the transgenic crop may involve pollen escape of the transgenic crop to wild crops, impact on nontarget organisms, impact on biodiversity, and development of resistant pests/pathogens. Use of a gene-edited crop without management, for example if the crop is highly resistant to pests, will allow for the development of resistance to these pests over time. Likewise, traits (alleles) which are edited for fitness could be introduced into other related wild species by cross-species pollination. The following are risks characteristic of gene editing, but need to be assessed, monitored and carefully managed in this context:

A lack of public acceptance is another key limitation. Some consumers seem not to make a clear distinction between genetic modification and the technique of gene editing. However, gene-edited crops could also be subject to backlash from the public because of food safety fears, concerns of naturalness, fears of monopolization by corporations, and fears of long-term environmental effects. Agbiotech is only possible in the context of public trust. Even technologies that are deemed scientifically safe can be rejected if a farmer or a consumer is not made fully aware of it. Transparent communication, appellation, public engagement and education are needed to foster understanding and acceptance of gene-edited crops.

Ethical issues are also associated with gene editing in agriculture. A challenge is whether modifications to the crop's genetic make-up are an appropriate or desirable direct goal, and what level of modification is considered appropriate and desirable. The second one has to do with ownership of the gene editing technologies. Several gene editing technologies have patents that could restrict access by such other researchers, small seed companies, and developing countries. In a situation that is largely controlled by large companies, some small farmers might not reap the same rewards. Proper use of gene editing must be based on equitable access, responsible research and transparent governance, and safeguard the rights of farmers.

8. FUTURE PROSPECTS OF GENE EDITING IN SUSTAINABLE AGRICULTURE

The future of gene editing in sustainable agriculture is highly promising because global agriculture is moving toward precision, resilience, and resource efficiency. As food demand increases and environmental challenges become more severe, crop improvement must become faster, more accurate, and more adaptable to changing conditions. Gene editing technologies such as CRISPR/Cas, base editing, prime editing, and multiplex genome editing are expected to play a major role in developing crops that can produce higher yields, tolerate climate stress, resist pests and diseases, improve nutritional quality, and reduce dependence on chemical inputs. Therefore, gene editing is likely to become an important component of future sustainable farming systems.

One major future prospect is the development of climate-smart crops. Climate change is expected to increase the frequency of drought, heat waves, salinity, flooding, and new disease outbreaks. Future gene editing research will focus more strongly on crops that can maintain productivity under these multiple stresses. Instead of improving tolerance to only one stress, scientists are likely to target combinations of genes responsible for drought tolerance, heat resistance, salt tolerance, root development, water-use efficiency, and reproductive stability. Such crops will be highly valuable for farmers in arid, semi-arid, coastal, and climate-vulnerable regions. This approach can help protect food security while reducing crop failure caused by unpredictable weather conditions.

Another important future direction is multiplex gene editing. Many important agricultural traits are controlled by several genes rather than a single gene. Yield, stress tolerance, nutrient-use efficiency, and disease resistance are complex traits that require coordinated genetic changes. Multiplex CRISPR systems allow researchers to edit several genes at the same time. This can accelerate the development of crops with combined traits such as high yield, disease resistance, and drought tolerance. In the future, multiplex editing may help produce crop varieties that are not only productive but also stable under diverse environmental conditions.

Base editing and prime editing are also expected to become more important in crop improvement. Traditional CRISPR/Cas editing often works by cutting DNA and allowing the plant to repair the break. However, base editing can change a single DNA base without creating a double-strand break, while prime editing can introduce more precise substitutions, insertions, and deletions. These advanced tools may allow scientists to reproduce beneficial natural mutations or correct harmful genetic variations with greater accuracy. In future sustainable agriculture, base and prime editing may be used to improve grain quality, plant architecture, nutritional composition, disease resistance, and stress tolerance with fewer unwanted changes.

Artificial intelligence and bioinformatics will further strengthen the future of gene editing. The success of gene editing depends on identifying the right genes and designing accurate guide RNAs. AI-based tools can help predict gene function, identify useful genetic variation, reduce off-target effects, and select the best editing targets. Integration of genomics, transcriptomics, proteomics, metabolomics, and phenomics will allow researchers to understand crop traits more deeply. In the future, gene editing combined with AI-assisted breeding can make crop improvement faster, more predictive, and more efficient.

Gene editing will also contribute to improving nutritional security. Future crops will not only need to produce more food but also healthier food. Gene editing can be used to increase vitamins, minerals, proteins, essential amino acids, antioxidants, and beneficial fatty acids in crops. It can also reduce anti-nutritional factors, allergens, toxins, and compounds responsible for poor food quality. Biofortified gene-edited crops may help reduce hidden hunger and micronutrient deficiencies, especially in developing countries where people depend heavily on staple crops such as rice, wheat, maize, and potatoes.

Another future prospect is the development of crops with improved resource-use efficiency. Sustainable agriculture requires reducing excessive use of fertilizers, pesticides, and irrigation water. Gene editing can improve nitrogen-use efficiency, phosphorus-use efficiency, water-use efficiency, and photosynthetic efficiency. Crops that produce more yield with fewer inputs can reduce production costs and environmental pollution. In the future, gene-edited crops may help farmers maintain productivity while lowering greenhouse gas emissions, nutrient runoff, soil degradation, and water consumption.

Gene editing may also support environmentally friendly pest and disease management. Future gene-edited crops can be developed with broad-spectrum resistance against fungi, bacteria, viruses, insects, and nematodes. This can reduce the use of chemical pesticides and protect beneficial organisms such as pollinators and natural enemies of pests. However, future crop protection strategies must combine gene-edited resistance with integrated pest management to prevent pests and pathogens from evolving resistance. Therefore, gene editing should be used as part of a wider ecological approach to pest and disease control.

The improvement of orphan crops and underutilized crops is another important future opportunity. Many local crops are highly nutritious and adapted to harsh environments, but they have received less attention from commercial breeding programs. Gene editing can help improve these crops by enhancing yield, reducing anti-nutritional factors, improving disease resistance, and increasing market value. This is especially important for food security in developing countries, where local crops support rural livelihoods and dietary diversity. Future gene editing research should therefore include not only major global crops but also regionally important crops.

Transgene-free gene-edited crops are expected to gain more importance in the future. In many cases, gene editing can create small changes in the plant genome without leaving foreign DNA in the final variety. Such crops may be more acceptable to regulators and consumers in some countries. If regulatory systems clearly distinguish between transgenic GMOs and transgene-free gene-edited crops, commercialization may become easier. This could encourage public and private investment in gene-edited crop development.

However, the future success of gene editing will depend on responsible governance. Clear biosafety assessment, transparent regulation, ethical research practices, and public engagement will be necessary. Farmers, consumers, scientists, policymakers, and industry stakeholders must be involved in decision-making. Public trust will be essential, especially because many people still confuse gene editing with older genetic modification technologies. Open communication about benefits, risks, limitations, and safety measures can improve acceptance.

Equitable access will also shape the future of gene editing. If gene editing technologies remain expensive or controlled only by large companies, smallholder farmers may not benefit fully. Public-sector research, international collaboration, open licensing, and capacity building are needed to make gene editing useful for developing countries. Future policies should ensure that gene-edited seeds are affordable, locally adapted, and available to farmers who need them most.

Overall, the future prospects of gene editing in sustainable agriculture are strong. Gene editing can support climate-smart crops, higher yield, improved nutrition, efficient resource use, reduced pesticide dependence, and better adaptation to environmental stress. Advanced tools such as multiplex editing, base editing, prime editing, and AI-assisted breeding will expand the possibilities of crop improvement. However, gene editing should not be treated as a single solution to all agricultural problems. It must be integrated with conventional breeding, agroecology, soil conservation, biodiversity protection, and sustainable farming practices. If applied responsibly and inclusively, gene editing can become a powerful tool for achieving food security and sustainable agricultural development in the future.

9. CONCLUSION

Gene editing has emerged as one of the most important scientific tools for advancing sustainable agriculture in the modern era. As global agriculture faces increasing pressure from population growth, climate change, land degradation, water scarcity, pest outbreaks, disease pressure, and nutritional insecurity, there is a strong need for crop improvement methods that are precise, efficient, and environmentally responsible. Gene editing technologies, particularly CRISPR/Cas systems, base editing, and prime editing, provide new opportunities to improve crop traits more rapidly than conventional breeding methods. These tools allow targeted modification of genes associated with yield, stress tolerance, disease resistance, nutritional quality, and resource-use efficiency, making them highly valuable for future food security.

The review shows that gene editing can significantly contribute to crop yield improvement by modifying genes involved in grain size, plant architecture, flowering time, photosynthesis, reproductive efficiency, and nutrient uptake. Since yield is a complex trait influenced by both genetic and environmental factors, gene editing offers a more direct way to improve specific yield-related pathways. In major crops such as rice, wheat, maize, soybean, potato, tomato, and cotton, gene editing has already demonstrated potential for improving productivity and crop quality. These improvements are especially important because increasing food production through land expansion is no longer a sustainable option.

FUNDING

This work was supported by the Deanship of Scientific Research, the Vice Presidency for Graduate Studies and Scientific Research, King Faisal University, Saudi Arabia, under Grant No. KFU263175.

REFERENCES

Ali, Q., Yu, C., Hussain, A., Ali, M., Ahmar, S., Sohail, M. A., Riaz, M., Ashraf, M. F., Abdalmegeed, D., Wang, X., Imran, M., Manghwar, H., & Zhou, L. (2022). Genome engineering technology for durable disease resistance: Recent progress and future outlooks for sustainable agriculture. *Frontiers in Plant Science*, *13*, 1025145. DOI: 10.3389/fpls.2022.860281 PMID: 35371164

Amoah, P., Oumarou Mahamane, A. R., Byiringiro, M. H., Mahula, N. J., Manneh, N., Oluwasegun, Y. R., Assfaw, A. T., Mukiti, H. M., Garba, A. D., Chiemeke, F. K., Bernard Ojuederie, O., & Olasanmi, B. (2024). Genome editing in Sub-Saharan Africa: A game-changing strategy for climate change mitigation and sustainable agriculture. *GM Crops & Food*, *15*(1), 233–248. DOI: 10.1080/21645698.2024.2411767 PMID: 39481911

Anyshchenko, A. (2022). Aligning policy design with science to achieve food security: The contribution of genome editing to sustainable agriculture. *Frontiers in Sustainable Food Systems*, *6*, 937742. DOI: 10.3389/fsufs.2022.897643

Brower-Toland, B., Stevens, J. L., Ralston, L., Kosola, K., & Slewinski, T. L. (2024). A crucial role for technology in sustainable agriculture. *ACS Agricultural Science & Technology*, *4*(2), 173–180. DOI: 10.1021/acsagscitech.3c00426

Camerlengo, F., Frittelli, A., & Pagliarello, R. (2022). CRISPR towards a sustainable agriculture. *Encyclopedia*, *2*(2), 650–664. DOI: 10.3390/encyclopedia2020044

Clapp, J., & Ruder, S. L. (2020). Precision technologies for agriculture: Digital farming, gene-edited crops, and the politics of sustainability. *Global Environmental Politics*, *20*(3), 49–69. DOI: 10.1162/glep_a_00566

Das, S., Ray, M. K., Panday, D., & Mishra, P. K. (2023). Role of biotechnology in creating sustainable agriculture. *PLOS Sustainability and Transformation*, *2*(5), e0000065. DOI: 10.1371/journal.pstr.0000069

Dima, O., & Inzé, D. (2021). The role of scientists in policy making for more sustainable agriculture. *Current Biology : CB*, *31*(20), R1235–R1238. DOI: 10.1016/j.cub.2021.01.090 PMID: 33689712

Fernandes, P. M. B., Favaratto, L., Fernandes, A. A. R., Vicien, C., Capalbo, D. M. F., & Zerbini, F. M. (2022). To become more sustainable organic agriculture needs genome editing technology. *Frontiers in Bioengineering and Biotechnology*, *10*, 860051. DOI: 10.3389/fbioe.2022.912793 PMID: 36061445

Gordon, D. R., Jaffe, G., Doane, M., Glaser, A., Gremillion, T. M., & Ho, M. D. (2021). Responsible governance of gene editing in agriculture and the environment. *Nature Biotechnology*, *39*(9), 1055–1057. DOI: 10.1038/s41587-021-01023-1 PMID: 34381209

Karavolias, N. G., Horner, W., & Abugu, M. N.. (2021). Gene editing applications for climate-resilient agriculture and sustainable food systems. *Frontiers in Sustainable Food Systems*, *5*, 685801. DOI: 10.3389/fsufs.2021.685801

Karavolias, N. G., Horner, W., Abugu, M. N., & Evanega, S. N. (2021). Application of gene editing for climate change in agriculture. *Frontiers in Sustainable Food Systems*, *5*, 685801. DOI: 10.3389/fsufs.2021.685801

Kitenge, S. Y., Nkhabindze, B., Ratemo, B. O., Dada, S. A., Ogbimi, E. R., Karikari, B., Nuaila, V. N. A., Toili, M. E. M., Ndudzo, A., Adetunji, C. O., Fotabong, E., & Akinbo, O. (2025). The role of young scientists in promoting genome editing for sustainable agriculture and food systems in Africa. *Frontiers in Sustainable Food Systems*, *9*, 1523546. DOI: 10.3389/fsufs.2025.1530800

Lalotra, S., Mukherjee, T., & Singh, A.. (2024). Unlocking sustainable agriculture: The crucial role of plant genetic engineering. In *Sustainable Agriculture and Biotechnology*. Springer. DOI: 10.1515/9783111369754-007

Mehta, S., Lal, S. K., Sahu, K. P., Venkatapuram, A. K., ... (2020). CRISPR/Cas9-edited rice: A new frontier for sustainable agriculture. In *CRISPR Crops for Durable Agriculture* (pp. 149–171). Springer.

Mestanza, M., Hernández-Amasifuen, A. D., Pineda-Lázaro, A. J., Eriksson, D., & Guerrero-Abad, J. C. (2025). Genome editing for sustainable agriculture in Peru: Advances, potential applications and regulation. *Frontiers in Genome Editing*, *7*, 1574488. DOI: 10.3389/fgeed.2025.1611040 PMID: 40661867

Nor A'azizam, N. M., Chopra, S., Guleria, P., Kumar, V., Abd Rahim, M. H., & Yaacob, J. S. (2024). Harnessing the potential of mutation breeding, CRISPR genome editing, and beyond for sustainable agriculture. *Functional & Integrative Genomics, 24*(1), 15. DOI: 10.1007/s10142-024-01325-y PMID: 38421529

Panigrahi, K. K., Mohanty, A., & Padhan, S. R.. (2023). Recent advances in genome editing towards sustainable agriculture. In *Genome Editing and Sustainable Crop Improvement*. CRC Press. DOI: 10.4324/9781003382102-7

Pattnaik, D., Avinash, S. P., Panda, S., & Bansal, K. C.. (2023). Accelerating crop domestication through genome editing for sustainable agriculture. *Journal of Plant Growth Regulation, 42*(7), 4524–4545. DOI: 10.1007/s00344-022-10765-7

Penna, S., Shirani Bidabadi, S., & Jain, S. M. (2023). Mutation breeding to promote sustainable agriculture and food security in the era of climate change. In S. M. Jain & S. Penna (Eds.), *Mutation Breeding for Sustainable Agriculture and Food Security* (pp. 1–24). Springer. DOI: 10.1007/978-981-16-9720-3_1

Qaim, M. (2020). Role of new plant breeding technologies for food security and sustainable agricultural development. *Applied Economic Perspectives and Policy, 42*(2), 129–150. DOI: 10.1002/aepp.13044

Rajput, M., Choudhary, K., Kumar, M., Vivekanand, V., Chawade, A., Ortiz, R., & Pareek, N. (2021). RNA interference and CRISPR/Cas gene editing for crop improvement: Paradigm shift towards sustainable agriculture. *Plants, 10*(9), 1914. DOI: 10.3390/plants10091914 PMID: 34579446

Rehman, R. S., Zafar, S. A., Ali, M., Pasha, A. N., Naveed, M. S., Waseem, M., Ahmad, M., & Raza, A. (2022). CRISPR-Cas mediated genome editing: A paradigm shift towards sustainable agriculture and biotechnology. *Asian Plant Research Journal, 9*(4), 32–48. DOI: 10.9734/aprj/2022/v9i130197

Roy, S., Barman, M., & Das, A.. (2026). Application of genome editing for sustainable development of agriculture: A perspective view. In *Advances in Sustainable Crop Production*. CRC Press. DOI: 10.1201/9781779643155-5

Sarfraz, S., Ali, F., Hameed, A., Ahmad, Z., & Riaz, K. (2023). Sustainable agriculture through technological innovations. In *Sustainable Agriculture in the Era of Climate Change*. Springer.

Singh, A., Rajput, V. D., Rawat, S., & Sharma, R.. (2021). Emerging tools for sustainable agriculture and food security: An introduction. In *Emerging Tools for Sustainable Agriculture and Food Security* (pp. 1–15). Springer., DOI: 10.1007/978-981-16-0917-6_1

Thankappan, S., Binodh, A. K., & Kumar, P. R.. (2024). Genome editing of plant growth-promoting microbes (PGPM) towards developing smart bio-formulations for sustainable agriculture: Current trends and perspectives. In *Genome Editing in Microbial Biotechnology*. Bentham Science Publishers. DOI: 10.2174/9789815223798124010006

Tripathi, L., Dhugga, K. S., Ntui, V. O., Runo, S., Syombua, E. D., Muiruri, S., Wen, Z., & Tripathi, J. N. (2022). Genome editing for sustainable agriculture in Africa. *Frontiers in Genome Editing*, *4*, 876674. Advance online publication. DOI: 10.3389/fgeed.2022.876697 PMID: 35647578

Tripathi, L., Dhugga, K. S., Ntui, V. O., Runo, S., Syombua, E. D., Muiruri, S., Wen, Z., & Tripathi, J. N. (2022). Genome editing for sustainable agriculture in Africa: Opportunities and challenges for crop improvement and food security. *Frontiers in Genome Editing*, *4*, 876674. DOI: 10.3389/fgeed.2022.876697

Waghmode, P., Pitha, C. C., & Banka, V. (2024). The role of genome editing technologies in sustainable agriculture. *Journal of Advances in Biology & Biotechnology*, *27*(4), 50–61. DOI: 10.9734/jabb/2024/v27i81168

Wan, X., Hou, Q., & McConnell, L. L. (2022). Advances in genome editing for sustainable agriculture. *ACS Agricultural Science & Technology*, *2*(12), 2390–2405. DOI: 10.1021/acsagscitech.2c00090

Zheng, H., Wang, X., Qiu, J., & Zhou, S. (2025). The application of gene editing technology in reducing the use of pesticides and chemical fertilizers and mitigating the environmental impact of agriculture. *Advances in Resources Research*, *3*(1), 45–58.

Chapter 2
Applications of Gene Editing in Major Crops

M. Vincy Mary
KPR Institute of Engineering and Technology, India

S. Kiruthika
http://orcid.org/0009-0007-5910-1892
KPR Institute of Engineering and Technology, India

S. Kavitha
KPR Institute of Engineering and Technology, India

C. Pooja
KPR Institute of Engineering and Technology, India

ABSTRACT

Gene editing has emerged as a new tool in crop biotechnology, allowing for precise, efficient, and targeted change of plant genomes. Advanced techniques like CRISPR-Cas9, CRISPR-Cas12a, TALENs, and Zinc Finger Nucleases (ZFNs) have allowed for quick advances in major crops such as rice, wheat, maize, soybean, potato, and tomato. Applications include improved yield, disease resistance, abiotic stress tolerance, nutritional quality, and post-harvest properties. Gene editing speeds up breeding programs by directly targeting genes responsible for important agronomic features, eliminating reliance on traditional, labor-intensive breeding cycles. Despite its benefits, problems exist in regulatory frameworks, ethical issues, and public acceptance. Ongoing research and technical improvements are projected to increase efficiency and expand applications. Gene editing has great promise for generating climate-resilient crops, promoting sustainable agriculture, and enhancing global food security in the face of population expansion and environmental change.

DOI: 10.4018/979-8-3373-9534-0.ch002

INTRODUCTION

A few of the major issues facing agriculture in the twenty-first century are the growing population, the changing climate, the scarcity of arable land, the emergence of pests and illnesses, and the rising need for wholesome food. By 2050, there will likely be more than 9 billion people on the planet, necessitating a significant rise in food production while preserving environmental sustainability. Crop development has benefited greatly from conventional plant breeding techniques, but these approaches are frequently labour- intensive and constrained by genetic variety found in nature (Sahu, 2024). In this regard, cutting-edge biotechnological instruments in particular, gene editing have become revolutionary methods for accurate and quick crop modification.

The term "gene editing" describes sophisticated molecular methods that allow for exact changes to an organism's genome (Puchta, 2022). Gene editing enables targeted modifications such insertions, deletions, or replacements inside particular genomic areas, in contrast to traditional genetic engineering, which usually entails the insertion of foreign DNA. The efficiency and precision of genome alteration have been greatly increased by the invention of programmable nucleases, such as Zinc Finger Nucleases (ZFNs), Transcription Activator-Like Effector Nucleases (TALENs), and most notably the CRISPR/Cas systems (Arora & Narula, 2017; K. Bhushan et al., 2016; Ochiai & Yamamoto, 2014). Due to its ease of use, affordability, and adaptability, CRISPR/Cas9 has become a popular technology for crop genome editing because of these factors (Dhugga, 2022).

New approaches to solving important agricultural issues have been made possible by the use of gene editing in important crops like rice, wheat, maize, soybean, and barley (Atia et al., 2024). These staple crops give billions of people the vital calories and nutrients they need, making them the foundation of global food security. Direct targeting of the genes causing agronomic features is made possible by gene editing, which speeds up the creation of superior varieties without requiring the drawn-out breeding cycles of traditional techniques (Aglawe et al., 2018).

One of the most prominent applications of gene editing in major crops is enhancing resistance to biotic stresses. Insect pests and bacterial, viral, and fungal infections constantly threaten crops, causing large output losses globally (Zhan, 2021). Researchers have created disease-resistant cultivars by altering susceptibility genes or making plant defence pathways active. For example, rice's resistance to bacterial blight has been improved by editing certain genes (Oliva et al., 2019), while wheat's resistance to powdery mildew has been improved by changes (Zou et al., 2023). These developments lessen the need for chemical pesticides, supporting ecologically friendly farming methods.

In addition to disease resistance, gene editing has been widely used to improve tolerance to abiotic stresses such as drought, salinity, heat, and cold (Kumar, 2023). Climate change has intensified environmental stresses, negatively impacting crop productivity and stability. Targeted modification of stress-responsive genes enables crops to better withstand adverse conditions, ensuring stable yields even under unpredictable climates. For example, editing genes involved in drought response pathways has resulted in improved water-use efficiency in several crops (Joshi, 2023). These innovations are crucial for sustaining agricultural production in regions vulnerable to climate variability.

Another important use of gene editing is yielding enhancement (Mishra & Zhao, 2018). Future food demands must be met by increasing yield potential without increasing agricultural land. Scientists have significantly increased yields by modifying genes related to plant design, flowering period, grain size, and nutrient utilization. Rice productivity has increased as a result of altering genes that control tillering and grain number (Tabassum, 2021). In a similar vein, maize and wheat genome editing has improved harvest index and optimized plant structure (Wang et al., 2020). These focused enhancements speed up breeding initiatives and boost agricultural productivity. Similarly, genome editing in maize and wheat has optimized plant structure and enhanced harvest index. Such targeted improvements accelerate breeding programs and increase agricultural efficiency (Kumlehn, 2018). One area of gene editing applications that is becoming more and more significant is nutritional enhancement, or biofortification (Sahu, 2024). Millions of people worldwide suffer from micronutrient deficiencies, sometimes known as "hidden hunger." Crops can be genetically modified to have higher concentrations of vital minerals including iron, zinc, and provitamin A. Gene editing has also been utilized to lessen allergens and anti-nutritional elements, enhancing food safety and consumer wellness (Dhugga, 2022). These developments show that gene editing has the potential to improve food quality as well as quantity.

Additionally, gene editing helps to enhance industrial quality and post-harvest characteristics. By altering the genes that cause enzymatic browning and fruit ripening, fruits and vegetables have a longer shelf life, which lowers post-harvest losses and increases marketability. Targeted genome alterations have changed the content of fatty acids in oilseed crops like canola and soybean, improving the quality of the oil for use in industry and nutrition (Dossa et al., 2017). These enhancements satisfy customer desires while giving agricultural produce more economic worth (Wang et al., 2014).

Despite its considerable potential, the application of gene editing in major crops is accompanied by regulatory, ethical, and socio-economic considerations. Regulatory frameworks vary across countries, particularly regarding whether gene-edited crops without foreign DNA are classified as genetically modified organisms (GMOs)

(Ravikiran, 2023). Public perception and acceptance significantly influence the adoption of these technologies. Transparent communication, rigorous risk assessment, and harmonized policies are essential to ensure responsible deployment and public trust (Puchta, 2022).

To sum up, gene editing is a ground-breaking development in crop biotechnology that provides accurate, effective, and adaptable ways to enhance important crops. Disease resistance, stress tolerance, yield increase, nutritional improvement, and quality attributes are among its uses. Gene-editing technologies have enormous potential to advance sustainable agriculture and guarantee global food security as they develop further. The various uses of gene editing in important crops are covered in this chapter, along with current developments, real-world applications, and potential future directions in this quickly developing subject.

Overview of Gene Editing in Crop Improvement

In agricultural science, gene editing has become a potent breakthrough that opens up new avenues for more precise and effective crop improvement. It enables scientists to precisely modify plant DNA to increase beneficial traits or decrease unfavourable ones. Gene editing allows scientists to directly alter the genes responsible for significant plant traits, in contrast to traditional breeding, which frequently relies on repeated crossing and selection over many years (Yıldırım, 2023). This focused strategy has sped up the creation of better crop types.

Researchers may now thoroughly examine plant genetics and find genes associated with yield, resistance, and quality thanks to modern gene-editing technologies. Once these genes are discovered, desired results can be obtained by introducing minor, controlled alterations (Zhang, 2017). Crop types that are more appropriate for the difficulties encountered in contemporary agriculture might thus be created (Zhan, 2021). These issues include the necessity for sustainable farming methods, the shifting environment, and the rising need for wholesome food.

Enhancing plant performance in challenging environmental settings is a key area where gene editing makes a substantial contribution(Tufail et al., 2026). Stronger resistance to drought, excessive salt, or temperature swings can be produced in crops, enabling them to flourish in areas with less ideal farming conditions (Joshi, 2023; Kumar, 2023). Farmers are able to maintain consistent production even in the face of environmental stressors thanks to this resilience.

The enhancement of crop health is a significant advantage of gene editing. Scientists can create plants that are inherently more resistant to illnesses brought on by bacteria, fungus, or viruses by altering genes linked to disease susceptibility (Zou et al., 2023). In addition to increasing yields, healthier plants also lessen the need for chemical pesticides, supporting ecologically friendly farming methods (Song et al., 2026).

Enhancing the nutritional value and quality of food crops is another benefit of gene editing. To boost advantageous substances like proteins, vitamins, or antioxidants, researchers can modify metabolic pathways in plants. While preserving the flavour and texture that consumers anticipate from their food, these advancements help to improve the dietary quality for consumers (Sahu, 2024). This method can enhance post-harvest characteristics in addition to its nutritional advantages. Crops can be kept fresher for longer by modifying genes that regulate ripening or storage behaviour. Reducing food losses during storage and transit is crucial for global food supply systems. All things considered, gene editing is turning into a crucial instrument in contemporary crop enhancement initiatives. It facilitates the production of crops that are more resilient, productive, nutritious, and sustainable by allowing for precise genetic modifications. Gene editing is anticipated to become more significant in bolstering global agriculture and guaranteeing food security for future generations as research advances (Zhou, 2020).

Gene Editing Technologies Used in Crops

Advanced molecular techniques known as gene editing technologies enable researchers to precisely and specifically alter plant DNA. By making it possible to quickly improve agricultural characteristics like yield, disease resistance, stress tolerance, and nutritional quality, these methods have completely changed modern plant breeding (Puchta, 2022). Gene editing is quicker, more precise, and capable of directly altering particular genes that give crops their desired traits when compared to traditional breeding techniques (Wang et al., 2020).

Agricultural biotechnology has created and effectively implemented a number of gene-editing techniques. The following describes the main gene-editing techniques utilized in crops.

The following describes the main gene-editing techniques utilized in crops.

CRISPR-Cas9

CRISPR-Cas9 is one of the most powerful and extensively used techniques in contemporary genetic engineering and biotechnology. Clustered Regularly Interspaced Short Palindromic Repeats, or CRISPR for short, are distinct DNA sequences that are naturally present in bacterial genomes. These sequences are essential to the bacterial immune system and are dotted with brief, repeating palindromic sequences. CRISPR-associated protein 9, or Cas9, is an endonuclease enzyme that can cut DNA at specific sites like molecular scissors. CRISPR and Cas9 work together to create a programmable system that can identify and alter particular DNA sequences, allowing for highly focused genetic changes (Scheben, 2017).

The CRISPR-Cas9 system was discovered as a result of research on bacterial adaptive immunity. Viruses, also known as bacteriophages, are continually in contact with bacteria and have the ability to infiltrate and kill them. Bacteria use the CRISPR locus to integrate viral DNA pieces into their own genomes as a defensive mechanism. In the event that the virus tries to re-infect the cell, the Cas9 protein is guided by these sequences, which are subsequently transcribed into short RNA molecules. Bacteria are able to identify and eliminate invasive viral DNA with exceptional specificity thanks to this natural "memory-based" defence mechanism (Song et al., 2026).

Scientists have developed CRISPR-Cas9 as a flexible genome-editing tool that can be used on a variety of creatures, such as bacteria, plants, and animals, by building on this natural mechanism. Researchers can direct Cas9 to a specific genomic site and create a double-stranded break by creating a synthetic guide RNA (gRNA) that matches a target DNA sequence. This break is subsequently repaired by the cell's own DNA repair processes, either by Homology-Directed Repair, which enables exact insertion or correction of DNA sequences using a repair template, or by Non-Homologous End Joining, which frequently introduces tiny insertions or deletions. Because of its versatility, CRISPR-Cas9 is a very effective and versatile gene editing technique that allows for both precise gene modification and gene knockout (Rasheed, 2022). The flow of the CRISPR-Cas9 were shown in below Figure 1.

Figure 1. Schematic overview of the CRISPR–Cas9 genome-editing workflow in plants

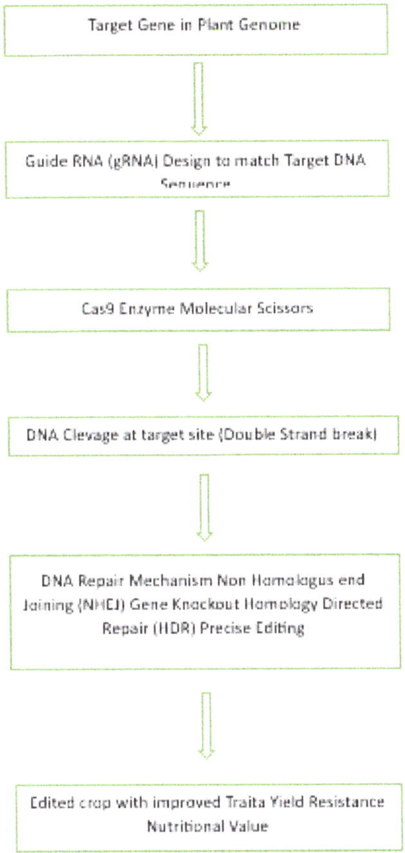

The process begins with the identification of a target gene associated with a desired trait, followed by the design of a complementary guide RNA (gRNA). The gRNA forms a ribonucleoprotein complex with the Cas9 nuclease, which recognizes the target DNA sequence adjacent to a protospacer adjacent motif (PAM) and introduces a site-specific double-strand break (DSB). The induced DNA break is repaired through the cellular non-homologous end joining (NHEJ) pathway, resulting in targeted insertions or deletions (indels) and gene knockout, or through homology-directed repair (HDR) using a donor DNA template for precise sequence modification. Successfully edited plants are subsequently selected to develop improved crop varieties with enhanced agronomic traits, including increased yield, improved nutritional quality, and greater resistance to biotic and abiotic stresses.

CRISPR-Cas9 has become a remarkable technique for crop improvement in agricultural biotechnology. Crossbreeding and selection are the mainstays of traditional plant breeding techniques, which might take many years to achieve desired features. On the other hand, CRISPR-Cas9 enables scientists to specifically target genes that are in charge of important agronomic traits including crop productivity, disease resistance, resilience to abiotic stressors like salinity and drought, and nutritional quality (Zhou, 2020). To create crops resistant to bacterial or fungal infections, for instance, scientists can alter genes related to plant immunology. They can also alter genes that control water usage efficiency to improve drought tolerance. The nutritional profile of staple crops like rice, wheat, and maize can also be enhanced by editing genes that affect the manufacture of vitamins, minerals, and other nutrients (Puchta, 2022; Tabassum, 2021).

CRISPR-Cas9 has benefits for crop development that go beyond accuracy and productivity. This technology is reasonably easy to use, reasonably priced, and suitable for a large range of plant species. In contrast to conventional genetic manipulation, which frequently entails introducing foreign DNA, CRISPR-Cas9 can produce alterations that are identical to mutations that occur naturally. Gene-edited crops may therefore encounter less regulatory obstacles and are becoming more widely used in both commercial and research agriculture (Mishra & Zhao, 2018; Negi, 2022).

The use of CRISPR-Cas9 has enormous potential, but there are some drawbacks as well, such as the possibility of off-target modifications, moral dilemmas, and unclear regulations in many nations. However, the precision and safety of this gene-editing system are constantly being improved by continuous advancements in guide RNA design, Cas9 variations, and delivery techniques (Wang et al., 2020; Xia, 2021; Xiong et al., 2015). In summary, CRISPR-Cas9 is a revolutionary technology in contemporary research that provides accurate, effective, and flexible ways to modify the genomes of microorganisms, plants, and mammals. It is a key technology for sustainable agriculture and global food security in the twenty-first century since its use in agriculture has created previously unheard-of chances to improve crop yield, quality, resilience, and nutritional content.

2. CRISPR-Cas12a

A advanced genome-editing tool, CRISPR-Cas12a (formerly Cpf1) belongs to the family of CRISPR-based gene-editing systems (Bandyopadhyay et al., 2020). Like CRISPR-Cas9, CRISPR-Cas12a comes from the bacterial adaptive immune system, where it serves as a protection against viral infections. By cutting viral DNA sequences, CRISPR-Cas systems enable bacteria to identify and kill invasive viruses, giving the organism immunity against further infections. In order to precisely modify the genomes of a wide range of creatures, including plants, animals, and microbes, scientists have adopted this natural bacterial technique (Yao et al., 2018).

The method CRISPR-Cas12a recognizes and breaks DNA is one of the key characteristics that sets it apart from CRISPR-Cas9. Cas12a recognizes a T-rich PAM (usually TTTV, where V can be A, C, or G) upstream of the target site, whereas Cas9 requires the presence of a specific Protospacer Adjacent Motif (PAM) sequence (usually NGG) immediately following the target DNA. Furthermore, as opposed to Cas9's blunt-end cut, Cas12a causes a staggered or "sticky-end" double-strand break in the DNA. When making focused gene alterations, this tool enables possibly more accurate insertions or modifications (Rasheed, 2022).

The use of a single, shorter CRISPR RNA (crRNA) instead of an extra trans-activating RNA (tracrRNA) simplifies the design and delivery of the editing components, which is another important benefit of CRISPR-Cas12a (Zhao et al., 2024). It is also easier to transfer into plant cells via viral vectors or other transformation techniques because of its lower size compared to Cas9. Because of these qualities, Cas12a is a useful and effective tool for accurate genome editing, particularly in plant biotechnology (Ma, 2016).

CRISPR-Cas12a is now a highly useful tool in crop science for enhancing agronomic features. Researchers can raise crop productivity, boost resistance to pests and diseases, and improve tolerance to abiotic stresses like drought, salinity, and severe temperatures by focusing on particular genes (Ma et al., 2023). Furthermore, Cas12a permits changes in genes related to nutritional pathways, enabling the production of crops with increased protein, vitamin, or mineral content. In contrast to traditional breeding techniques, which frequently need several generations to produce desired features, CRISPR-Cas12a enables direct, quick, and highly specific genome alteration in plants (Bandyopadhyay et al., 2020). The flow of the CRISPR-Cas12 a were shown in below Figure 2

Figure 2. Schematic representation of the CRISPR–Cas12a (Cpf1) genome-editing workflow for crop improvement

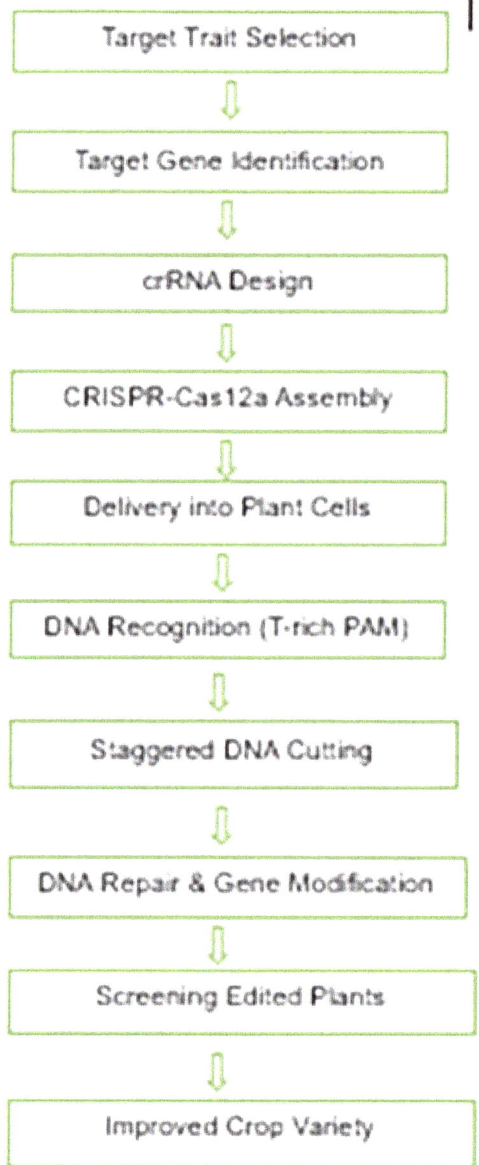

The process begins with the selection of the target trait and identification of the corresponding gene, followed by the design of a CRISPR RNA (crRNA) and assembly of the Cas12a editing complex. The CRISPR–Cas12a construct is delivered into plant cells, where Cas12a recognizes a T-rich protospacer adjacent motif (PAM) and binds the target DNA. Cas12a introduces staggered double-stranded DNA breaks, which are subsequently repaired through the cellular DNA repair pathways, including non-homologous end joining (NHEJ) or homology-directed repair (HDR), resulting in targeted genetic modifications. Successfully edited plants are screened and selected to develop improved crop varieties with desirable agronomic traits, such as enhanced yield, stress tolerance, disease resistance, and improved quality.

CRISPR-Cas12a offers an opportunity to perform multiplex genome editing in addition to its technical benefits. It is feasible to target numerous genes at once in a single experiment as Cas12a may process multiple crRNAs from a single transcript (Zhao et al., 2024). This capacity is especially useful for crops when several genes regulate complicated qualities like yield or stress tolerance. Scientists can expedite the creation of better crop types by altering several targets at once.

Like other genome-editing techniques, CRISPR-Cas12a has drawbacks despite its numerous benefits (Zhao et al., 2024). When using Cas12a in crop development initiatives, it's crucial to take into account potential off-target impacts, efficiency variability based on plant species, and ethical and regulatory issues. The specificity, effectiveness, and utility of this technique are nevertheless being improved by continuous study into guide RNA design, delivery strategies, and enzyme variations (Kaniganti et al., 2026).

In summary, CRISPR-Cas12a is an advanced and extremely effective genome-editing mechanism that is derived from bacterial immunity. It is a useful tool in plant biotechnology because of its special characteristics, which include T-rich PAM recognition, staggered DNA cleavage, and multiplexed editing capability. CRISPR-Cas12a has enormous potential to improve agronomic attributes including yield, stress tolerance, disease resistance, and nutritional quality by allowing accurate, effective, and targeted changes of crop genomes (Zhang et al., 2021). Because of this, it is quickly emerging as a key technology for crop development and sustainable agriculture in the twenty-first century.

3. TALENs (TRANSCRIPTION ACTIVATOR-LIKE EFFECTOR NUCLEASES)

TALENs, or transcription activator-like effector nucleases, are powerful and extremely accurate gene-editing instruments that enable researchers to alter certain DNA sequences in plants, animals, and other living things (Sun & Zhao, 2013). This method is crucial to contemporary biotechnology since it allows scientists to modify specific genes to enhance crops, investigate gene functions, and create novel biological solutions for agriculture and health.

Transcription Activator-Like Effectors (TALEs), which are naturally occurring proteins, provide the basis for the creation of TALEN technology. Some plant-pathogenic bacteria, especially those in the genus Xanthomonas, are the original producers of these proteins. TALE proteins are released into plant cells by these bacteria when they infect plants (Moore et al., 2014). Once inside, the proteins attach to particular DNA sequences in the plant genome to trigger the expression of genes that aid in the bacteria's infection and survival within the host plant.

Transcription Activator-Like Effectors (TALEs), which are naturally occurring proteins, provide the basis for the creation of TALEN technology (Sun & Zhao, 2013). Some plant-pathogenic bacteria, especially those in the genus Xanthomonas, are the original producers of these proteins. TALE proteins are released into plant cells by these bacteria when they infect plants (Erkes et al., 2017). Once inside, the proteins attach to particular DNA sequences in the plant genome to trigger the expression of genes that aid in the bacteria's infection and survival within the host plant. After researching this natural process, scientists discovered that TALE proteins' capacity to bind DNA could be modified to identify nearly any DNA sequence (Ji et al., 2016).

In order to design TALE proteins to attach to specific DNA regions in a genome, researchers used this knowledge. Every TALE protein has a sequence of repeating units, each of which is capable of identifying a particular DNA nucleotide. Scientists can design a unique TALE protein that targets a certain DNA sequence by assembling these repetitions in various ways. TALENs are incredibly versatile and adaptable for many gene-editing applications because of their modular nature (Ousterout & Gersbach, 2016).

Scientists united these modified DNA-binding proteins with a nuclease, an enzyme that cuts DNA, to make them useful tools for editing genes. The FokI nuclease is the most often utilized enzyme. The connected nuclease cleaves the DNA at that precise point when the tailored TALE protein attaches to its target DNA sequence. TALENs typically function in pairs, with each protein attaching itself to the target DNA sequence's opposing side. The FokI nucleases unite and cause a double-strand break in the DNA when both TALEN proteins bind to their corresponding locations (Sun & Zhao, 2013).

The cell's inherent repair systems are triggered when the DNA is damaged. The desired genetic changes can be introduced by scientists throughout the repair process. Adding new genes, removing undesirable genes, or fixing mutations are a few examples of these changes (K. Bhushan et al., 2016; Ochiai & Yamamoto, 2014; Ousterout & Gersbach, 2016). TALENs enable accurate and regulated genome editing because they target particular sequences.

TALENs have been extensively employed in plant biotechnology to enhance crops (Zaman et al., 2025). Using this technology, scientists have created plants with improved characteristics like greater productivity, better tolerance to environmental challenges, and increased disease resistance. Apart from plants, TALENs have been used in genetic research, animal studies, and even early medical research to comprehend genetic illnesses.

Figure 3. Schematic overview of the TALEN (Transcription Activator-Like Effector Nuclease) genome-editing workflow

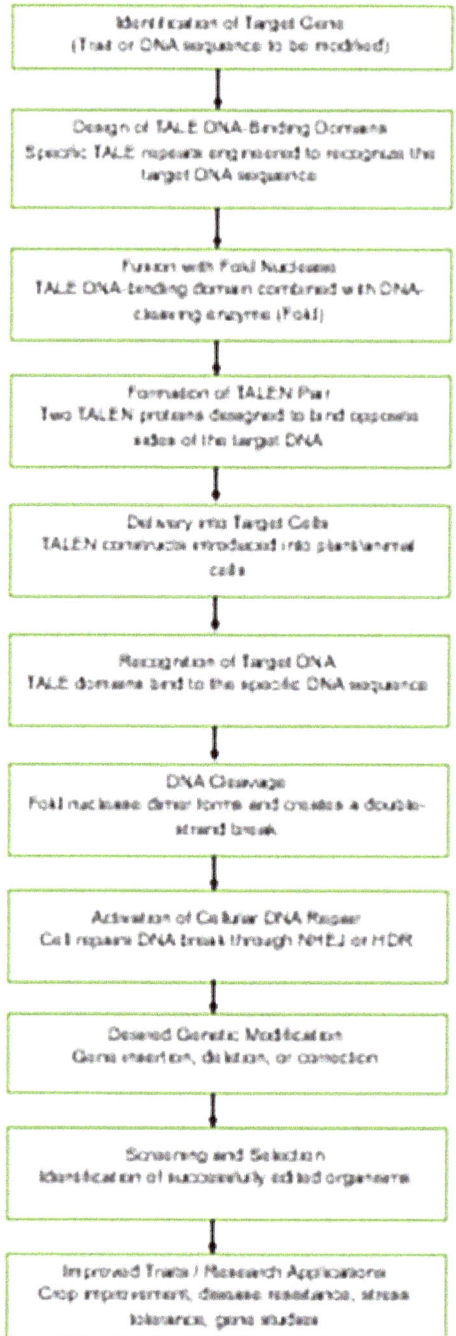

The process begins with identification of the target gene, followed by the design of sequence-specific TALE DNA-binding domains and their fusion with the FokI nuclease to generate TALEN constructs. Paired TALENs are introduced into target cells, where they bind opposite strands of the target DNA sequence, enabling FokI dimerization and induction of a site-specific double-strand break (DSB). The resulting DNA damage is repaired by the cell through either non-homologous end joining (NHEJ) or homology-directed repair (HDR), leading to targeted gene knockout, insertion, deletion, or sequence correction. Successfully edited organisms are subsequently screened and selected for downstream applications, including crop improvement, enhanced disease resistance, abiotic stress tolerance, and functional genomics research.

Before the widespread adoption of CRISPR-based technologies, TALENs were considered one of the most advanced and reliable genome-editing tools available. They offered greater precision compared to earlier methods like zinc finger nucleases. Although newer techniques such as CRISPR have become more popular due to their simplicity and efficiency, TALENs are still valuable in many research areas because they provide highly specific DNA targeting with relatively low off-target effects (Scheben, 2017; Zaman et al., 2025). The flow of the TALENs were shown in the above Figure 3.

TALENs were regarded as one of the most sophisticated and dependable genome-editing methods available prior to the widespread acceptance of CRISPR-based technologies. Compared to previous techniques like zinc finger nucleases, they provided higher precision. TALENs are still useful in many scientific fields because they offer extremely specific DNA targeting with comparatively little off-target effects, even though more recent methods like CRISPR have gained popularity because to their ease of use and effectiveness (Nemudryi et al., 2014). All things considered, TALEN technology is a significant advancement in genetic engineering. Researchers have developed a flexible technique that permits accurate genome editing by using naturally occurring bacterial proteins and altering them for scientific purposes. This invention has greatly aided in the development of genetic research, agriculture, and biotechnology.

Structure of TALENs (Transcription Activator-Like Effector Nucleases)

TALENs, or transcription activator-like effector nucleases, are created proteins that are intended to identify particular DNA sequences and cause targeted double-strand breaks in the genome. Because they enable scientists to precisely alter genes within an organism, they are widely utilized in genetic engineering and biotechnology. TALENs work by fusing an enzyme that cleaves DNA with a component that recognizes DNA. With this combination, scientists may target a specific gene and precisely cut the DNA. Scientists can inject desired genetic modifications as the cell's natural repair systems work to mend the broken DNA (Khalid et al., 2025).

TALENs have a modular structure, which means that the protein is made up of multiple functional domains or sections that cooperate to recognize and cut DNA. Every domain has a distinct function, and TALENs operate efficiently when these elements are properly coordinated. The nuclease domain, linker region, and DNA-binding domain are the three primary structural elements of TALENs. Together, these three components find the target DNA sequence and precisely break the DNA molecule (K. Bhushan et al., 2016; Boch et al., 2009).

a. DNA-Binding Domain (TALE Repeats)

One of the most crucial elements of the TALEN structure is the DNA-binding domain. Transcription Activator-Like Effector (TALE) proteins, which are produced naturally by several plant-pathogenic bacteria, are the source of this domain. TALE proteins are released into plant cells by these bacteria when they infect plants. Once within the cell, the proteins attach to particular plant genome DNA sequences and trigger the expression of genes that aid in the bacteria's survival and growth (Kul Bhushan et al., 2016; Ochiai & Yamamoto, 2014).

These proteins have a special capacity to identify particular DNA sequences, according to research done on them. These characteristic allowed researchers to tweak and create TALE proteins for use as genome editing tools. TALE repeats are a sequence of repeating protein units found in the DNA-binding domain of TALENs. Each repeat is in charge of identifying a single nucleotide in the DNA sequence and is made up of roughly 33 to 35 amino acids (Carroll, 2011).

Two crucial amino acids called Repeat Variable Di-residues (RVDs) are present in every repeat. Which DNA base—adenine (A), thymine (T), cytosine (C), or guanine (G)—the repeat will identify is determined by these RVDs. Scientists may create TALENs that attach to nearly any desired DNA sequence by placing these repetitions in a particular order. Researchers can tailor TALENs to target specific genes in the genome thanks to its modular design (Cermak et al., 2011). The DNA-binding domain is essential to ensuring that TALENs only cut DNA at the desired spot because of its accurate DNA-recognition capability. The likelihood of undesirable alterations in other regions of the genome is decreased by this selectivity (Christian et al., 2010).

b. Linker Region

The TALEN structure's linker region is what joins the nuclease and DNA-binding domains. Despite having a straightforward connecting segment appearance, it serves a crucial structural and functional purpose. The various components of the TALEN protein can move and align correctly during the gene-editing process thanks to this region's flexibility (Carroll, 2011). The linker region aids in preserving the proper spatial orientation between the nuclease and DNA-binding domains when the DNA-binding domain binds to the target DNA sequence. For the nuclease domain to reach the right location on the DNA and carry out the cleavage effectively, this alignment is required (Christian et al., 2010).

Additionally, the linker region guarantees that the TALEN protein can adjust to the three-dimensional DNA structure. DNA molecules are twisted double helices rather than straight lines. The linker region's flexibility enables the TALEN protein to self-adjust so that the nuclease domain can precisely reach the DNA strands. Without this area, the TALEN structure could stiffen and be unable to properly situate the cutting enzyme (Miller et al., 2011).

c. Nuclease Domain (FokI)

The TALEN protein's nuclease domain is in charge of cutting DNA. This domain is generated from the restriction enzyme FokI, which is frequently employed in molecular biology to manipulate DNA. Proteins known as restriction enzymes are capable of identifying and cutting particular DNA sequences at precise sites. The FokI nuclease domain is altered in TALEN technology so that it cannot cut DNA unless two TALEN molecules bind near one another on the target DNA region. TALENs typically operate in pairs, according to this. Each TALEN protein attaches itself to the target DNA region's opposing edges. The FokI nuclease domains unite to produce an active enzyme complex when both TALENs bind to their corresponding binding sites (Ochiai & Yamamoto, 2014; Sun & Zhao, 2013; Urnov et al., 2010).

A double strand break in the DNA between the two binding sites is produced when the two FokI domains dimerize. The cell's built-in DNA repair mechanisms are activated by this break. Scientists can make desired genetic modifications during the repair process, like adding new DNA sequences, eliminating undesirable genes, or fixing mutations (Miller et al., 2011).

4. ZINC FINGER NUCLEASES (ZFNs)

Zinc Finger Nucleases (ZFNs) are among the first genome-editing tools designed to precisely alter DNA sequences (Bibikova et al., 2003). These are modified proteins that are intended to identify and cut DNA at particular genomic regions. ZFNs were crucial to the development of contemporary genetic engineering methods because of their capacity to target specific genes (Curtin et al., 2011). Prior to the development of more recent genome-editing technologies like CRISPR-based systems and TALENs, scientists frequently employed ZFNs to alter the genetic makeup of different creatures and investigate gene function (Ochiai & Yamamoto, 2014).

A DNA-binding domain and a DNA-cleaving domain are two crucial parts of the structure of ZFNs (Urnov et al., 2010). Zinc finger proteins, which can identify and bind to particular DNA sequences, make up the DNA-binding domain. Usually, each zinc finger unit can identify a brief sequence consisting of three base pairs of DNA. Scientists can produce a protein that can identify longer, more precise DNA sequences in the genome by joining multiple zinc finger units. ZFNs are an effective tool for targeted genome editing because of their capacity to create proteins that bind to specific DNA regions.

The DNA-cleaving enzyme, which is often generated from a restriction enzyme known as FokI, is the second part of ZFNs (Carroll, 2011). The FokI enzyme can cleave DNA strands since it is a nuclease. Nevertheless, this enzyme only functions well when two copies of the enzyme combine to create a dimer. ZFNs are therefore made to operate in pairs. Every ZFN molecule attaches itself to the target DNA sequence's opposing sides. The FokI nuclease domains unite and produce a double-strand break in the DNA at the desired spot when both ZFNs bind to their individual binding sites.

The cell initiates its own DNA repair processes after the DNA is damaged. The integrity of the genome depends on these repair mechanisms. These repair mechanisms are exploited by scientists to introduce particular genetic modifications. Researchers can use homology-directed repair (HDR), which enables them to insert or replace particular genes using a template DNA sequence, or non-homologous end joining (NHEJ), which can result in minor insertions or deletions in the DNA sequence. ZFNs enable precise alterations such gene disruption, gene insertion, or genetic mutation correction through these repair processes (Ochiai & Yamamoto, 2014).

Before more modern genome-editing technologies were developed, ZFNs were widely employed in genetic engineering research (Townsend et al., 2009). They made a substantial contribution to the understanding of gene activities in a wide range of taxa, including microbes, plants, and mammals. ZFNs were used in plant biotechnology to alter crop genomes to enhance critical agricultural characteristics. Using this method, researchers were able to create plants that were more productive, more resilient to environmental challenges like salinity or drought, and more resistant to diseases (Shukla et al., 2009).

The advancement of molecular biology and biotechnology studies was also significantly aided by the technology. Scientists were able to gain a better understanding of how particular genes govern biological processes by using ZFNs to enable targeted modifications in the genome. Innovations in biotechnology, medical research, and improved crops all benefited from this understanding. Zinc finger proteins (ZFNs) were revolutionary when they were originally developed, but creating ZFNs for particular DNA sequences can be difficult and time-consuming (Townsend et al., 2009). These difficulties have led to an increase in the adoption of more contemporary genome-editing tools, including as TALENs and CRISPR systems. While maintaining great levels of precision, these contemporary instruments are frequently simpler to create and use.

The contribution of ZFNs to the field of genome editing is still very important, even with the advent of newer technologies (Urnov et al., 2010). They proved that targeted DNA modification is feasible and set the stage for many of the current gene-editing techniques. In conclusion, one of the first genome-editing instruments that allowed researchers to carry out precise genetic alterations is zinc finger nucleases (Carroll, 2011; Urnov et al., 2005). The advancement of contemporary genome-editing technologies and the enhancement of crops and other species were significantly aided by their application in early genetic engineering research and plant biotechnology.

Applications of Gene Editing in Major Crops:

With the ability to precisely alter plant genomes to improve desired features, gene editing has emerged as a revolutionary method to modern crop modification. Gene editing directly modifies certain genes responsible for significant traits, in contrast to traditional breeding techniques that depend on crossing plants across multiple generations. The procedure is quicker, more precise, and more effective because to this focused approach. CRISPR systems, TALENs, and Zinc Finger Nucleases (ZFNs) are examples of advanced technologies that have enabled the introduction of advantageous genetic modifications in numerous significant crops (Altpeter et al., 2016).

Increasing crop yield is one of the main uses of gene editing in large crops. Through the manipulation of genes linked to plant growth, yield potential, and nutrient usage efficiency, scientists can create crop types that yield more even in harsh environments. Creating crops that are resistant to disease is another important application of gene editing. Scientists can develop crops that are more resistant to bacterial, viral, and fungal infections by focusing on genes involved in plant–pathogen interactions. This will decrease crop losses and the need for chemical pesticides (Aglawe et al., 2018; Atia et al., 2024; Erkes et al., 2017).

Enhancing resistance to abiotic stressors such drought, salinity, heat, and cold is another important use. These pressures have become more frequent and severe due to climate change, and gene editing aids plants in adjusting to these environmental difficulties. Additionally, gene editing can result in crops that are resistant to herbicides, which aids farmers in better controlling weeds without sacrificing agricultural yield(Joshi, 2023; Kumar, 2023; Sahu, 2024).

Additionally, gene editing also known as "biofortification" is utilized to improve nutrition. Higher levels of vitamins, minerals, and other vital nutrients can be added to crops by specific genetic modifications. In addition, scientists are employing gene editing to enhance crop quality, post-harvest traits, and shelf life, which helps improve food supply chains and decrease food waste (Sahu, 2024). In general, the uses of gene editing in important crops are numerous and still growing. Gene editing technologies are contributing to the development of crops that are more resilient, prolific, and nutritious by permitting accurate and effective genetic alterations (Ochiai & Yamamoto, 2014). It is anticipated that these developments will be essential to maintaining sustainable agriculture and satisfying the world's food needs in the future.

Increasing Crop Productivity:

Increasing crop production is a top issue in agriculture, particularly as farmers need to produce more food to feed the world's expanding population. By precisely altering a plant's genetic composition, gene editing has emerged as a significant technology that helps scientists increase crop productivity (Aglawe et al., 2018). Gene editing enables researchers to directly target genes that control significant plant features rather than depending solely on conventional breeding techniques, which can take several years to generate detectable improvements. Scientists can alter genes that control plant growth, blooming period, grain development, and general plant structure by using gene editing (Dhugga, 2022). These characteristics have a direct bearing on how well a plant can yield grains or seeds. Researchers can create crop varieties that grow more effectively and yield more harvestable product by modifying these genetic characteristics (Cui, 2023).

For instance, gene editing has been utilized to improve grain formation and plant architecture in major crops like rice and maize. Stronger plants that use resources more effectively may arise from these modifications, increasing yields. Researchers have employed gene editing to boost grain weight and spikelet count in wheat, which directly increases plant output. All things considered, gene editing opens up new possibilities for producing crop kinds with excellent yields. This technology promotes sustainable farming and helps guarantee a steady food supply for the future by increasing output in an accurate and efficient manner (Abdelrahman, 2021; Sahu, 2024; Schaart et al., 2021).

Creation of Crops with Disease Resistance:

Plant diseases are one of the main issues influencing global agricultural productivity. Crop damage from pathogens including bacteria, fungus, and viruses can result in large output losses and lower food quality. In the past, farmers have used traditional breeding techniques and chemical pesticides to control plant diseases. Nevertheless, these methods can be expensive, time-consuming, and occasionally detrimental to the environment. Gene editing has made it possible to create crops that are inherently disease-resistant in a new and more accurate manner (Song et al., 2026; Zou et al., 2023).

Scientists can change or eliminate particular plant genes that make crops susceptible to disease attacks by using gene editing methods. These genes, also known as susceptibility genes, can be changed to prevent disease-causing organisms from efficiently infecting the plant. This increases the plant's resistance to some illnesses without impairing its regular development or yield. For example, scientists have successfully created rice cultivars that are resistant to bacterial blight, a disease that severely damages crops in many areas where rice is grown. Similarly, wheat types resistant to powdery mildew, a prevalent fungal disease, have been created by genome editing. Scientists have modified some genes in tomatoes to increase resilience to a number of viral illnesses that typically harm the crop. Gene editing encourages environmentally friendly farming methods and lessens the need for chemical pesticides by producing crops that are resistant to disease. This strategy preserves crop quality and output while promoting sustainable agriculture(Ji et al., 2016; Song et al., 2026) .

Strengthening Abiotic Stress Resistance:

Crop development and productivity are greatly impacted by abiotic stresses like drought, salt, severe temperatures, and other adverse climatic conditions. These pressures have increased in frequency and severity due to the growing effects of climate change, creating significant problems for agricultural systems across the globe. Therefore, creating crop varieties that can withstand adverse weather circumstances has emerged as a key objective in contemporary plant science. A useful technique for enhancing crops' resistance to various challenges is gene editing (Joshi, 2023).

Genes that control how plants react to their surroundings can be accurately altered by scientists using gene editing. Certain genes regulate stress signaling pathways, heat resistance, salt tolerance, and water use efficiency. Researchers can improve plants' inherent defense mechanisms by modifying these genes, allowing them to endure and flourish even in harsh environmental circumstances (Kumar, 2023).

For specific examples, rice varieties that can withstand high soil salinity levels—a significant issue in many agricultural regions—have been created using genome editing. Similarly, genes linked to stress response and water regulation have been altered to create drought-tolerant cultivars of maize. Additionally, researchers are trying to develop wheat varieties that can withstand high temperatures, which is crucial in regions where temperatures are rising. In general, enhancing abiotic stress tolerance via gene editing contributes to the development of resilient crop types that can sustain productivity even in harsh environmental circumstances (Joshi, 2023; Kumar, 2023).

Nutritional Improvement (Biofortification):

Improving food crops' nutritional value is a major objective in contemporary agriculture, especially in areas where people rely mostly on a small number of staple foods. Proteins, vitamins, and minerals are among the vital nutrients that are deficient in many people worldwide. By directly altering genes involved in nutrient production and metabolic pathways, gene editing provides a precise and effective method for increasing the nutritional content of crops. The goal of this procedure, called "biofortification," is to create crop varieties with naturally increased concentrations of advantageous nutrients (Negi, 2022; Sahu, 2024).

Scientists can modify the function of particular genes that control the synthesis and accumulation of significant chemicals in plants using gene editing technology. By focusing on these genes, scientists can raise nutrient concentrations without adversely affecting plant growth or production. By using this technique, crops can retain their agricultural performance while offering greater nutritional benefits. As an example, soybeans are now a more nutrient-dense source of plant-based protein because to gene editing. Scientists have found that tomatoes have higher quantities of antioxidants, which are crucial for preventing harm to human cells and promoting general health. In a similar vein, gene editing has been used to alter the starch composition of potatoes, enhancing their nutritional value and processing quality. All things considered, gene editing-based biofortification offers a viable approach to creating healthier food crops and assisting in addressing the world's nutritional problems (Haque, 2018; Kaniganti et al., 2026; Zaman et al., 2025; Zhan, 2021; Zhang, 2017).

Tolerance to Herbicides:

Because they compete with crops for vital resources like sunshine, water, space, and soil nutrients, weeds are one of the biggest issues facing farmers. Crop development and overall productivity may be severely hampered by this competition. Farmers have historically used chemical pesticides, mechanical techniques, or manual weed eradication to control weeds. Herbicides, however, can occasionally harm crops in addition to weeds, particularly if the plants are susceptible to these substances. By making it possible to create crops that can withstand particular herbicides, gene editing has offered a practical option(Karavolias, 2021).

Researchers can change specific genes that regulate a plant's susceptibility to herbicides by using gene editing. It is possible to make crops immune to the negative effects of specific herbicides while leaving weeds vulnerable by modifying these genes. This enables farmers to manage weed growth in their fields without endangering the primary crop by using herbicides. Weed control consequently becomes simpler, more effective, and requires less work. (Chavhan, 2025; Dhugga, 2022; Farinati, 2023)In this regard, herbicide-tolerant cultivars of significant crops like maize and rice have been created using gene editing. In order to sustain healthy crop development and increase yields, farmers must be able to successfully manage weeds in the early phases of crop growth, which is made possible by these improved cultivars. In general, herbicide-tolerant crops created via gene editing encourage more effective agricultural output, enhance farm management techniques, and lessen crop losses brought on by weeds (Ma, 2016).

Enhancement of Shelf Life and Crop Quality:

In today's agriculture and food production, increasing the shelf life and general quality of crops is a key objective. Many fruits and vegetables immediately experience changes like softening, discolouration, and nutrient loss after harvest. These modifications may lower the produce's market value and result in a large amount of food waste during marketing, shipping, and storage. By altering particular genes that regulate ripening, texture, and other post-harvest traits, gene editing provides a practical solution to these issues (Ravikiran, 2023).

The research community can change the genes in charge of processes like fruit ripening and spoilage-causing enzymatic activities by using gene editing techniques. Crops can be kept fresher for longer without losing their nutritional content or flavor by slowing down these processes. This contributes to the preservation of product quality throughout storage and transit, which is crucial for crops that are shipped across great distances (Guo, 2023).

Thus, scientists have altered ripening-related genes in tomatoes to postpone the softening of the fruit. By making this change, tomatoes' flavour and texture are preserved while their shelf life is increased. Gene editing has also been used to lessen enzymatic browning in potatoes, a reaction that results in discolouration during processing and storage. This enhancement improves potatoes' look and storage quality. In all cases, gene editing benefits farmers and consumers by increasing crop quality, extending shelf life, and decreasing food waste (Ku & Ha, 2020).

CONCLUSION

In agricultural science, gene editing has become an effective innovation that opens up new avenues for more precise and effective crop improvement. It enables scientists to precisely modify a plant's DNA to increase beneficial qualities or decrease unfavourable ones. With gene editing, scientists can directly alter the genes responsible for significant plant traits, in contrast to traditional breeding, which frequently relies on repeated crossing and selection over many years. The development of better crop types has been accelerated by this focused strategy.

With the use of modern gene-editing techniques, scientists may thoroughly examine plant genetics and pinpoint genes associated with quality, yield, and resistance. Once these genes are discovered, desired results can be attained by introducing minor, controlled alterations. Crop types that are more appropriate for the difficulties of contemporary agriculture might thus be created. Climate change, the necessity for sustainable farming methods, and the rising demand for wholesome food are some of these concerns.

Enhancing plant performance in challenging environmental settings is a key area where gene editing makes a substantial contribution. Stronger resistance to drought, excessive salt, or temperature swings can be produced in crops, enabling them to flourish in areas with less ideal farming conditions. Farmers are able to maintain consistent production even in the face of environmental stressors thanks to this resilience. The improved level of crop health is a significant advantage of gene editing. Scientists can create plants that are inherently more resistant to illnesses brought on by bacteria, fungus, or viruses by altering genes linked to disease susceptibility. In addition to increasing yields, healthier plants also lessen the need for chemical pesticides, supporting ecologically friendly farming methods.

Optimizing the nutritional value and quality of food crops is another benefit of gene editing. To boost advantageous substances like proteins, vitamins, or antioxidants, researchers can modify metabolic pathways in plants. While preserving the flavor and texture that consumers anticipate from their food, these advancements help to improve the dietary quality for consumers. This method can improve post-harvest characteristics in addition to its nutritional advantages. Crops can be kept fresher for longer by modifying genes that regulate ripening or storage behaviour. Reducing food losses during storage and transit is crucial for global food supply systems.

In general, gene editing is turning into a crucial tool for contemporary crop development initiatives. It facilitates the production of crops that are more resilient, nutritious, productive, and sustainable by allowing for precise genetic modifications. Gene editing is anticipated to become a more significant factor in bolstering global agriculture and guaranteeing food security for future generations as research advances.

REFERENCES

Abdelrahman, M., Wei, Z., Rohila, J. S., & Zhao, K. (2021). Multiplex genome-editing technologies for revolutionizing plant biology and crop improvement. *Frontiers in Plant Science*, *12*, 721203. DOI: 10.3389/fpls.2021.721203

Aglawe, S. B., Barbadikar, K. M., Mangrauthia, S. K., & Madhav, M. S. (2018). New breeding technique "genome editing" for crop improvement: Applications, potentials and challenges. *3 Biotech*, *8*(8), 336. https://doi.org/DOI: 10.1007/s13205-018-1355-3

Altpeter, F., Springer, N. M., Bartley, L. E., Blechl, A. E., Brutnell, T. P., Citovsky, V., Conrad, L., Gelvin, S. B., Jackson, D., Kausch, A. P., Lemaux, P. G., Medford, J. I., Orozo-Cardenas, M., Tricoli, D., VanEck, J., Voytas, D. F., Walbot, V., Wang, K., Zhang, Z. J., & Stewart, C. N.Jr. (2016). Advancing crop transformation in the era of genome editing. *The Plant Cell*, *28*(7), 1510–1520. DOI: 10.1105/tpc.16.00196

Arora, L., & Narula, A. (2017). Gene editing and crop improvement using CRISPR-Cas9 system. *Frontiers in Plant Science*, *8*, 1932. DOI: 10.3389/fpls.2017.01932

Atia, M., Jiang, W., Sedeek, K., Butt, H., & Mahfouz, M. (2024). Crop bioengineering via gene editing: Reshaping the future of agriculture. *Plant Cell Reports*, *43*(4), 98. DOI: 10.1007/s00299-024-03183-1

Bandyopadhyay, A., Kancharla, N., Javalkote, V. S., Dasgupta, S., & Brutnell, T. P. (2020). CRISPR-Cas12a (Cpf1): A Versatile Tool in the Plant Genome Editing Tool Box for Agricultural Advancement [Review]. *Volume 11 - 2020*. https://doi.org/DOI: 10.3389/fpls.2020.584151

Bhushan, K., Pratap, D., & Sharma, P. K. (2016). Transcription activator-like effector nucleases (TALENs): An efficient tool for plant genome editing. *Engineering in Life Sciences*, *16*(4), 330–337. DOI: 10.1002/elsc.201500126

Bhushan, K., Pratap, D., & Sharma, P. K. (2016). Transcription activator-like effector nucleases (TALENs): An efficient tool for plant genome editing. *Engineering in Life Sciences*, *16*(4), 330–337. DOI: 10.1002/elsc.201500126

Bibikova, M., Beumer, K., Trautman, J. K., & Carroll, D. (2003). Enhancing gene targeting with designed zinc finger nucleases. *Science*, *300*(5620), 764. DOI: 10.1126/science.1079512

Boch, J., Scholze, H., Schornack, S., Landgraf, A., Hahn, S., Kay, S., Lahaye, T., Nickstadt, A., & Bonas, U. (2009). Breaking the code of DNA binding specificity of TAL-type III effectors. *Science*, *326*(5959), 1509–1512. DOI: 10.1126/science.1178811

Carroll, D. (2011). Genome engineering with zinc-finger nucleases. *Genetics, 188*(4), 773–782. DOI: 10.1534/genetics.111.131433

Cermak, T., Doyle, E. L., Christian, M., Wang, L., Zhang, Y., Schmidt, C., Baller, J. A., Somia, N. V., Bogdanove, A. J., & Voytas, D. F. (2011). Efficient design and assembly of custom TALEN and other TAL effector-based constructs for DNA targeting. *Nucleic Acids Research, 39*(12), e82. DOI: 10.1093/nar/gkr218

Chavhan, R. L., Jaybhaye, S. G., Hinge, V. R., Deshmukh, A. S., Shaikh, U. S., Jadhav, P. K., Kadam, U. S., & Hong, J. C. (2025). Emerging applications of gene editing technologies for the development of climate-resilient crops. *Frontiers in Genome Editing, 7*, 1524767. DOI: 10.3389/fgeed.2025.1524767

Christian, M., Cermak, T., Doyle, E. L., Schmidt, C., Zhang, F., Hummel, A., Bogdanove, A. J., & Voytas, D. F. (2010). Targeting DNA double-strand breaks with TAL effector nucleases. *Genetics, 186*(2), 757–761. DOI: 10.1534/genetics.110.120717

Cui, Y., Cao, Q., Li, Y., He, M., & Liu, X. (2023). Advances in cis-element- and natural variation-mediated transcriptional regulation and applications in gene editing of major crops. *Journal of Experimental Botany, 74*(18), 5441–5457. DOI: 10.1093/jxb/erad248

Curtin, S. J., Zhang, F., Sander, J. D., Haun, W. J., Starker, C., Baltes, N. J., Reyon, D., Dahlborg, E. J., Goodwin, M. J., Coffman, A. P., Dobbs, D., Joung, J. K., Voytas, D. F., & Stupar, R. M. (2011). Targeted mutagenesis of duplicated genes in soybean with zinc-finger nucleases. *Plant Physiology, 156*(2), 466–473. DOI: 10.1104/pp.111.172981

Dhugga, K. S. (2022). Gene editing to accelerate crop breeding. *Frontiers in Plant Science, 13*, 889995. DOI: 10.3389/fpls.2022.889995

Dossa, K., Diouf, D., Wang, L., Wei, X., Zhang, Y., Niang, M., & Yehouessi, L. W. (2017). The emerging oilseed crop Sesamum indicum enters the "Omics" era. *Frontiers in Plant Science, 8*, 1154. DOI: 10.3389/fpls.2017.01154

Erkes, A., Reschke, M., Boch, J., & Grau, J. (2017). Evolution of Transcription Activator-Like Effectors in Xanthomonas oryzae. *Genome Biology and Evolution, 9*(6), 1599–1615. DOI: 10.1093/gbe/evx108

Farinati, S., Draga, S., Betto, A., Palumbo, F., Vannozzi, A., Lucchin, M., & Barcaccia, G. (2023). Current insights and advances into plant male sterility: New precision breeding technology based on genome editing applications. *Frontiers in Plant Science, 14*, 1223861. DOI: 10.3389/fpls.2023.1223861

Guo, Y., Zhao, G., Gao, X., Zhang, L., Zhang, Y., Cai, X., Yuan, X., & Guo, X. (2023). CRISPR/Cas9 gene editing technology: A precise and efficient tool for crop quality improvement. *Planta*, *258*(2), 36. DOI: 10.1007/s00425-023-04187-z

Haque, E., Taniguchi, H., Hassan, M. M., Bhowmik, P., Karim, M. R., Śmiech, M., Zhao, K., Rahman, M., & Islam, T. (2018). Application of CRISPR/Cas9 genome editing technology for the improvement of crops cultivated in tropical climates: Recent progress, prospects, and challenges. *Frontiers in Plant Science*, *9*, 617. DOI: 10.3389/fpls.2018.00617

Ji, Z., Ji, C., Liu, B., Zou, L., Chen, G., & Yang, B. (2016). Interfering TAL effectors of Xanthomonas oryzae neutralize R-gene-mediated plant disease resistance. *Nature Communications*, *7*(1), 13435. DOI: 10.1038/ncomms13435

Joshi, A., Yang, S.-Y., Song, H.-G., Min, J., & Lee, J.-H. (2023). Genetic databases and gene editing tools for enhancing crop resistance against abiotic stress. *Biology (Basel)*, *12*(11), 1400. DOI: 10.3390/biology12111400

Kaniganti, S., Saini, H., Chaitanya, A. K., Hegde, N., Shah, P., Magar, N. D., Rijal, R., Kaushik, J. J., Nanda, D., Sachan, S., Kumar, A., Bhoite, R., & Jamedar, H. R. (2026). CRISPR/Cas Genome Editing and Its Applications in Cereal Crop Improvement. *Plant-Environment Interactions*, *7*(2), e70133. DOI: 10.1002/pei3.70133

Karavolias, N. G., Horner, W., Abugu, M. N., & Evanega, S. N. (2021). Application of gene editing for climate change in agriculture. *Frontiers in Sustainable Food Systems*, *5*, 685801. DOI: 10.3389/fsufs.2021.685801

Khalid, S., Shahid, M. N., Nawaz, K., Naeem, M., & Jabeen, K. (2025). Structure and Mechanism of Transcription Activator-like Effector Nuclease-mediated Genome Editing in Plants. 47–55. https://doi.org/DOI: 10.1079/9781800622517.0004

Ku, H. K., & Ha, S. H. (2020). Improving nutritional and functional quality by genome editing of crops: Status and perspectives. *Frontiers in Plant Science*, *11*, 577313. DOI: 10.3389/fpls.2020.577313

Kumar, M., Prusty, M. R., Pandey, M. K., Singh, P. K., Bohra, A., Guo, B., & Varshney, R. K. (2023). Application of CRISPR/Cas9-mediated gene editing for abiotic stress management in crop plants. *Frontiers in Plant Science*, *14*, 1157678. DOI: 10.3389/fpls.2023.1157678

Kumlehn, J., Pietralla, J., Hensel, G., Pacher, M., & Puchta, H. (2018). The CRISPR/Cas revolution continues: From efficient gene editing for crop breeding to plant synthetic biology. *Journal of Integrative Plant Biology*, *60*(12), 1127–1153. DOI: 10.1111/jipb.12734

Ma, X., Zhu, Q., Chen, Y., & Liu, Y.-G. (2016). CRISPR/Cas9 platforms for genome editing in plants: Developments and applications. *Molecular Plant*, *9*(7), 961–974. DOI: 10.1016/j.molp.2016.04.009

Miller, J. C., Tan, S., Qiao, G., Barlow, K. A., Wang, J., Xia, D. F., & Hinkley, S. J. (2011). A TALE nuclease architecture for efficient genome editing. *Nature Biotechnology*, *29*(2), 143–148. DOI: 10.1038/nbt.1755

Mishra, R., & Zhao, K. (2018). Genome editing technologies and their applications in crop improvement. *Plant Biotechnology Reports*, *12*(2), 57–68. DOI: 10.1007/s11816-018-0472-0

Moore, R., Chandrahas, A., & Bleris, L. (2014). Transcription activator-like effectors: A toolkit for synthetic biology. *ACS Synthetic Biology*, *3*(10), 708–716. DOI: 10.1021/sb400137b

Negi, C., Vasistha, N. K., Singh, D., Vyas, P., & Dhaliwal, H. S. (2022). Application of CRISPR-mediated gene editing for crop improvement. *Molecular Biotechnology*, *64*(11), 1198–1217. DOI: 10.1007/s12033-022-00507-y

Nemudryi, A. A., Valetdinova, K. R., Medvedev, S. P., & Zakian, S. M. (2014). TALEN and CRISPR/Cas Genome Editing Systems: Tools of Discovery. *Acta Naturae*, *6*(3), 19–40. DOI: 10.32607/20758251-2014-6-3-19-40

Ochiai, H., & Yamamoto, T. (2014). Genome editing using zinc-finger nucleases (ZFNs) and transcription activator-like effector nucleases (TALENs). In *Targeted Genome Editing Using Site-Specific Nucleases: ZFNs, TALENs, and the CRISPR/Cas9 System* (pp. 3–24). Springer Japan., DOI: 10.1007/978-4-431-55200-4_1

Oliva, R., Ji, C., Atienza-Grande, G., Huguet-Tapia, J. C., Perez-Quintero, A., Li, T., Eom, J.-S., Li, C., Nguyen, H., Liu, B., Auguy, F., Sciallano, C., Luu, V. T., Dossa, G. S., Cunnac, S., Schmidt, S. M., Slamet-Loedin, I. H., Vera Cruz, C., Szurek, B., & Yang, B. (2019). Broad-spectrum resistance to bacterial blight in rice using genome editing. *Nature Biotechnology*, *37*(11), 1344–1350. DOI: 10.1038/s41587-019-0267-z

Ousterout, D. G., & Gersbach, C. A. (2016). The Development of TALE Nucleases for Biotechnology. *Methods in Molecular Biology (Clifton, N.J.)*, *1338*, 27–42. DOI: 10.1007/978-1-4939-2932-0_3

Puchta, H., Jiang, J., Wang, K., & Zhao, Y. (2022). Updates on gene editing and its applications. *Plant Physiology*, *188*(4), 1725–1730. DOI: 10.1093/plphys/kiac032

Rasheed, A., Barqawi, A. A., Mahmood, A., Nawaz, M., Shah, A. N., Bay, D. H., Alahdal, M. A., Hassan, M. U., & Qari, S. H. (2022). CRISPR/Cas9 is a powerful tool for precise genome editing of legume crops: A review. *Molecular Biology Reports*, *49*(6), 5595–5609. DOI: 10.1007/s11033-022-07529-4

Ravikiran, K. T., Thribhuvan, R., Sheoran, S., Kumar, S., Kushwaha, A. K., Vineeth, T. V., & Saini, M. (2023). Tailoring crops with superior product quality through genome editing: An update. *Planta*, *257*(5), 86. DOI: 10.1007/s00425-023-04112-4

Sahu, A., Verma, R., Gupta, U., Kashyap, S., & Sanyal, I. (2024). An overview of targeted genome editing strategies for reducing the biosynthesis of phytic acid: An anti-nutrient in crop plants. *Molecular Biotechnology*, *66*(1), 11–25. DOI: 10.1007/s12033-023-00722-1

Schaart, J. G., van de Wiel, C. C. M., & Smulders, M. J. M. (2021). Genome editing of polyploid crops: Prospects, achievements and bottlenecks. *Transgenic Research*, *30*(4), 337–351. DOI: 10.1007/s11248-021-00251-0

Scheben, A., Wolter, F., Batley, J., Puchta, H., & Edwards, D. (2017). Towards CRISPR/Cas crops–bringing together genomics and genome editing. *The New Phytologist*, *216*(3), 682–698. DOI: 10.1111/nph.14702

Shukla, V. K., Doyon, Y., Miller, J. C., DeKelver, R. C., Moehle, E. A., Worden, S. E., Mitchell, J. C., Arnold, N. L., Gopalan, S., Meng, X., Choi, V. M., Rock, J. M., Wu, Y.-Y., Katibah, G. E., Zhifang, G., McCaskill, D., Simpson, M. A., Blakeslee, B., Greenwalt, S. A., & Urnov, F. D. (2009). Precise genome modification in the crop plant maize using zinc-finger nucleases. *Nature*, *459*(7245), 437–441. DOI: 10.1038/nature07992

Song, X., Dang, C., Wang, F., Yao, H., Song, Q., & Ye, G. (2026). *CRISPR/Cas gene editing for plant resistance to pests, diseases, and herbicide: application, risk assessment, and safety evaluation*. GM Crops Food. 2026 Mar 16;17(1):2641893. . eCollection 2026.DOI: 10.1080/21645698.2026.2641893

Sun, N., & Zhao, H. (2013). Transcription activator-like effector nucleases (TALENs): A highly efficient and versatile tool for genome editing. *Biotechnology and Bioengineering*, *110*(7), 1811–1821. DOI: 10.1002/bit.24890

Tabassum, J., Ahmad, S., Hussain, B., Mawia, A. M., Zeb, A., & Ju, L. (2021). Applications and potential of genome-editing systems in rice improvement: Current and future perspectives. *Agronomy (Basel)*, *11*(7), 1359. DOI: 10.3390/agronomy11071359

Townsend, J. A., Wright, D. A., Winfrey, R. J., Fu, F., Maeder, M. L., Joung, J. K., & Voytas, D. F. (2009). High-frequency modification of plant genes using engineered zinc-finger nucleases. *Nature*, *459*(7245), 442–445. DOI: 10.1038/nature07845

Tufail, U., Kanwal, R., Asghar, A., Ali, B., Shabbir, F., Rehman, K., & Raza, A. (2026). The Role of Calcium in Salinity Stress Alleviation: Mechanisms and Implications. *Plant Science Horizons*, *1*(1), 41–58. DOI: 10.66835/psh.2026.e0007

Urnov, F. D., Miller, J. C., Lee, Y. L., Beausejour, C. M., Rock, J. M., Gregory, P. D., & Holmes, M. C. (2005). Highly efficient endogenous human gene correction using designed zinc-finger nucleases. *Nature*, *435*(7042), 646–651. DOI: 10.1038/nature03556

Urnov, F. D., Rebar, E. J., Holmes, M. C., Zhang, H. S., & Gregory, P. D. (2010). Genome editing with engineered zinc finger nucleases. *Nature Reviews. Genetics*, *11*(9), 636–646. DOI: 10.1038/nrg2842

Wang, K., Gong, Q., & Ye, X. (2020). Recent developments and applications of genetic transformation and genome editing technologies in wheat. *Theoretical and Applied Genetics*, *133*(5), 1603–1622. DOI: 10.1007/s00122-019-03464-4

Wang, L., Yu, S., Tong, C., Zhao, Y., Liu, Y., Song, C., & Hua, W. (2014). Genome sequencing of the high oil crop sesame provides insight into oil biosynthesis. *Genome Biology*, *15*(2), R39. DOI: 10.1186/gb-2014-15-2-r39

Xia, X., Cheng, X., Li, R., Yao, J., Li, Z., & Cheng, Y. (2021). Advances in application of genome editing in tomato and recent development of genome editing technology. *Theoretical and Applied Genetics*, *134*(9), 2727–2747. DOI: 10.1007/s00122-021-03874-3

Xiong, J. S., Ding, J., & Li, Y. (2015). Genome-editing technologies and their potential application in horticultural crop breeding. *Horticulture Research*, *2*(1), 15019. DOI: 10.1038/hortres.2015.19

Yao, R., Liu, D., Jia, X., Zheng, Y., Liu, W., & Xiao, Y. (2018). CRISPR-Cas9/Cas12a biotechnology and application in bacteria. *Synthetic and Systems Biotechnology*, *3*(3), 135–149. DOI: 10.1016/j.synbio.2018.09.004

Yıldırım, K., Miladinović, D., Sweet, J., Akin, M., Galović, V., Kavas, M., Zlatković, M., & de Andrade, E. (2023). Genome editing for healthy crops: Traits, tools and impacts. *Frontiers in Plant Science*, *14*, 1231013. DOI: 10.3389/fpls.2023.1231013

Zaman, W., Khalil, A. A., & Amin, A. (2025). TALEN-Interceded Genome Editing in Plants: Unveiling New Frontiers in Secondary Metabolite Improvement and Genetic Diversity. *Plants*, *14*(19), 3024. DOI: 10.3390/plants14193024

Zhan, X., Lu, Y., Zhu, J.-K., & Botella, J. R. (2021). Genome editing for plant research and crop improvement. *Journal of Integrative Plant Biology*, *63*(1), 3–33. DOI: 10.1111/jipb.13063

Zhang, H., Zhang, J., Lang, Z., Botella, J. R., & Zhu, J.-K. (2017). Genome editing—Principles and applications for functional genomics research and crop improvement. *Critical Reviews in Plant Sciences*, *36*(4), 291–309. DOI: 10.1080/07352689.2017.1402989

Zhang, Y., Ren, Q., Tang, X., Liu, S., Malzahn, A. A., Zhou, J., Wang, J., Yin, D., Pan, C., Yuan, M., Huang, L., Yang, H., Zhao, Y., Fang, Q., Zheng, X., Tian, L., Cheng, Y., Le, Y., McCoy, B., & Qi, Y. (2021). Expanding the scope of plant genome engineering with Cas12a orthologs and highly multiplexable editing systems. *Nature Communications*, *12*(1), 1944. DOI: 10.1038/s41467-021-22330-w

Zhao, P., Feng, Z., Cai, L., Phurbu, D., Duan, W., Xie, F., Li, X., & Liu, F. (2024). Development of an RPA-CRISPR/Cas12a Assay for Rapid and Sensitive Diagnosis of Plant Quarantine Fungus Setophoma terrestris. *Journal of Fungi (Basel, Switzerland)*, *10*(10), 716. DOI: 10.3390/jof10100716

Zhou, J., Li, D., Wang, G., Wang, F., Kunjal, M., Joldersma, D., & Liu, Z. (2020). Application and future perspective of CRISPR/Cas9 genome editing in fruit crops. *Journal of Integrative Plant Biology*, *62*(3), 269–286. DOI: 10.1111/jipb.12793

Zou, S., Xu, Y., Li, Q., Wei, Y., Zhang, Y., & Tang, D. (2023). Wheat powdery mildew resistance: From gene identification to immunity deployment. *Frontiers in Plant Science*, *14*, 1269498. DOI: 10.3389/fpls.2023.1269498

Chapter 3
Enhancing Crop Yield Through Gene Editing

Waseem Ahmed Khattak
http://orcid.org/0009-0004-9525-5627
Quaid-i-Azam University, Islamabad, Pakistan

Muhammad Anas
http://orcid.org/0000-0001-5127-8043
University of Selcuk, Konya, Turkey

Amjid Khan
http://orcid.org/0009-0001-5475-5156
University of South Africa, South Africa

Banzeer Ahsan Abbasi
http://orcid.org/0000-0003-3036-4481
Rawalpindi Women University, Pakistan

Zahid Ullah
University of Swat, Charbagh, Pakistan

Kashaf Khan
Bacha Khan University, Charsadda, Pakistan

ABSTRACT

Gene editing can substantially enhance crop yield, strengthening global food security where traditional breeding is often too slow to match climate change, pests, and diseases. Technologies such as CRISPR-Cas9, TALENs, and ZFNs enable precise, targeted changes that improve drought and pest resistance, nutrient-use efficiency, and growth rate. By identifying and modifying genes controlling key traits, researchers have developed higher-performing crops, including rice, wheat, maize, and tomato. Major applications include boosting stress tolerance, improving pest and disease resistance, increasing nutrient uptake, and shortening growth cycles. Key constraints remain—editing efficiency, off-target effects, cost, and regulatory barriers—alongside ethical concerns about ecological impacts and public perceptions. Future progress will likely depend on multi-gene editing, climate-adaptive trait stacking, and integration with precision agriculture and AI-driven decision tools, supporting higher yields alongside sustainability and resilience.

DOI: 10.4018/979-8-3373-9534-0.ch003

INTRODUCTION

Brief Overview of the Importance of Crop Yield in Global Food Security

As will be highlighted throughout this chapter, crop yield is central to food security worldwide. With this consideration present, it therefore suggests that in the next couple of years, by 20501 the global population will increase to over nine billion people (Anas et al., 2025; Awulachew & Weldehawaria, 2024). This is because we find ourselves struggling in agriculture to attend to the food needs of this growing population. Due to the scarcity of land for cultivation and other conditions such as climate change, water rationing, and decline in soil fertility, it is vital to increase the yield of crops (Nikolaou et al., 2020). The demand for increasing yields per unit area is another imperative in the modern world. According to the FAO, global food production must increase by 70% by 2050 to cope with the growing population. Nevertheless, achieving this goal using only conventional farming techniques is a daunting task (Roy et al., 2024; van Dijk et al., 2021). Therefore, this subject presents a need for change through innovation, as the consequences cannot be compared to anything other than hunger and malnutrition. Yield, or the number of crops per unit area, depends on factors such as the quality of seeds that have been planted, climate conditions, methods used to control pests, and techniques used during growth (Pandey & Mishra, 2024; Tandzi & Mutengwa, 2020). Conventional farming techniques have been practiced for thousands of years, but due to increasing challenges on farms, these practices have proven to be very risky. Therefore, the incorporation of current technologies in farm activities has become relevant for boosting food production without exerting more pressure on the environment (Muhie, 2022).

The Role of Traditional Breeding Methods and Their Limitations

In the past, selection has played an important role in enhancing crop productivity. It involves planting the seeds of two different plants that have the respective genes of the desirable characteristics of each parent plant and growing them into the next generation. Through selective breeding, where farmers grow crops of the best quality, they have been able to make minor enhancements to the variety (Kaiser et al., 2020). It has contributed to the development of some of the crop types we use today, such as high-yield varieties of wheat, maize, and rice. However, it also has some drawbacks that make it less effective than modern breeding methods (Benavente & Giménez, 2021). First, it is a slow process. It may literally take a number of years or even decades to increase yield significantly because it involves conventional methods where cross-pollination is used, and, as breeders know, it is a rather unpredictable method. In addition, conventional breeding methods require large companics to breed the population, which is tedious and costly. Furthermore, selective breeding is constrained within a certain level of variation in the range of a specific crop variety. As such genes are fixed in a population, it becomes challenging to locate suitable parent plants for the crossbreeding process (Begna, 2021). This restriction is almost rigorous for crops with low genetic variability or for crops that grow under extremely hot atmospheric conditions defined by climatic factors or diseases. Therefore, there is a need to develop more accurate and faster methods to increase yield to meet the challenges introduced by environmental factors (Anas et al., 2024; Pixley et al., 2023).

Introduction to Gene Editing as a Revolutionary Tool in Agriculture

These techniques have emerged as promising innovations in agriculture, representing a new method for controlling crop genomes. Unlike selective breeding, which involves a combination of two different individuals, gene editing enables the modification of an organism's DNA at particular locations in its genetic material (Ahmadikhah et al., 2025). This provides an opportunity to improve or add characteristics to crops in a manner that was not possible using normal breeding techniques. CRISPR-Cas9 or (Clustered Regularly Interspaced Short Palindromic Repeats with CRISPR-associated protein 9) has become particularly popular in making desired changes in the DNA of living organisms, including plants. Thus, speaking of CRISPR-Cas9, the specificity allows researchers to "cut out" the gene at the desired position and either add, remove, or change genes responsible for such traits as drought tolerance, pest resistance, or faster growth (Hamdan & Tan, 2024; Kafle, 2023; Ndudzo et al., 2024). This technique is much more accurate and quicker than other methods of genetic modification, whereby a gene is inserted into the DNA of a plant. Gene editing has several advantages over traditional breeding technique as listed below; The first one is performance gain, it is specifically more responsive than traditional approaches, and here are its benefits. This would have required extensive breeding processes and years, but is now accomplishable within a short span of time (Pandey et al., 2022). For instance, gene editing can produce plants with favorable characteristics in only one generation, rather than through several generations, as practiced in traditional breeding. In addition, gene editing has the advantage of inserting genes that already exist within crops; thus, there is a possibility of achieving desirable characteristics that may not exist naturally within certain species of crops. This offers the opportunity to enhance the varieties of plants whose alteration is normally challenging, such as some grains or beans (Shah et al., 2018).

Gene editing is also highly precise, and one may distinguish between wet and dry lab gene editing, the former being relatively more accurate than the latter. In our case, we can avoid other parts of the DNA getting altered in the process, since this is a major issue when it comes to other forms of genetic modification. This is because the potential for increasing the level of certain genes is high enough to increase yield-related traits, although undesirable characteristics may also appear. For instance, CRISPR-Cas9 has been applied to modify genes in crops to enhance their susceptibility to diseases, stress, and efficient utilization of nutrients (Rasheed et al., 2021). However, there are several points that show that gene editing can fully overcome some of the critical issues of the contemporary agricultural industry. There are major challenges, such as climatic and environmental changes, which are causing unpredictable weather conditions, resulting in droughts, heat, and floods. These crops can be developed using biotechnology to create resistance to stress factors with a view to preserving crop output in the face of changing climatic conditions (Alotaibi, 2023; Benitez-Alfonso et al., 2023). In addition, enhanced gene editing can improve the tolerance of crops to pests and diseases, thereby reducing the use of chemical pesticides and their negative impact on the environment (KhokharVoytas et al., 2023).

Fundamentals of Gene Editing

Researchers use this technology to edit targeted areas of genetic material in living organisms through precise DNA changes. Scientists can now edit DNA material in organismal genomes with superior accuracy. Genetic modification occurs through natural processes and laboratory-generated techniques using different editing tools (Kopczyńska et al., 2025). In contrast to regular genetic alteration methods, scientists can adjust particular genes without introducing foreign DNA into the organism's genome using this technique. Scientists apply gene-editing techniques across different applications to make DNA modifications for medical, agricultural, and biological needs. Scientists use gene-editing tools to improve crop farming by strengthening agricultural value, fighting plant diseases and drought conditions, and making plants work better with nutrients (Ansori et al., 2023). Gene-editing tools can transform agriculture by solving problems of damaged soil and crops, climate change, and feeding the world's population (Fig 1).

Figure 1. Transforming agriculture through gene editing.

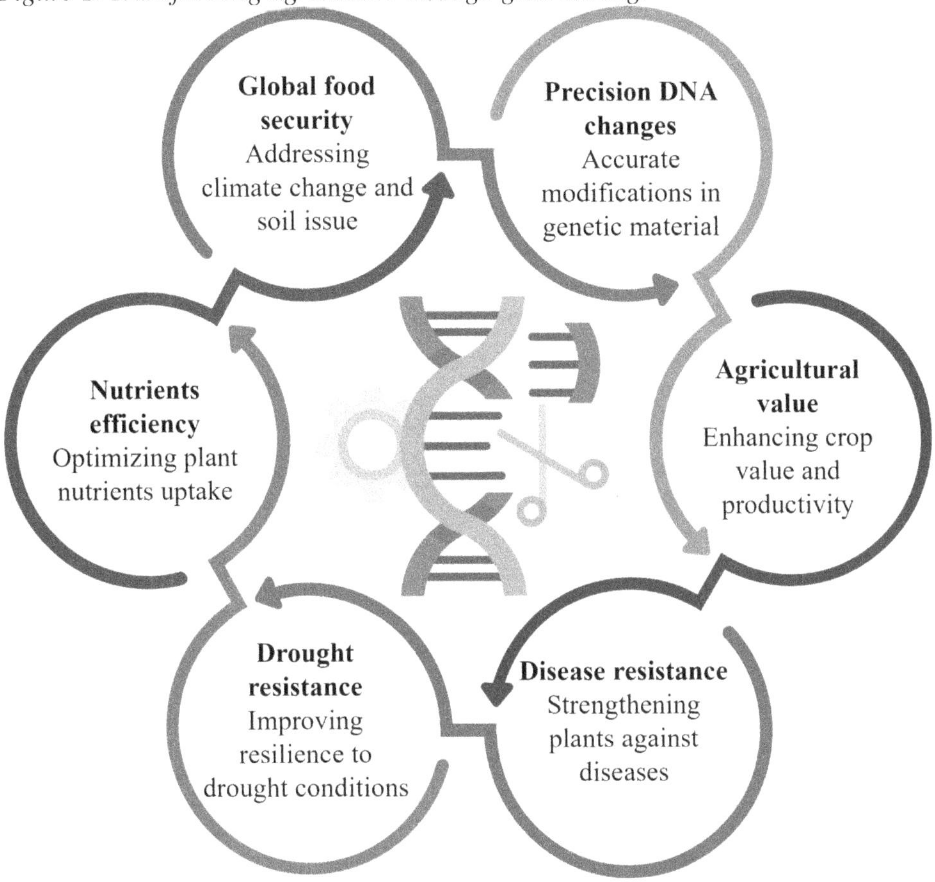

Key Technologies: CRISPR-Cas9, TALENs, and ZFNs

Three dominant gene-editing techniques, CRISPR-Cas9, TALENs, and ZFNs, currently provide effective solutions in molecular biology. These advanced tools enable researchers to make accurate, targeted changes to genes within plants and other life forms (Shamshirgaran et al., 2022). CRISPR-Cas9 is the modern base technology for biotechnology gene editing. Bacteria originally used CRISPR as an adaptive immune response to fight viruses. A guide RNA directs the CRISPR-Cas9 system to cut DNA at a specific location. Following guide RNA instructions, Cas9 cuts through DNA at the chosen site to enable precise changes (Akram et al., 2023). Scientists favor this system because it works well in many situations owing to its basic design and exact targeting abilities. The system allows scientists to advance crop research through specific modifications that improve drought tolerance, protect against pests, and boost growth efficiency (Kumar et al., 2023).

Transcription Activator-Like Effector Nucleases represent an effective method for gene editing. TALENs function as engineered proteins that identify particular genetic areas and produce breaks at desired locations. These proteins have two major features: a DNA-binding area that attaches to precise parts of the genetic material and a nuclease area that produces DNA damage (Shamshirgaran et al., 2022). TALENs can edit specific genes effectively, but developers must expend extra resources to create and use them compared to CRISPR technology. Scientists have applied TALEN technology to enhance crop species traits, such as immunity against diseases and other crop features (Bhardwaj & Nain, 2021).

Zinc Finger Nucleases are the earliest tools for gene editing. These enzymes contain zinc finger regions that attach to specific DNA parts, alongside an enzymatic segment that cuts double-stranded DNA. Unlike other gene-editing tools, the precision of ZFNs is inferior to that of innovations such as CRISPR (Bhagtaney & Sundarrajan, 2023). Scientists have applied ZFNs to numerous plant species for GM development and have achieved successful results, such as resistance to pests and illnesses. Since CRISPR-Cas9 led gene-editing methods are used today, TALENs, ZFNs, and CRISPR offer special benefits and address specific agricultural needs (Rafiq, 2024).

How Gene Editing Differs from Genetic Modification (GMOs)

Although gene editing is related to genetic modification of GMOs, both processes have unique functions. The two methods use genetic transformation to add desired features, but they differ in their mechanisms and precisions. Pioneers modify an organism's genetic material through targeted processes. The Gene Modification process involves inserting foreign DNA from various species into the genetics of target living organisms. Scientists use Agrobacterium-mediated transformation or gene gun methods to introduce foreign DNA into plant genomes (Rozas et al., 2022). Introducing foreign genes into GMOs yields useful outcomes but sparks fears related to safety problems, environmental pressure, and official permit challenges. Using gene-editing methods, developers can make precise DNA edits to an organism (Subejano et al., 2024). Gene editing modifies DNA internally and leaves any new genetic material. This method alters the natural order of genetic material in living organisms. Scientists can introduce minor modifications to DNA by deleting one base pair or making small changes in gene function. Gene-editing systems produce precise results that minimize unwanted effects during DNA manipulation compared to traditional GMO methods. Several governmental bodies do not treat gene-edited life forms as GMOs because they operate without including external DNA (Nadakuduti & Enciso-Rodríguez, 2021). Gene editing is regarded as a better choice than GMO methods because it offers precise manipulation that receives easier regulatory approval and public acceptance. Organisms with beneficial production traits can be produced through gene modification without foreign DNA, which helps avoid many GMO-related controversies (Gupta et al., 2021).

The Science behind Gene Editing: How It Works at the Molecular Level

To change living cells, scientists use tools that correct DNA structures at the basic molecular level. Scientists have developed a guide sequence that identifies the CRISPR-Cas9 tool required to access the target DNA (Akram et al., 2023).

DNA Targeting and Cleavage:

Gene editing begins with the selection of an exact location in the genome for this process. Scientists design guide RNA sequences to find and match the targeted DNA regions when using CRISPR-Cas9 technology. The Cas9 protein and gRNA mixture locate and cut the DNA target region. Cas9 reaches the DNA inside the target cell, where it locates its gRNA target sequence to cut the genetic material at that site (Newman et al., 2020).

DNA Repair and Modification

When DNA is cut, the cell starts its regular system of repair operations. Cells use two repair pathways, NHEJ and HR, to mend broken DNA. The cell uses NHEJ to repair DNA breaks by connecting cut ends, which tend to generate small alterations that damage genetic instructions. Scientists typically use this method to delete specific genes from the genome. Instead of NHEJ, HR enables cells to repair DNA breaks with the assistance of a template to accurately change the genetic makeup (Oh & Myung, 2022). Scientists use DNA repair systems to introduce desired changes or replace existing genes for specific purposes. By targeting genes through gene-editing technology, scientists can create better-tolerated crop varieties that grow faster and absorb more nutrients. Genetic editing permits exact precision in modifying animal and crop DNA for better farming practices (Sathee et al., 2022).

KEY GENES INVOLVED IN CROP YIELD

Overview of Plant Traits That Contribute to Yield

Plants that yield crops suffer from a wide range of influencing elements that move from natural conditions to inherent genetic aspects. Improvement of plant traits linked to growth production and survivorship under climate variations represents a strategy to enhance yield performance. The important characteristics that lead to higher crop yields include resistance to drought conditions, protection from pests, and accelerated plant development (Burgess et al., 2023).

Drought Resistance:

Water shortages are a major threat to global crop cultivation. Plant growth becomes restricted, and seed production decreases, while flowering efficiency declines in drought-stricken environments, leading to very low yield counts (Seleiman et al., 2021). Plants that survive in dry zones develop adaptive processes that enable water conservation and resistance to dry conditions (Wang et al., 2025). Plants have developed three mechanisms of drought resistance, which consist of controlled stomatal function combined with osmotic mechanisms and specific root adaptations. Crops can become more resistant to future drought patterns caused by climate change through the improvement of genes that create drought-tolerant characteristics (Liao et al., 2025).

Pest Resistance:

Worldwide crop loss is primarily caused by pests and pathogens. Plants incorporate two types of protection against enemies, including physical elements such as tough cell walls and defensive compounds that produce poisonous substances (Tlak Gajger & Dar, 2021; Yactayo-Chang et al., 2020). Currently, agricultural systems maintain most crops with reduced natural pest protection features after domestication. Gene-editing techniques enhance crop defenses to make them resistant to damage and protect them from requiring chemical pesticides (Abdul Aziz et al., 2022).

Faster Growth:

Plants that grow at faster rates mature sooner, which is advantageous when operating in locations with limited growing windows or culturing annually (Roy et al., 2025). The combination of growth hormone gene regulation by gene editing produces accelerated plant development, as well as better photosynthetic efficiency and auxin or gibberellin hormone regulation. The accelerated maturation rate of modified crops enhances both productivity from each field unit and the number of harvests per agricultural year (Fudjoe et al., 2025; Wang et al., 2025).

Identification of Target Genes for Improvement in Various Crops

Researchers use modern gene-editing methods to precisely change the genes that regulate vital yield-related characteristics. Modern crop enhancement programs require the identification of major genes that would lead to improvements. Several vital target genes in various crops confer resistance against diseases, heat tolerance, and improved nutritional efficiency (Chachar et al., 2025).

Genes for Disease Resistance

Plants that grow in the fields are at risk of contracting fungal, bacterial, viral, and nematode diseases, which substantially diminish crop yields. The implementation of disease-resistant crops is vital because it guarantees food stability. The disease resistance trait in plants depends on the well-known pathogenesis-related gene 1 (PR-1), which functions as an immune response regulator (Manzoor et al., 2024). Modern agricultural methods involve modifying wheat rust defense and related resistance genes to upgrade plant immunity against wheat rust, which threatens wheat cultivation worldwide. The addition of resistance genes, known as R genes, in crops such as tomatoes and rice enables immunity against bacterial blight and Fusarium wilt infections (Limenie & Alehegn, 2025).

Genes for Heat Tolerance

Increased heat tolerance enables various crops to remain viable because climate change increases global temperatures. Heat stress causes physiological issues in plants because it negatively affects protein synthesis, membrane integrity, and enzyme activity (Yadav et al., 2022). The key genes that mediate heat tolerance are heat shock proteins (HSPs), which stabilize protein structures and defend cellular frameworks from heat damage. Researchers have used gene editing methods to increase heat shock factor expression to enhance crop heat tolerance in rice, wheat, and maize (Lang et al., 2021).

Genes Involved in Nutrient Efficiency:

Plants demonstrate nutrient-use efficiency when they effectively utilize both nitrogen and phosphorus. Efficient nutrient management in plants is important in agricultural production because it advances both yield production strength and fertilizer pollution reduction effectiveness. Scientists are now using the rice gene NRT1.1B for targeting efforts to enhance NUE. Crop efficiency in nutrient absorption improves in nutrient-poor soils once scientists edit genes that regulate nutrient sensing and transport mechanisms (Ju et al., 2024).

Examples of Successfully Edited Genes in Crop Species

The applications of gene editing have resulted in multiple improved crop varieties that exhibit stronger disease resistance, better nutrient utilization, and adaptable environmental characteristics.

Drought-Tolerant Crops:

Gene-edited rice crops exemplify success in the development of drought-tolerant varieties. Researchers have directed their efforts toward manipulating the OsDREB1A gene in rice varieties to improve drought tolerance. The introduction of extra copies of this genetic sequence allowed rice plants to grow optimally under limited water availability conditions (Panda et al., 2021). The P5CS gene responsible for proline synthesis (a water retention compound) has been edited through various biotechnology applications in wheat and maize crops to boost drought tolerance. Drought-resistant plant varieties exhibit enhanced yield output when water conditions become scarce (Kiran, 2022).

Pest-Resistant Crops:

Bt cotton plants produced through genetic editing have additional advantages over traditionally engineered crops because the process enhances the natural insect-killing capabilities of the plants. This modification enables farmers to decrease their dependence on chemical pesticides, making cotton cultivation more sustainable (Bhandari & Paudel, 2024). Gene-editing techniques have led to higher fungal pathogen resistance in maize by modifying genes that recognize pathogens and activate immune responses (Pathi, 2021).

Improved Nutrient Efficiency in Crops

Rice plants exhibit enhanced nitrogen utilization and improved nutrient uptake because scientists edited the NRT1.1B gene to achieve these increases. Scientists have altered rice plants to improve their development and production when exposed to sparse nitrogen levels to decrease their dependency on nitrogen fertilizer applications (Hu et al., 2023). Gene editing in wheat has involved TaPDR1 management because it controls phosphorus uptake, making wheat crops sustainable in low-phosphorus areas (Maharajan et al., 2021).

Table 1. Summary of Studies Utilizing CRISPR/Cas9 for Crop Improvement.

Crop	Target Gene(s)	Trait Improved	Outcome	Reference
Rice	OsCKX2	Grain yield	Increased grain yield by enhancing cytokinin levels	
Wheat	TaCKX1	Grain yield	Enhanced grain yield through cytokinin accumulation	(Jabłoński et al., 2020)
Maize	ZmCKX3	Grain yield	Improved grain yield by reducing cytokinin degradation	(Vyroubalová et al., 2009)
Tomato	SFT	Fruit size and yield	Increased fruit size and overall yield	(Cui et al., 2022)
Soybean	GmFT2a	Flowering time	Altered flowering time for improved adaptation	(Su et al., 2024)
Potato	StGBSS	Starch composition	Produced amylose-free starch for industrial use	(Mali, Dutta, & Zinta, 2023)
Barley	HvHina	Grain hardness	Increased grain hardness suitable for malting	(Aubert et al., 2018)
Cassava	CYP79D1	Cyanogenic glycosides	Reduced cyanide content for safer consumption	(Juma et al., 2022)
Banana	PPO	Browning	Developed non-browning bananas to reduce waste	(Tripathi, Ntui, & Tripathi, 2024)
Tomato	Unknown	Plant architecture	Created compact plants suitable for vertical farming	(Ndudzo et al., 2024)
Rice	OsBADH2	Aroma	Enhanced fragrance by increasing 2-acetyl-1-pyrroline	(Tian et al., 2023)
Wheat	TaMLO	Powdery mildew resistance	Conferred resistance by knocking out susceptibility gene	(S. Li et al., 2022)
Rice	OsCOP1	UV-B tolerance	Improved tolerance to UV-B radiation	(F. Chen et al., 2024)
Barley	HvARE1	Nitrogen use efficiency	Enhanced nitrogen uptake and utilization	(Karunarathne, Han, Zhang, & Li, 2022)
Tomato	Unknown	Sugar content	Increased sweetness by targeting sugar-regulating genes	(L. Yuan et al., 2024)
Rice	OsAAP3	Tiller number	Increased tiller number leading to higher grain yield	(J. Wang et al., 2019)
Wheat	TaDREB2	Drought tolerance	Enhanced drought resistance through gene editing	(Mei et al., 2022)
Maize	ZmARGOS8	Stress tolerance	Improved stress tolerance and yield stability	(Shi et al., 2015)
Tomato	SlFSR	Shelf life	Extended shelf life by delaying fruit ripening	(L. Zhang et al., 2018)

continued on following page

Table 1. Continued

Crop	Target Gene(s)	Trait Improved	Outcome	Reference
Soybean	GmFAD2-1A/B	Fatty acid composition	Increased oleic acid content for healthier oil	(Song, Taylor, & Zhang, 2023)
Rice	OsSWEET14	Bacterial blight resistance	Enhanced resistance to bacterial blight	(Zhong, Xu, & Rao, 2024)
Wheat	TaERF3	Drought tolerance	Improved tolerance to drought temperatures	(Rong et al., 2014)
Maize	ZmPSY1	Provitamin A content	Increased provitamin A levels in kernels	(Maqbool, Aslam, Beshir, & Khan, 2018)
Tomato	SlGLK2	Fruit color and quality	Enhanced fruit coloration and nutritional quality	(Lupi et al., 2019)

Faster Growth and Higher Yields:

Gene editing has achieved significant progress in enhancing corn (maize) growth rates and improving yield production. Scientists have enhanced maize production by successfully modifying the GPC-1 gene, which controls growth and grain development, through genetic editing. Gene editing has produced corn varieties that display faster growth patterns and better performance in various environmental settings, resulting in higher yield capacity (Hernandes-Lopes et al., 2023; Raza et al., 2025). Gene editing has unlimited capacity to enhance agricultural yields by controlling fundamental genetic components that affect drought defense, pest protection, nutrient optimization, and accelerated growth (KhokharVoytas et al., 2023). Precise genetic editing technology delivers an effective instrument to tackle current agricultural demands because it enables experts to develop resilient plant varieties which adapt to climate change pressures and pest threats alongside nutrient shortage conditions (Fig 2)

Figure 2. Corn can be improved through gene editing.

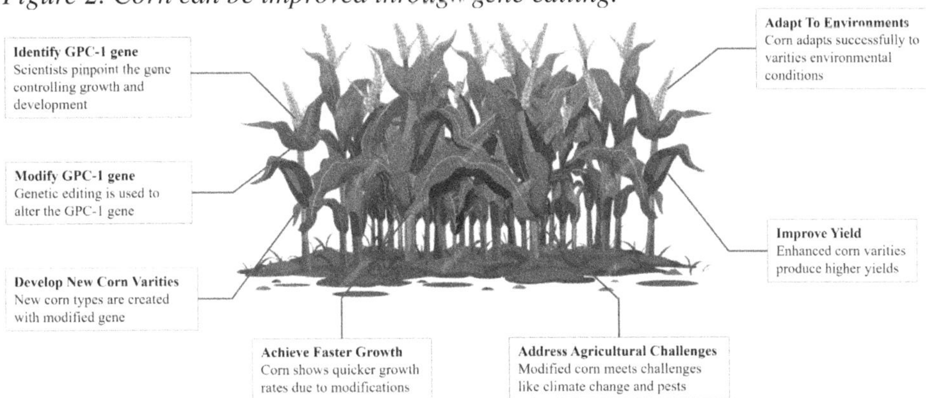

Applications of Gene Editing in Enhancing Crop Yield

Gene editing has introduced a new agricultural biotechnology period through genome sequencing precision that enhances crop trait characteristics. This section details genetic engineering applications that optimize agricultural yields by fortifying plants against pests and diseases while also providing resistance against stress and increasing nutrient acquisition rates as well as growth acceleration (Yuan et al., 2024).

Improving Pest and Disease Resistance

Crop yield faces major agricultural challenges because pests and pathogens severely damage crops. The use of traditional pest and disease control methods through chemical pesticides has become unsustainable because of environmental issues and resistant pest development (Tudi et al., 2021). The use of gene-editing techniques provides crop breeders with a performance-driven method to achieve better pest and disease protection in agricultural developments (Kumar et al., 2024).

Editing Genes to Enhance Natural Resistance

Using CRISPR-Cas9 gene editing techniques, scientists can make precise genetic changes to plants, enhancing their natural defenses against pests and pathogens. Scientists have developed crops with reinforced defense mechanisms by altering genes that determine plant immunity, thereby making the crops resistant to diverse pests and diseases. Studying and modifying plant PR-1 genes enhances resistance to fungal diseases, whereas gene changes in secondary metabolite production genes lead to plants becoming less appealing or toxic to natural predators (Rato et al., 2021).

Case Studies of Crops with Improved Resistance

Multiple studies have shown that gene-editing applications successfully improve crop resistance to pests and diseases. Gene editing has enabled Bt cotton manufacturers to develop this pest-resistant variety that produces poisonous substances effective against cotton bollworms. The implementation of Bt cotton reduces the demand for chemical pesticides because it decreases both production expenses and environmental disruption (Nagaraj et al., 2024). CRISPR/Cas9-edited rice varieties are resistant to bacterial blight due to the implementation of Xa21 resistance genes. The modified plants show improved resistance to destructive organisms, resulting in greater output and more environmentally friendly cultivation methods (Tyagi et al., 2021).

Enhancing Stress Tolerance

The combination of drought, heat, and salinity stress creates immense obstacles for agriculture, as climate change is intensifying their frequency and severity (Dos Santos et al., 2022). Through gene editing, scientists can develop plant varieties that demonstrate enhanced adaptability to stressful environmental conditions, as this adaptation benefits yields when facing adverse environmental conditions. Developing plants through genetic alteration to gain tolerance to drought conditions and increased heat resistance and salt resistance tolerance is important. Through gene editing, researchers have been able to find and redesign genes that manage stress responses, leading to greater crop resistance (KhokharVoytas et al., 2023). The OsDREB1A gene and similar genetic material from rice were modified using systematic editing techniques to improve drought tolerance by enhancing stomatal water conservation (Usman et al., 2020). Through targeted approaches, researchers have identified HSP101 protein, among others, to enhance crop heat resistance (Ritesh Kumar et al., 2020). The NHX1 gene has been successfully used to enhance the tolerance of wheat and rice to saline conditions through its protein production for maintaining cellular osmotic balance (Solis et al., 2022).

Examples of Drought-Tolerant and Heat-Resistant Crops

Gene editing of ZmDREB2A regulates the drought response mechanism in resistant maize plants. Scientists have achieved improved drought resistance and superior production performance under limited water availability, resulting in elevated maize harvest yields (Filyushin et al., 2022). Research has shown that modified wheat strains with altered HSP101 genes have better heat tolerance, which improves yield sustainability in hot climate zones where wheat production occurs (Arif et al., 2025).

Boosting Nutrient Efficiency

Improving crop yield requires proper nutrient efficiency in both nitrogen and phosphorus uptake and utilization. The process of applying too many fertilizer products in agricultural fields creates two major problems: environmental contamination and significantly increased farm expenses (Umar et al., 2020). Through genetic modification, scientists have developed means to improve the nutrient uptake capabilities of crops, thus eliminating the need for excessive fertilizers (Sadikiel Mmbando & Ngongolo, 2024).

GENE EDITING TO INCREASE NUTRIENT UPTAKE AND UTILIZATION EFFICIENCY

Scientists use gene-editing techniques to modify genes that operate nucleotide transport along with nutrient metabolic pathways in plants. NRT1.1B gene from rice, which controls nitrate intake to increase plant efficiency in nitrogen absorption and utilization (Yawen Ju et al., 2024). The nutrient efficiency of rice plants is improved by edited versions of genes such as OsPSTOL1, which aids phosphate uptake, thereby enhancing growth in mineral-deficient soil conditions (Kettenburg et al., 2023).

Potential for Reducing Fertilizer Dependency

Gene-editing technologies enhance nutrient efficiency and help farmers decrease their dependence on synthetic fertilizers while protecting environmental interests. The reduction of fertilizers enables farm owners to save costs while creating fewer adverse ecological effects from excessive fertilizer distribution, resulting in water pollution and greenhouse gas emissions. Enhanced nitrogen and phosphorus efficiency in edited crops allows farmers to achieve sustainably high yields across different soil environments while minimizing agricultural environmental impacts (Qaim, 2020; Smyth, 2022).

Speeding Up Crop Growth

Crop productivity grows considerably when farmers use plant growth acceleration techniques, especially when their growing seasons are short or if quick harvest cycles become necessary. Through gene-editing techniques, scientists can regulate growth factors that reduce plant maturity duration (Bhatta et al., 2021).

Modification of Growth Regulators to Accelerate Growth and Reduce Time to Harvest

Growth regulators composed of gibberellins, auxins, and cytokinins serve as critical controllers that direct plant growth and developmental advancement. Gene editing technology manipulates specific hormonal genes, allowing farms to achieve faster crop development with reduced growth times. The modification of maize GPC1 helps increase growth speed and raises grain output levels through advanced photosynthetic activity and enhanced nutrient usage (Singh et al., 2023).

Impact on Crop Productivity and Food Supply Chains

Accelerating crop growth rates provide essential benefits to agricultural regions that cultivate more than one harvest each year and those experiencing rapid population expansion. The application of gene editing produces the benefits of frequent crop harvesting, in addition to higher total crop productivity levels. The accelerated growth speed of crops creates supply chain stability by decreasing the threats that destructive environmental factors like droughts and heatwaves would normally cause (Ahmar et al., 2020)

Case Studies of Gene Edited Crops

Improving various crops through gene editing promises to make food crops more resistant to environmental challenges, more economical in terms of nutrient use, and more productive (Gao, 2021). Here, we summarize case studies of several key crops in which gene editing has been applied for a range of purposes, resulting in a variety of outcomes.

Rice: Improvements in Drought Resistance and Yield Enhancement

There are billions of people around the world who eat rice as a staple food, and its production is greatly threatened by environmental stresses, particularly drought.

Drought-resistant varieties are crucial for maintaining global food security because of increasing concerns about water scarcity (Sarma et al., 2023). The modification of the OsDREB1A gene is one of the notable achievements in gene-edited rice, since it improved drought resistance. This gene is involved in the plant response to water stress. Once this gene is activated, researchers have been able to successfully produce rice plants that can hold water better, feature better root architecture, and have higher survival rates during droughts. Gene editing in rice has also improved its yield and drought tolerance (Usman et al., 2020). Researchers have edited genes in plants related to photosynthesis, such as OsSUT1, to increase the conversion of photosynthetic solar energy. As part of this, these improvements resulted in a higher grain yield per plant, especially under suboptimal growing conditions (Nazari et al., 2024). Generations of starving people in arid regions may look forward to food security as a result (Fig 3).

Figure 3. Enhancement of gene editing in rice.

Wheat: Resistance to Fungal Diseases and Environmental Stressors

Another globally important crop, wheat, is also susceptible to fungal diseases, such as wheat rust and Fusarium head blight, which can devastate entire crops. Wheat production is also affected by environmental stresses, such as heat and drought (Bhavani et al., 2021). The Yr10 gene, which codes for a resistance protein, has been inserted into wheat by editing the gene to produce a genetically modified plant that resists wheat rust. This gene marks wheat rust pathogens so that they cannot infect the crop (Dibley et al., 2024). CRISPR-Cas9 has also been used to boost the resistance of wheat against fungal diseases to decrease losses in crops from wheat rust (Waites et al., 2025). Heat and drought stress also affect wheat crops. Researchers have succeeded in modifying heat-shock protein genes, such as HSP101, and developed wheat varieties that are more tolerant to high temperatures, even in warmer growing seasons. As global temperatures continue to rise, these advancements are important (Yadav et al., 2022).

Corn: Enhancing Pest Resistance and Nitrogen Use Efficiency

One of the most widely grown crops in the world, corn is heavily plagued by pests, such as the corn rootworm and corn borer, and inefficiencies surrounding nitrogen use. The most successful application of gene editing in corn is the development of Bt corn, which produces a protein toxic to certain pests, such as the corn borer (Li et al., 2024). Therefore, this decreases the need for chemical pesticides to cultivate corn, which is a more environmentally friendly approach. CRISPR/Cas9 has been used to improve the pest resistance of corn by expressing a protein that kills itself if it is consumed (F. Chen et al., 2024). Insect-resistant corn has greatly reduced crop loss from insect damage. Corn requires high amounts of nitrogen for growth; however, the application of excess nutrients can cause environmental pollution. Enhancing nitrogen use efficiency in corn through gene editing has been achieved in the past by modifying genes such as ZmNRT1.1, which regulates nitrate uptake'. In terms of nitrogen, edited corn plants have improved nitrogen uptake and performance under nitrogen-limiting conditions, resulting in reduced fertilizer use and enhanced sustainability (Mahboob et al., 2023).

Tomatoes: Faster Growth and Disease Resistance

Although tomatoes are an important crop in global agriculture, they are vulnerable to a host of diseases and pests, and significant time is needed for tomatoes to grow and reach maturity. The ability to grow faster and with greater disease resistance can have major benefits for growers and consumers. To speed up tomato growth, genes that control plant hormones, including gibberellins, have been modified using gene editing (Gatahi, 2020). The germination of the edited tomatoes is faster, early growth is enhanced, and time to harvest is shortened, attributes relevant to markets in which speed to market is desired. Thus, productivity is greater, and the food supply chain efficiency is improved. Bacterial wilt and late blight are diseases that affect tomatoes (Nie et al., 2024). Researchers have successfully improved disease resistance in tomatoes using CRISPR-based gene editing. An example is the modification of the SlHSP70 gene, which encodes a heat shock protein involved in stress tolerance, to improve the plant's resistance to fungal pathogens, such as late blight. This allows tomatoes to be more immune to disease outbreaks, requires fewer chemical treatments, and helps make the crop more sustainable (Ilyas, 2025).

Outcomes, Challenges, and Lessons Learned

These case studies show the great potential of gene editing in increasing crop yield and resilience. However, these advancements come with challenges in terms of outcomes. Improvements have been made in practical traits such as drought resistance, pest resistance, and nutrient efficiency in all of these crops, where gene editing has played a role. Gene editing has directly led to increased yield, improved resilience to environmental stress, and reduced dependence on chemical inputs (for example, fertilizers and pesticides) in rice, wheat, corn, and tomatoes. These outcomes show that this gene-editing strategy can provide solutions to several major agricultural problems. Although successes have been achieved, crop gene editing still faces challenges that need to be overcome (Shelake et al., 2022).

Ethical Considerations and Public Perceptions

The use of gene-editing technology, particularly where the technology is being used in agriculture, has been a source of great ethical debate and concerns regarding long-term implications. The potential of gene editing to increase crop yield and resilience is without doubt, but questions about the ethical problems it poses and the public perception, regulatory, and educational issues surrounding it must be addressed (Clapp & Ruder, 2020; Munir et al., 2024).

Concerns Surrounding Gene Editing in Agriculture

Although gene editing in agriculture is a method with potential, it has also provoked concerns about its safety, effects on the environment, and unintended consequences. The main concern is the risk of unintended genetic changes (Gordon et al., 2021). Although gene-editing technologies such as CRISPR-Cas9 are precise, off-target effects on the unintended part of the genome could have unknown consequences for the plant or its consumer. All of these could change inadvertently and potentially upset ecological balances or cause unintended health effects in humans or animals that eat the crops. A concern is the monopolization of agricultural biotechnology (Zhang et al., 2020). Gene editing is mostly used in crops by private biotechnology companies, raising the specter of these companies having the power to consolidate the market and small farmers dependent on these crops and other resources. In this regard, access challenges could be created with regard to ethics; poorer regions or developing countries may not have access to these technologies and may further increase global inequalities (Borg & Policante, 2024).

Ethical Dilemmas: Gene Editing vs. Natural Evolution

A central ethical debate in agriculture regarding gene editing is between human interference and natural evolution of species. Others criticize gene editing because it involves altering a crop's genome in ways that would never happen in nature and therefore interferes with nature's own evolutionary process. This intervention is unnatural or playing God for some people, as it may lead to unforeseen consequences (Idris et al., 2023). However, as gene-editing techniques become more of a fact than fiction, proponents say it is just an extension of such favored traditional breeding techniques that have been practiced for years but with vastly more precision. Previously, plant breeders would have crossbred plants to introduce certain traits, but this could take many years of trial and error. In contrast, gene editing can be more specific and targeted to changes that need to be made in the genome, leading to shorter durations and more predictable results. From this perspective, gene editing is not viewed as a break from the practice of agricultural science but as an evolution of it (Bullock et al., 2021).

Regulatory Frameworks and Global Perspectives on Gene-Edited Crops

The regulation of gene-edited crops varies depending on the country and region. Gene-edited crops must be safely and sustainably developed, and regulatory frameworks are crucial in this regard (Turnbull et al., 2021). With regard to gene-edited crops, the U.S. regulatory framework is more permissive. However, gene-edited crops are generally regulated by the U.S. Department of Agriculture (USDA) and Food and Drug Administration (FDA) on a case-by-case basis, based on the substance of the product and not on the process by which the genes are edited (Tachikawa & Matsuo, 2024). In contrast, the European Union has adopted a more conservative position. Consequently, gene-edited crops are viewed by the EU as equivalent to GMOs and are subject to the same rules and extensive safety testing. Regulation of gene-edited crops: A global affair. Although countries like Argentina, Brazil, China, and India are yet to formulate their regulatory framework (but in progress), there are instances where countries like Argentina and Brazil have taken a more flexible approach towards gene editing (Jiang, 2020).

Public Acceptance and the Role of Consumer Education

Gene-edited crops are subject to the general public's perception for their implementation and acceptance. The potential benefits are still on the table, but many consumers are skeptical about gene-edited foods, often because they do not understand the technology or the implications. The perceived safety of gene-edited crops has been recognized as one of the major factors of public acceptance. According to the studies, consumers are more open to accepting gene-edited foods since they believe that gene-editing technology is safe and presents no risk to human health or the environment (Kato-Nitta et al., 2022). Nevertheless, media portrayals of biotechnology and the general view of it often affect the public's attitudes towards gene-edited crops. For example, negative press coverage of GMOs has resulted in a general perception of genetic modification by the public, even when referring to gene editing. Thus, if consumer education is not involved, acceptance is not easy (Bašinskienė & Šeinauskienė, 2021). Education about gene editing should not be left to the general public through news media, where it could, quite frankly, potentially become quite scary; rather, it should be accomplished through effective education campaigns that demystify the process, explain the benefits, and address safety and ethical concerns. Involving consumers in decision-making processes, being transparent, and maintaining available information on how gene-edited crops are developed and regulated can build public trust (Ebeling & Gebhard, 2022).

Challenges in Implementing Gene Editing for Crop Yield

Gene editing has demonstrated superior potential to enhance crop yield, alongside improving plant resilience and sustainability. The reliable technological deployment of this technology encounters multiple substantial obstacles. The massive implementation of gene-edited crops encounters significant barriers based on technical, economic, environmental, and regulatory categories. The introduction section demonstrates the central hurdles that block the execution of gene editing for crop yield improvement (Ansari et al., 2020).

Technical and Scientific Challenges: Precision and Off-Target Effects

The gene-editing approach using CRISPR-Cas9 shows significant potential for performing specific DNA alterations in plant genome sequences. Achieving exact precision is the main technical requirement for implementing gene-editing strategies. The greater accuracy of CRISPR-Cas9 compared to that of prior tools does not eliminate all potential errors during modifications (Manghwar et al., 2020).

CRISPR-Cas9 technology continues to pose major risks because it can modify unintended regions of the genome. Small, unpredictable genomic alterations pose a risk of triggering detrimental genetic effects or damaging crucial control mechanisms in plants. Such unintentional genetic alterations can result in crop health complications and yield decreases when gene-edited plants are released into the environment before undertaking proper testing procedures (Ayanoğlu et al., 2020).

Economic Barriers: Cost of Research and Technology Dissemination

The necessary expenses for conducting gene-editing studies, along with technology distribution, create obstacles that hinder the complete adoption of gene-edited crop cultivation. The creation of gene-edited crops demands substantial initial expenses because it entails funding research along with the procurement of gene-editing equipment and the execution of safety and effectiveness tests (Marone et al., 2023). Gene editing features a significant capital requirement for both specialized research equipment and laboratories, as well as personnel with the necessary knowledge and expertise. The expenses associated with gene-edited crop development limit access to technology, mainly because they burden the financial capacity of small-scale agricultural producers operating in low-income areas or developing countries. The high investment in gene-edited seeds and technology distribution methods acts as a barrier that prevents smallholder farmers from accessing the technology. Market control over gene-edited seeds rests primarily with patent owners from private organizations because they possess ownership rights to gene-editing technology. The high economic costs present a significant barrier for small-scale farmers who want to benefit from gene-edited crops, thus worsening global food system inequalities (Wei, 2024).

Environmental Risks: Potential Ecological Impacts of Edited Crops

Gene-editing technology provides enhanced crop production capabilities but generates critical environmental hazards. The main environmental concern involves the possible adverse effects of allowing gene-edited crops to escape into open ecosystems. Genetically modified organisms from edited crops could interbreed with their wild relatives and non-modified crops, thus dispersing edited genes into natural ecosystems. Gene-edited crops with pest resistance have detrimental effects on native species biodiversity because they modify ecological pest communities and impact unidentified species (Sabat & Tripathy, 2024). Proper evaluation of environmental risks must precede the commercial licensing of gene-edited crops owing to safety concerns identified by researchers and regulatory bodies. The regular monitoring of gene-edited crops under natural conditions must continue indefinitely to detect unknown environmental effects that require proper management (Brookes & Smyth, 2024).

Regulatory Challenges: Policies, Testing, and Approval Processes

Gene-edited crops encounter multiple regulatory obstacles during the implementation process for yield improvement. Different regulatory systems regarding gene-edited crops in various nations lead to research and production uncertainties for scientists and crop producers. Gene-edited plant varieties in the United States fall under current GMO oversight regulations, but other governments, including European Union states, regulate such organisms as GMOs (Turnbull et al., 2021). Gene-edited crop assessments follow a protracted multi-step assessment procedure that researchers and producers experience. Gene-edited crops in the EU must comply with extensive testing to assess both their safety aspects for human health and environmental impacts. Public safety testing is frequently required by the industry; however, it slows down the commercial release of gene-edited crops, delaying their beneficial application. Gene-edited crops receive quicker approval periods in the U.S. than in the EU, which promotes faster market releases of these varieties (Eckerstorfer et al., 2023).

Future Directions

The agricultural field will experience a revolutionary transformation through gene editing technology, as it enables solutions for food security issues and climate resilience combined with improved crop yields. Multiple emerging technology research areas, alongside potential applications, provide ways to optimize sustainable crop production through future developments in technology (Khan et al., 2021). This section discusses the major prospective applications of gene editing in agriculture, which combines multiple genes with climate change mitigation strategies alongside other agricultural technologies. Multi-gene editing represents an exciting research field in gene editing because it simultaneously edits various genes to create crops that combine desired traits (D. Li et al., 2024). Since the advent of CRISPR-Cas9 gene editing technology, scientists have the capability to edit several genes simultaneously, whereas traditional approaches provided mostly single-gene editing functions. Multi-gene editing techniques offer prospects for developing crop varieties that demonstrate improved resistance to multiple pests and diseases, better stress adaptation, and enhanced nutrient management capability, which collectively drive increased crop production and environmental stability (Wang et al., 2022). Future research on gene editing should focus on transforming multi-gene editing capabilities and developing strategies for climate change adaptation. Gene editing is an essential tool for producing crops that survive extreme droughts, heatwaves, and salty conditions because climate change threatens worldwide farming processes. Scientists use genetic alterations to create plants that resist drought by controlling the gene expression associated with water conservation and stress tolerance (Khokha-rVoytas et al., 2023).

Gene editing provides extensive capabilities for solving global issues related to food security. Human society faces growing demands for increased food production with diminished resource requirements because of global population expansion. Gene editing involves the creation of new crops that produce higher yields while using decreased water and fewer chemicals and show enhanced resistance to pathogens (Karavolias et al., 2021). Through genetic editing technology, researchers can develop crops that enhance their ability to use nutrients more efficiently to decrease reliance on manufactured fertilizers containing harmful environmental effects. The combination of enhanced nutrient absorption and reduced chemical fertilization in gene-edited crops creates economic benefits in food production through lower costs and enhanced sustainability standards for worldwide food accessibility (Sathee et al., 2022). Precision farming, along with artificial intelligence (AI), becomes more effective through collaboration with gene-editing techniques. Precision farming implements data-based approaches to boost agricultural efficiency by allowing operators to check and modify water usage, nutrition, and pesticide applications. When integrating gene editing with precision farming, it becomes possible to create crops that match the distinctive ecological characteristics of every farming area, providing improved resource efficiency alongside enhanced agricultural output (Chen et al., 2025).

CONCLUSION

Gene editing is now recognized as a revolutionary prospect in the agricultural field, as it provides the possibility to breed better, more productive, and more sustainable crops. Gene editing through newer techniques, such as CRISPR-Cas 9, allows plants to be genetically modified at the root DNA level to provide desirable qualities, such as pest resistance, drought tolerance, and nutrient uptake. These could help meet some of the challenges in food provisioning, especially in the world today, where climate change and population average life expectancy have deteriorated. Therefore, it is prudent to expect that ethical considerations also feature prominently in gene editing in agriculture. Although technology seems to have much potential for a positive impact from an ecological, equity, and supply perspective, potential consequences in these areas must also be deliberated. Thus, further clear rules to govern this new technology, more studies on different plants and crops, and more knowledge shared with the public will be important in making gene editing improve agricultural practices to feed the entire world.

REFERENCES

Abdul Aziz, M., Brini, F., Rouached, H., & Masmoudi, K. (2022). Genetically engineered crops for sustainably enhanced food production systems. *Frontiers in Plant Science, 13*, 1027828. DOI: 10.3389/fpls.2022.1027828

Ahmadikhah, A., Zarabizadeh, H., Nayeri, S., & Abbasi, M. S. (2025). Advancements in genome editing tools for genetic studies and crop improvement. *Frontiers in Plant Science, 15*, 1370675. DOI: 10.3389/fpls.2024.1370675

Ahmar, S., Saeed, S., Khan, M. H. U., Ullah Khan, S., Mora-Poblete, F., Kamran, M., & Saleem, S. (2020). A revolution toward gene-editing technology and its application to crop improvement. *International Journal of Molecular Sciences, 21*(16), 5665. DOI: 10.3390/ijms21165665

Akram, F., Sahreen, S., Aamir, F., Haq, I., Malik, K., Imtiaz, M., Naseem, W., Nasir, N., & Waheed, H. M. (2023). An insight into modern targeted genome-editing technologies with a special focus on CRISPR/Cas9 and its applications. *Molecular Biotechnology, 65*(2), 227–242. DOI: 10.1007/s12033-022-00501-4

Alotaibi, M. (2023). Climate change, its impact on crop production, challenges, and possible solutions. *Notulae Botanicae Horti Agrobotanici Cluj-Napoca, 51*(1), 13020. DOI: 10.15835/nbha51113020

Anas, M., Bibi, S., Farwa, U., Khan, A., & Quraishi, U. M. (2024). Manganese oxide nanoparticles elicit cadmium tolerance in wheat seedlings by ionomic and phenomic adjustment, regulation of AsA–GSH pathway, cellular thickness recovery, and antioxidant modulation. *Environmental Science. Nano, 11*(11), 4588–4605. DOI: 10.1039/D4EN00490F

Anas, M., Khattak, W. A., Fahad, S., Alrawiq, N., Alrawiq, H. S., Abdelsalam, N. R., Jaremko, M., & Quraishi, U. M. (2025). Mitigating heavy metal pollution in agriculture: A multi-omics and nanotechnology approach to safeguard global wheat production. *Journal of Hazardous Materials Advances, 17*, 100584. DOI: 10.1016/j.hazadv.2024.100584

Ansari, W. A., Chandanshive, S. U., Bhatt, V., Nadaf, A. B., Vats, S., Katara, J. L., Sonah, H., & Deshmukh, R. (2020). Genome editing in cereals: Approaches, applications and challenges. *International Journal of Molecular Sciences, 21*(11), 4040. DOI: 10.3390/ijms21114040

Ansori, A. N., Antonius, Y., Susilo, R. J., Hayaza, S., Kharisma, V. D., Parikesit, A. A., Zainul, R., Jakhmola, V., Saklani, T., Rebezov, M., Ullah, M. E., Maksimiuk, N., Derkho, M., & Burkov, P. (2023). Application of CRISPR-Cas9 genome editing technology in various fields: A review. *Narra J*, *3*(2), e184. DOI: 10.52225/narra.v3i2.184

Arif, M., Haroon, M., Nawaz, A. F., Abbas, H., Xu, R., & Li, L. (2025). Enhancing wheat resilience: Biotechnological advances in combating heat stress and environmental challenges. *Plant Molecular Biology*, *115*(2), 41. DOI: 10.1007/s11103-025-01569-7

Aubert, M. K., Coventry, S., Shirley, N. J., Betts, N. S., Würschum, T., Burton, R. A., & Tucker, M. R. (2018). Differences in hydrolytic enzyme activity accompany natural variation in mature aleurone morphology in barley (Hordeum vulgare L.). *Scientific Reports*, *8*(1), 11025. DOI: 10.1038/s41598-018-29068-4

Awulachew, M. T., & Weldehawaria, A. G. (2024). Food security and safety in light of global megatrends: A review. *Asian J. Biol. Sci*, *17*(4), 666–677.

Ayanoğlu, F. B., Elçin, A. E., & Elçin, Y. M. (2020). Bioethical issues in genome editing by CRISPR-Cas9 technology. *Turkish Journal of Biology = Türk Biyoloji Dergisi*, *44*(2), 110–120. DOI: 10.3906/biy-1912-52

Bašinskienė, L., & Šeinauskienė, B. (2021). Gene Editing versus Gene Modification: Awareness, Attitudes and Behavioural Intentions of Lithuanian Consumers, Producers and Farmers. *CET Journal-Chemical Engineering Transactions, 87.*

Begna, T. (2021). Conventional Breeding Methods Widely used to Improve Self-Pollinated Crops. *International Journal of Research Studies in Agricultural Sciences*, *7*(1), 2454–6224. DOI: 10.20431/2454-6224.0701001

Benavente, E., & Giménez, E. (2021). Modern approaches for the genetic improvement of rice, wheat and maize for abiotic constraints-related traits: A comparative overview. *Agronomy (Basel)*, *11*(2), 376. DOI: 10.3390/agronomy11020376

Benitez-Alfonso, Y., Soanes, B. K., Zimba, S., Sinanaj, B., German, L., Sharma, V., Bohra, A., Kolesnikova, A., Dunn, J. A., Martin, A. C., Khashi u Rahman, M., Saati-Santamaría, Z., García-Fraile, P., Ferreira, E. A., Frazão, L. A., Cowling, W. A., Siddique, K. H. M., Pandey, M. K., Farooq, M., & Foyer, C. H. (2023). Enhancing climate change resilience in agricultural crops. *Current Biology : CB*, *33*(23), R1246–R1261. DOI: 10.1016/j.cub.2023.10.028

Bhagtaney, L., & Sundarrajan, P. (2023). An overview of tools for genome editing: ZFNs, mega nucleases, and TALENs. *CRISPR/Cas-mediated genome editing in plants*, 37-64.

Bhandari, M. K., & Paudel, M. (2024). Genetic, biological and sterile insect techniques: Insect pest management strategies: A review. *Agricultural Reviews (Karnal)*, *45*(3), 390–399. DOI: 10.18805/ag.RF-311

Bhardwaj, A., & Nain, V. (2021). TALENs—an indispensable tool in the era of CRISPR: A mini review. *Journal of Genetic Engineering and Biotechnology*, *19*(1), 125. DOI: 10.1186/s43141-021-00225-z

Bhatta, M., Sandro, P., Smith, M. R., Delaney, O., Voss-Fels, K. P., Gutierrez, L., & Hickey, L. T. (2021). Need for speed: Manipulating plant growth to accelerate breeding cycles. *Current Opinion in Plant Biology*, *60*, 101986. DOI: 10.1016/j.pbi.2020.101986

Bhavani, S., Singh, P., Qureshi, N., He, X., Biswal, A. K., Juliana, P., . . . Mourad, A. M. (2021). Globally important wheat diseases: status, challenges, breeding and genomic tools to enhance resistance durability. *Genomic designing for biotic stress resistant cereal crops*, 59-128.

Borg, E., & Policante, A. (2024). The Gene Editing Business: Rent Extraction in the Biotech Industry. *Review of Political Economy*, •••, 1–36.

Brookes, G., & Smyth, S. J. (2024). Risk-appropriate regulations for gene-editing technologies. *GM Crops & Food*, *15*(1), 1–14. DOI: 10.1080/21645698.2023.2293510

Bullock, D., Wilson, W., & Neadeau, J. (2021). Gene Editing Versus Genetic Modification in the Research and Development of New Crop Traits: An Economic Comparison. *American Journal of Agricultural Economics*, *103*(5), 1700–1719. Advance online publication. DOI: 10.1111/ajae.12201

Burgess, A. J., Masclaux-Daubresse, C., Strittmatter, G., Weber, A. P., Taylor, S. H., Harbinson, J., & Westhoff, P. (2023). Improving crop yield potential: Underlying biological processes and future prospects. *Food and Energy Security*, *12*(1), e435. DOI: 10.1002/fes3.435

Chachar, Z., Xue, X., Junteng, F., Chen, M., Chen, W., Hui, L. X., & Liang Chen, Z. (2025). Genetic and Molecular Insights into Tiller Development and Approaches for Crop Yield Improvement. *Frontiers in Plant Science*, *16*, 1532180. DOI: 10.3389/fpls.2025.1532180

Chen, F., Chen, L., Yan, Z., Xu, J., Feng, L., He, N., & Chen, H. (2024). Recent advances of CRISPR-based genome editing for enhancing staple crops. *Frontiers in Plant Science*, *15*, 1478398. DOI: 10.3389/fpls.2024.1478398

Chen, X., Cui, X., & Ma, H. (2025). The application of artificial intelligence and gene editing technologies in high-yield crop breeding: Innovations and prospects for future agriculture. *Advances in Resources Research*, *5*(1), 146–166.

Clapp, J., & Ruder, S.-L. (2020). Precision technologies for agriculture: Digital farming, gene-edited crops, and the politics of sustainability. *Global Environmental Politics*, *20*(3), 49–69. DOI: 10.1162/glep_a_00566

Cui, L., Zheng, F., Wang, J., Zhang, C., Zhang, D., Gao, S., & Ouyang, B. (2022). The tomato CONSTANS-LIKE protein SlCOL1 regulates fruit yield by repressing SFT gene expression. *BMC Plant Biology*, *22*(1), 429. DOI: 10.1186/s12870-022-03813-4

Dibley, K., Jost, M., McIntosh, R., Lagudah, E., & Zhang, P. (2024). The wheat stripe rust resistance gene YrNAM is Yr10. *Nature Communications*, *15*(1), 3291. DOI: 10.1038/s41467-024-47513-z

Dos Santos, T. B., Ribas, A. F., de Souza, S. G. H., Budzinski, I. G. F., & Domingues, D. S. (2022). Physiological responses to drought, salinity, and heat stress in plants: A review. *Stresses*, *2*(1), 113–135. DOI: 10.3390/stresses2010009

Ebeling, S., & Gebhard, U. (2022). Ideas, hopes, and fears: What young adults think about genome editing, nature, and society. *Cultural Studies of Science Education*, *17*(3), 745–764. DOI: 10.1007/s11422-022-10102-x

Eckerstorfer, M. F., Dolezel, M., Engelhard, M., Giovannelli, V., Grabowski, M., Heissenberger, A., & Staiano, G. (2023). Recommendations for the assessment of potential environmental effects of genome-editing applications in plants in the EU. *Plants*, *12*(9), 1764. DOI: 10.3390/plants12091764

Filyushin, M. A., Kochieva, E. Z., & Shchennikova, A. V. (2022). ZmDREB2.9 gene in maize (Zea mays L.): Genome-wide identification, characterization, expression, and stress response. *Plants*, *11*(22), 3060. DOI: 10.3390/plants11223060

Fudjoe, S. K., Shi, S., Li, L., Anwar, S., Xie, J., Chimsah, F., & Wang, L. (2025). Gene expression through comparative transcriptome analysis unravels the molecular mechanisms of fertilizer-induced hormonal regulation and leaf senescence in maize for enhanced yield in semiarid regions. *BMC Plant Biology*, *25*(1), 327. DOI: 10.1186/s12870-025-06303-5

Gao, C. (2021). Genome engineering for crop improvement and future agriculture. *Cell, 184*(6), 1621–1635. DOI: 10.1016/j.cell.2021.01.005

Gatahi, D. M. (2020). Challenges and opportunities in tomato production chain and sustainable standards. *International Journal of Horticultural Science and Technology, 7*(3), 235–262.

Gordon, D. R., Jaffe, G., Doane, M., Glaser, A., Gremillion, T. M., & Ho, M. D. (2021). Responsible governance of gene editing in agriculture and the environment. *Nature Biotechnology, 39*(9), 1055–1057. DOI: 10.1038/s41587-021-01023-1

Gupta, S., Kumar, A., Patel, R., & Kumar, V. (2021). Genetically modified crop regulations: Scope and opportunity using the CRISPR-Cas9 genome editing approach. *Molecular Biology Reports, 48*(5), 4851–4863. DOI: 10.1007/s11033-021-06477-9

Hamdan, M. F., & Tan, B. C. (2024). Genetic modification techniques in plant breeding: A comparative review of CRISPR/Cas and GM technologies. *Horticultural Plant Journal.* Advance online publication. DOI: 10.1016/j.hpj.2024.02.012

Hernandes-Lopes, J., Yassitepe, J., Koltun, A., Pauwels, L., Silva, V., Dante, R. A., Gerhardt, I. R., & Arruda, P. (2023). Genome editing in maize: Toward improving complex traits in a global crop. *Genetics and Molecular Biology, 46*(1, Suppl 1), e20220217. DOI: 10.1590/1678-4685-gmb-2022-0217

Hu, B., Wang, W., Chen, J., Liu, Y., & Chu, C. (2023). Genetic improvement toward nitrogen-use efficiency in rice: Lessons and perspectives. *Molecular Plant, 16*(1), 64–74. DOI: 10.1016/j.molp.2022.11.007

Idris, S. H., Mat Jalaluddin, N. S., & Chang, L. W. (2023). Ethical and legal implications of gene editing in plant breeding: A systematic literature review. *Journal of Zhejiang University. Science. B, 24*(12), 1093–1105. DOI: 10.1631/jzus.B2200601

Ilyas, M. (2025). Improving resistance of traditional Italian tomato cultivars against biotic stress.

Jabłoński, B., Ogonowska, H., Szala, K., Bajguz, A., Orczyk, W., & Nadolska-Orczyk, A. (2020). Silencing of TaCKX1 Mediates Expression of Other TaCKX Genes to Increase Yield Parameters in Wheat. *International Journal of Molecular Sciences, 21*(13), 4809. Advance online publication. DOI: 10.3390/ijms21134809

Jiang, L. (2020). Commercialization of the gene-edited crop and morality: Challenges from the liberal patent law and the strict GMO law in the EU. *New Genetics and Society, 39*(2), 191–218. DOI: 10.1080/14636778.2019.1686968

Ju, Y., Jia, Y., Cheng, B., Wang, D., Gu, D., Jing, W., Zhang, H., Chen, X., & Li, G. (2024). NRT1.1B mediates rice plant growth and soil microbial diversity under different nitrogen conditions. *AMB Express, 14*(1), 39. DOI: 10.1186/s13568-024-01683-7

Ju, Y., Jia, Y., Cheng, B., Wang, D., Gu, D., Jing, W., Zhang, H., Chen, X., & Li, G. (2024). NRT1. 1B mediates rice plant growth and soil microbial diversity under different nitrogen conditions. *AMB Express, 14*(1), 39. DOI: 10.1186/s13568-024-01683-7

Juma, B. S., Mukami, A., Mweu, C., Ngugi, M. P., & Mbinda, W. (2022). Targeted mutagenesis of the CYP79D1 gene via CRISPR/Cas9-mediated genome editing results in lower levels of cyanide in cassava. *Frontiers in Plant Science, 13*, 1009860. DOI: 10.3389/fpls.2022.1009860

Kafle, S. (2023). CRISPR/CAS9: A new paradigm for crop improvement revolutionizing agriculture. *Journal of Agriculture and Food Research, 11*, 100484. DOI: 10.1016/j.jafr.2022.100484

Kaiser, N., Douches, D., Dhingra, A., Glenn, K. C., Herzig, P. R., Stowe, E. C., & Swarup, S. (2020). The role of conventional plant breeding in ensuring safe levels of naturally occurring toxins in food crops. *Trends in Food Science & Technology, 100*, 51–66. DOI: 10.1016/j.tifs.2020.03.042

Karavolias, N. G., Horner, W., Abugu, M. N., & Evanega, S. N. (2021). Application of gene editing for climate change in agriculture. *Frontiers in Sustainable Food Systems, 5*, 685801. DOI: 10.3389/fsufs.2021.685801

Karunarathne, S. D., Han, Y., Zhang, X. Q., & Li, C. (2022). CRISPR/Cas9 gene editing and natural variation analysis demonstrate the potential for HvARE1 in improvement of nitrogen use efficiency in barley. *Journal of Integrative Plant Biology, 64*(3), 756–770. DOI: 10.1111/jipb.13214

Kato-Nitta, N., Tachikawa, M., Inagaki, Y., & Maeda, T. (2022). Public Perceptions of Risks and Benefits of Gene-edited Food Crops: An International Comparative Study between the US, Japan, and Germany. *Science, Technology & Human Values, 48*(6), 1360–1392. DOI: 10.1177/01622439221123830

Kettenburg, A. T., Lopez, M. A., Yogendra, K., Prior, M. J., Rose, T., Bimson, S., Heuer, S., Roy, S. J., & Bailey-Serres, J. (2023). PHOSPHORUS-STARVATION TOLERANCE 1 (OsPSTOL1) is prevalent in upland rice and enhances root growth and hastens low phosphate signaling in wheat. *Plant, Cell & Environment, 46*(7), 2187–2205. DOI: 10.1111/pce.14588

Khan, N., Ray, R. L., Sargani, G. R., Ihtisham, M., Khayyam, M., & Ismail, S. (2021). Current progress and future prospects of agriculture technology: Gateway to sustainable agriculture. *Sustainability (Basel)*, *13*(9), 4883. DOI: 10.3390/su13094883

KhokharVoytas, A., Shahbaz, M., Maqsood, M. F., Zulfiqar, U., Naz, N., Iqbal, U. Z., Sara, M., Aqeel, M., Khalid, N., Noman, A., Zulfiqar, F., Al Syaad, K. M., & AlShaqhaa, M. A.KhokharVoytas. (2023). Genetic modification strategies for enhancing plant resilience to abiotic stresses in the context of climate change. *Functional & Integrative Genomics*, *23*(3), 283. DOI: 10.1007/s10142-023-01202-0

Kiran, F. (2022). Proline and Plant Stress Tolerance: Mechanisms, Applications, and Future Prospects. *Plant and Environment*, *3*(01), 23–36.

Kopczyńska, M., Saha, U., Romanenko, A., Nojima, T., Gdula, M. R., & Kamieniarz-Gdula, K. (2025). Defining gene ends: RNA polymerase II CTD threonine 4 phosphorylation marks transcription termination regions genome-wide. *Nucleic Acids Research*, *53*(2), gkae1240. DOI: 10.1093/nar/gkae1240

Kumar, M., Prusty, M. R., Pandey, M. K., Singh, P. K., Bohra, A., Guo, B., & Varshney, R. K. (2023). Application of CRISPR/Cas9-mediated gene editing for abiotic stress management in crop plants. *Frontiers in Plant Science*, *14*, 1157678. DOI: 10.3389/fpls.2023.1157678

Kumar, R., Khungar, L., Shimphrui, R., Tiwari, L. D., Tripathi, G., Sarkar, N. K., Agarwal, S.-K., Agarwal, M., & Grover, A. (2020). AtHsp101 research sets course of action for the genetic improvement of crops against heat stress. *Journal of Plant Biochemistry and Biotechnology*, *29*(4), 715–732. DOI: 10.1007/s13562-020-00624-2

Kumar, R., Shanker, R., Singh, P., Yadav, M. K., Chaudhary, V., & Kumar, M. (2024). Genome editing towards pests and disease management in agricultural crops: Recent developments, challenges and future prospects. *Physiological and Molecular Plant Pathology*, *134*, 102402. DOI: 10.1016/j.pmpp.2024.102402

Lang, B. J., Guerrero, M. E., Prince, T. L., Okusha, Y., Bonorino, C., & Calderwood, S. K. (2021). The functions and regulation of heat shock proteins; key orchestrators of proteostasis and the heat shock response. *Archives of Toxicology*, *95*(6), 1943–1970. DOI: 10.1007/s00204-021-03070-8

Li, D., Liu, X., Wang, H., Yuan, S., & Jiang, L. (2024). The application and prospects of gene editing technology in enhancing crop yields: A review. *Advances in Resources Research*, *4*(4), 654–667.

Li, H., Wang, W., Yang, X., Kang, G., Zhang, Z., & Wu, K. (2024). Toxic effects of Bt-(Cry1Ab + Vip3Aa) maize ("DBN3601T"event) on the Asian corn borer Ostrinia furnacalis (Guenée) in Southwestern China. *Agronomy (Basel)*, *14*(9), 1906. DOI: 10.3390/agronomy14091906

Li, S., Lin, D., Zhang, Y., Deng, M., Chen, Y., Lv, B., & Zhao, L. (2022). Genome-edited powdery mildew resistance in wheat without growth penalties. *Nature*, *602*(7897), 455–460. DOI: 10.1038/s41586-022-04395-9

Liao, Z., Chen, B., Boubakri, H., Farooq, M., Mur, L. A. J., Urano, D., & Aslam, M. M. (2025). The regulatory role of phytohormones in plant drought tolerance. *Planta*, *261*(5), 1–22. DOI: 10.1007/s00425-025-04671-8

Limenie, A. D., & Alehegn, M. (2025). Genetic Engineering for Cereal Crop Yield Improvement and Disease Resistant Breeding. *TheScientificWorldJournal*, *2025*(1), 6743917. DOI: 10.1155/tswj/6743917

Lupi, A. C. D., Lira, B. S., Gramegna, G., Trench, B., Alves, F. R. R., Demarco, D., Peres, L. E. P., Purgatto, E., Freschi, L., & Rossi, M. (2019). Solanum lycopersicum GOLDEN 2-LIKE 2 transcription factor affects fruit quality in a light-and auxin-dependent manner. *PLoS One*, *14*(2), e0212224. DOI: 10.1371/journal.pone.0212224

Maharajan, T., Roch, G. V., & Ceasar, S. A. (2021). Recent advancements of molecular breeding and functional genomics for improving nitrogen-, phosphorus-and potassium-use efficiencies in wheat. In *Molecular breeding in wheat, maize and sorghum: Strategies for improving abiotic stress tolerance and yield* (pp. 170–196). CABI Wallingford UK. DOI: 10.1079/9781789245431.0009

Mahboob, W., Yang, G., & Irfan, M. (2023). Crop nitrogen (N) utilization mechanism and strategies to improve N use efficiency. *Acta Physiologiae Plantarum*, *45*(4), 52. DOI: 10.1007/s11738-023-03527-6

Mali, S., Dutta, M., & Zinta, G. (2023). Genome editing advancements in potato (Solanum tuberosum L.): Operational challenges and solutions. *Journal of Plant Biochemistry and Biotechnology*, *32*(4), 730–742. DOI: 10.1007/s13562-022-00812-2

Manghwar, H., Li, B., Ding, X., Hussain, A., Lindsey, K., Zhang, X., & Jin, S. (2020). CRISPR/Cas systems in genome editing: Methodologies and tools for sgRNA design, off-target evaluation, and strategies to mitigate off-target effects. *Advanced Science (Weinheim, Baden-Württemberg, Germany)*, *7*(6), 1902312. DOI: 10.1002/advs.201902312

Manzoor, S., Nabi, S. U., Rather, T. R., Gani, G., Mir, Z. A., Wani, A. W., Ali, S., Tyagi, A., & Manzar, N. (2024). Advancing crop disease resistance through genome editing: A promising approach for enhancing agricultural production. *Frontiers in Genome Editing*, *6*, 1399051. DOI: 10.3389/fgeed.2024.1399051

Maqbool, M. A., Aslam, M., Beshir, A., & Khan, M. S. (2018). Breeding for provitamin A biofortification of maize (Zea mays L.). *Plant Breeding*, *137*(4), 451–469. DOI: 10.1111/pbr.12618

Marone, D., Mastrangelo, A. M., & Borrelli, G. M. (2023). From transgenesis to genome editing in crop improvement: Applications, marketing, and legal issues. *International Journal of Molecular Sciences*, *24*(8), 7122. DOI: 10.3390/ijms24087122

Mei, F., Chen, B., Du, L., Li, S., Zhu, D., Chen, N., & Cheng, X. (2022). A gain-of-function allele of a DREB transcription factor gene ameliorates drought tolerance in wheat. *The Plant Cell*, *34*(11), 4472–4494. DOI: 10.1093/plcell/koac248

Muhie, S. H. (2022). Novel approaches and practices to sustainable agriculture. *Journal of Agriculture and Food Research*, *10*, 100446. DOI: 10.1016/j.jafr.2022.100446

Munir, A., Salah, M. A., Ali, M., Ali, B., Saleem, M. H., Samarasinghe, K. G. B. A., De Silva, S. I. S., Ercisli, S., Iqbal, N., & Anas, M. (2024). Advancing Agriculture: Harnessing Smart Nanoparticles for Precision Fertilization. *BioNanoScience*, *14*(4), 3846–3863. DOI: 10.1007/s12668-024-01597-5

Nadakuduti, S. S., & Enciso-Rodríguez, F. (2021). Advances in genome editing with CRISPR systems and transformation technologies for plant DNA manipulation. *Frontiers in Plant Science*, *11*, 637159. DOI: 10.3389/fpls.2020.637159

Nagaraj, S., Rajasekaran, R., Palaniappan, J., Rangasamy, S., Narayanasamy, C., & Narayanan, M. B. (2024). Emerging technological developments to address pest resistance in Bt cotton. *Journal of Cotton Research*, *7*(1), 30. DOI: 10.1186/s42397-024-00192-z

Nazari, M., Kordrostami, M., Ghasemi-Soloklui, A. A., Eaton-Rye, J. J., Pashkovskiy, P., Kuznetsov, V., & Allakhverdiev, S. I. (2024). Enhancing photosynthesis and plant productivity through genetic modification. *Cells*, *13*(16), 1319. DOI: 10.3390/cells13161319

Ndudzo, A., Sibanda Makuvise, A., Moyo, S., & Bobo, E. D. (2024). CRISPR-Cas9 genome editing in crop breeding for climate change resilience: Implications for smallholder farmers in Africa. *Journal of Agriculture and Food Research*, *16*, 101132. DOI: 10.1016/j.jafr.2024.101132

Newman, A., Starrs, L., & Burgio, G. (2020). Cas9 Cuts and Consequences; Detecting, Predicting, and Mitigating CRISPR/Cas9 On-and Off-Target Damage: Techniques for Detecting, Predicting, and Mitigating the On-and off-target Effects of Cas9 Editing. *BioEssays : News and Reviews in Molecular, Cellular and Developmental Biology*, *42*(9), 2000047. DOI: 10.1002/bies.202000047

Nie, H., Yang, X., Zheng, S., & Hou, L. (2024). Gene-Based Developments in Improving Quality of Tomato: Focus on Firmness, Shelf Life, and Pre-and Post-Harvest Stress Adaptations. *Horticulturae*, *10*(6), 641. DOI: 10.3390/horticulturae10060641

Nikolaou, G., Neocleous, D., Christou, A., Kitta, E., & Katsoulas, N. (2020). Implementing sustainable irrigation in water-scarce regions under the impact of climate change. *Agronomy (Basel)*, *10*(8), 1120. DOI: 10.3390/agronomy10081120

Oh, J.-M., & Myung, K. (2022). Crosstalk between different DNA repair pathways for DNA double strand break repairs. *Mutation Research. Genetic Toxicology and Environmental Mutagenesis*, *873*, 503438. DOI: 10.1016/j.mrgentox.2021.503438

Panda, D., Mishra, S. S., & Behera, P. K. (2021). Drought Tolerance in Rice: Focus on Recent Mechanisms and Approaches. *Rice Science*, *28*(2), 119–132. DOI: 10.1016/j.rsci.2021.01.002

Pandey, D. K., & Mishra, R. (2024). Towards sustainable agriculture: Harnessing AI for global food security. *Artificial Intelligence in Agriculture*, *12*, 72–84. DOI: 10.1016/j.aiia.2024.04.003

Pandey, S., Singh, A., Parida, S. K., & Prasad, M. (2022). Combining speed breeding with traditional and genomics-assisted breeding for crop improvement. *Plant Breeding*, *141*(3), 301–313. DOI: 10.1111/pbr.13012

Pathi, K. (2021). Establishment of maize resistance to fungal diseases by host-induced gene silencing and site-directed mutagenesis.

Pixley, K. V., Cairns, J. E., Lopez-Ridaura, S., Ojiewo, C. O., Dawud, M. A., Drabo, I., Mindaye, T., Nebie, B., Asea, G., Das, B., Daudi, H., Desmae, H., Batieno, B. J., Boukar, O., Mukankusi, C. T. M., Nkalubo, S. T., Hearne, S. J., Dhugga, K. S., Gandhi, H., & Zepeda-Villarreal, E. A. (2023). Redesigning crop varieties to win the race between climate change and food security. *Molecular Plant*, *16*(10), 1590–1611. DOI: 10.1016/j.molp.2023.09.003

Qaim, M. (2020). Role of new plant breeding technologies for food security and sustainable agricultural development. *Applied Economic Perspectives and Policy*, *42*(2), 129–150. DOI: 10.1002/aepp.13044

Rafiq, A. R. (2024). Genomic editing techniques for ensuring food security: CRISPR Cas, TALEN, ZFN, RNAi and mutagenesis. *Journal of Pure and Applied Agriculture, 9*(2).

Rasheed, A., Gill, R. A., Hassan, M. U., Mahmood, A., Qari, S., Zaman, Q. U., & Li, H. (2021). A critical review: Recent advancements in the use of CRISPR/Cas9 technology to enhance crops and alleviate global food crises. *Current Issues in Molecular Biology, 43*(3), 1950–1976. DOI: 10.3390/cimb43030135

Rato, C., Carvalho, M. F., Azevedo, C., & Oblessuc, P. R. (2021). Genome editing for resistance against plant pests and pathogens. *Transgenic Research, 30*(4), 427–459. DOI: 10.1007/s11248-021-00262-x

Raza, A., Anas, M., Bhardwaj, S., Mir, R. A., Charagh, S., Elahi, M., Zhang, X., Mir, R. R., Weckwerth, W., Fernie, A. R., Siddique, K. H. M., Hu, Z., & Varshney, R. K. (2025). Harnessing metabolomics for enhanced crop drought tolerance. *The Crop Journal, 13*(2), 311–327. Advance online publication. DOI: 10.1016/j.cj.2025.01.001

Rong, W., Qi, L., Wang, A., Ye, X., Du, L., Liang, H., Xin, Z., & Zhang, Z. (2014). The ERF transcription factor TaERF3 promotes tolerance to salt and drought stresses in wheat. *Plant Biotechnology Journal, 12*(4), 468–479. DOI: 10.1111/pbi.12153

Roy, A., Moradkhani, H., Mekonnen, M., Moftakhari, H., & Magliocca, N. (2024). Towards strategic interventions for global food security in 2050. *The Science of the Total Environment, 954*, 176811. DOI: 10.1016/j.scitotenv.2024.176811

Roy, D., Deo, I., Kushwaha, U. K. S., & Singh, P. K. (2025). Speed Breeding: Space Inspired Plant Breeding for Crop Improvement. In *Smart Technologies in Sustainable Agriculture* (pp. 147-161): Apple Academic Press.

Rozas, P., Kessi-Pérez, E. I., & Martínez, C. (2022). Genetically modified organisms: Adapting regulatory frameworks for evolving genome editing technologies. *Biological Research, 55*(1), 55. DOI: 10.1186/s40659-022-00399-x

Sabat, M., & Tripathy, A. (2024). Genetically Modified and Gene-Edited Food Crops: Recent Status and Future Prospects. *Food Production, Diversity, and Safety Under Climate Change*, 211-222.

Sadikiel Mmbando, G., & Ngongolo, K. (2024). The recent genetic modification techniques for improve soil conservation, nutrient uptake and utilization. *GM Crops & Food, 15*(1), 233–247. DOI: 10.1080/21645698.2024.2377408

Sarma, B., Kashtoh, H., Lama Tamang, T., Bhattacharyya, P. N., Mohanta, Y. K., & Baek, K. H. (2023). Abiotic Stress in Rice: Visiting the Physiological Response and Its Tolerance Mechanisms. *Plants, 12*(23), 3948. Advance online publication. DOI: 10.3390/plants12233948

Sathee, L., Jagadhesan, B., Pandesha, P. H., Barman, D., & Adavi, B. (2022). Genome editing targets for improving nutrient use efficiency and nutrient stress adaptation. *Frontiers in Genetics, 13*, 900897. DOI: 10.3389/fgene.2022.900897

Seleiman, M. F., Al-Suhaibani, N., Ali, N., Akmal, M., Alotaibi, M., Refay, Y., Dindaroglu, T., Abdul-Wajid, H. H., & Battaglia, M. L. (2021). Drought Stress Impacts on Plants and Different Approaches to Alleviate Its Adverse Effects. *Plants, 10*(2), 259. Advance online publication. DOI: 10.3390/plants10020259

Shah, T., Andleeb, T., Lateef, S., & Noor, M. A. (2018). Genome editing in plants: Advancing crop transformation and overview of tools. *Plant Physiology and Biochemistry : PPB, 131*, 12–21. DOI: 10.1016/j.plaphy.2018.05.009

Shamshirgaran, Y., Liu, J., Sumer, H., Verma, P. J., & Taheri-Ghahfarokhi, A. (2022). Tools for efficient genome editing; ZFN, TALEN, and CRISPR. *Applications of genome modulation and editing*, 29-46.

Shelake, R. M., Kadam, U. S., Kumar, R., Pramanik, D., Singh, A. K., & Kim, J.-Y. (2022). Engineering drought and salinity tolerance traits in crops through CRISPR-mediated genome editing: Targets, tools, challenges, and perspectives. *Plant Communications, 3*(6), 100417. DOI: 10.1016/j.xplc.2022.100417

Shi, J., Habben, J. E., Archibald, R. L., Drummond, B. J., Chamberlin, M. A., Williams, R. W., Lafitte, H. R., & Weers, B. P. (2015). Overexpression of ARGOS genes modifies plant sensitivity to ethylene, leading to improved drought tolerance in both Arabidopsis and maize. *Plant Physiology, 169*(1), 266–282. DOI: 10.1104/pp.15.00780

Singh, V. K., Pundir, S., Chaturvedi, D., Kaur, A., Pandey, A., Mandal, S., . . . Dhanda, P. S. (2023). Enhancing Maize (Zea mays L.) Crop through Advanced Techniques: A Comprehensive Approach. In *New Prospects of Maize*: IntechOpen.

Smyth, S. J. (2022). Contributions of genome editing technologies towards improved nutrition, environmental sustainability and poverty reduction. *Frontiers in Genome Editing, 4*, 863193. DOI: 10.3389/fgeed.2022.863193

Solis, C. A., Yong, M.-T., Zhou, M., Venkataraman, G., Shabala, L., Holford, P., Shabala, S., & Chen, Z.-H. (2022). Evolutionary significance of NHX family and NHX1 in salinity stress adaptation in the genus oryza. *International Journal of Molecular Sciences*, *23*(4), 2092. DOI: 10.3390/ijms23042092

Song, H., Taylor, D. C., & Zhang, M. (2023). Bioengineering of Soybean Oil and Its Impact on Agronomic Traits. *International Journal of Molecular Sciences*, *24*(3), 2256. Advance online publication. DOI: 10.3390/ijms24032256

Su, Q., Chen, L., Cai, Y., Wang, L., Chen, Y., Zhang, J., & Gao, Y. (2024). The FLOWERING LOCUS T 5b positively regulates photoperiodic flowering and improves the geographical adaptation of soybean. *Plant, Cell & Environment*, *47*(1), 246–258. DOI: 10.1111/pce.14739

Subejano, M. S. E. P., Daquioag, J. E. L., & Penuliar, G. M. (2024). Genetically Modified Organisms in Research and Biotechnology. In *Biosafety and Biosecurity* (pp. 235–258). CRC Press. DOI: 10.1201/9781003426219-12

Tachikawa, M., & Matsuo, M. (2024). Global regulatory trends of genome editing technology in agriculture and food. *Breeding Science*, *74*(1), 3–10. DOI: 10.1270/jsbbs.23046

Tandzi, L., & Mutengwa, C. (2020). Factors Affecting Yield of Crops. In.

Tian, Y., Zhou, Y., Gao, G., Zhang, Q., Li, Y., Lou, G., & He, Y. (2023). Creation of two-line fragrant glutinous hybrid rice by editing the Wx and OsBADH2 genes via the CRISPR/Cas9 system. *International Journal of Molecular Sciences*, *24*(1), 849. DOI: 10.3390/ijms24010849

Tlak Gajger, I., & Dar, S. A. (2021). Plant Allelochemicals as Sources of Insecticides. *Insects*, *12*(3), 189. Advance online publication. DOI: 10.3390/insects12030189

Tripathi, L., Ntui, V. O., & Tripathi, J. N. (2024). Application of CRISPR/Cas-based gene-editing for developing better banana. *Frontiers in Bioengineering and Biotechnology*, *12*, 1395772. DOI: 10.3389/fbioe.2024.1395772

Tudi, M., Daniel Ruan, H., Wang, L., Lyu, J., Sadler, R., Connell, D., Chu, C., & Phung, D. T. (2021). Agriculture Development, Pesticide Application and Its Impact on the Environment. *International Journal of Environmental Research and Public Health*, *18*(3), 1112. Advance online publication. DOI: 10.3390/ijerph18031112

Turnbull, C., Lillemo, M., & Hvoslef-Eide, T. A. (2021). Global regulation of genetically modified crops amid the gene edited crop boom–a review. *Frontiers in Plant Science*, *12*, 630396. DOI: 10.3389/fpls.2021.630396

Tyagi, S., Kumar, R., Kumar, V., Won, S. Y., & Shukla, P. (2021). Engineering disease resistant plants through CRISPR-Cas9 technology. *GM Crops & Food*, *12*(1), 125–144. DOI: 10.1080/21645698.2020.1831729

Umar, W., Ayub, M. A., Rehman, M. Z. u., Ahmad, H. R., Farooqi, Z. U. R., Shahzad, A., . . . Nadeem, M. (2020). Nitrogen and phosphorus use efficiency in agroecosystems. *Resources use efficiency in agriculture*, 213-257.

Usman, B., Nawaz, G., Zhao, N., Liao, S., Liu, Y., & Li, R. (2020). Precise editing of the OsPYL9 gene by RNA-guided Cas9 nuclease confers enhanced drought tolerance and grain yield in rice (Oryza sativa L.) by regulating circadian rhythm and abiotic stress responsive proteins. *International Journal of Molecular Sciences*, *21*(21), 7854. DOI: 10.3390/ijms21217854

van Dijk, M., Morley, T., Rau, M. L., & Saghai, Y. (2021). A meta-analysis of projected global food demand and population at risk of hunger for the period 2010–2050. *Nature Food*, *2*(7), 494–501. DOI: 10.1038/s43016-021-00322-9

Vyroubalová, S., Václavíková, K. i., Turecková, V., Novák, O., Šmehilová, M., Hluska, T., Ohnoutková, L., Frébort, I., & Galuszka, P. (2009). Characterization of new maize genes putatively involved in cytokinin metabolism and their expression during osmotic stress in relation to cytokinin levels. *Plant Physiology*, *151*(1), 433–447. DOI: 10.1104/pp.109.142489

Waites, J., Achary, V. M. M., Syombua, E. D., Hearne, S. J., & Bandyopadhyay, A. (2025). CRISPR-mediated genome editing of wheat for enhancing disease resistance. *Frontiers in Genome Editing*, *7*, 1542487. DOI: 10.3389/fgeed.2025.1542487

Wang, J., Wu, B., Lu, K., Wei, Q., Qian, J., Chen, Y., & Fang, Z. (2019). The amino acid permease 5 (OsAAP5) regulates tiller number and grain yield in rice. *Plant Physiology*, *180*(2), 1031–1045. DOI: 10.1104/pp.19.00034

Wang, P., Si, H., Li, C., Xu, Z., Guo, H., Jin, S., & Cheng, H. (2025). Plant genetic transformation: Achievements, current status and future prospects. *Plant Biotechnology Journal*, *23*(6), 2034–2058. DOI: 10.1111/pbi.70028

Wang, R., Yang, Q., Deng, Z., & Nian, W. (2025). The research on soil-plant-climate interactions: An integrated assessment of water management and drought resilience. *Advances in Resources Research*, *5*(1), 456–476.

Wang, S.-W., Gao, C., Zheng, Y.-M., Yi, L., Lu, J.-C., Huang, X.-Y., Cai, J.-B., Zhang, P.-F., Cui, Y.-H., & Ke, A.-W. (2022). Current applications and future perspective of CRISPR/Cas9 gene editing in cancer. *Molecular Cancer*, *21*(1), 57. DOI: 10.1186/s12943-022-01518-8

Wei, Z. (2024). Analysis Of Publicly Funded Intellectual Property Management To Support Agricultural Development And Food Security In Low-And Middle-income Countries.

Yactayo-Chang, J. P., Tang, H. V., Mendoza, J., Christensen, S. A., & Block, A. K. (2020). Plant Defense Chemicals against Insect Pests. *Agronomy (Basel)*, *10*(8), 1156. Advance online publication. DOI: 10.3390/agronomy10081156

Yadav, M. R., Choudhary, M., Singh, J., Lal, M. K., Jha, P. K., Udawat, P., & Maheshwari, C. (2022). Impacts, tolerance, adaptation, and mitigation of heat stress on wheat under changing climates. *International Journal of Molecular Sciences*, *23*(5), 2838. DOI: 10.3390/ijms23052838

Yuan, L., Gai, W., Xuan, X., Ahiakpa, J. K., Li, F., Ge, P., & Zhang, Y. (2024). Advances in improving tomato fruit quality by gene editing. *Horticultural Plant Journal*. Advance online publication. DOI: 10.1016/j.hpj.2024.04.008

Yuan, P., Usman, M., Liu, W., Adhikari, A., Zhang, C., Njiti, V., & Xia, Y. (2024). Advancements in Plant Gene Editing Technology: From Construct Design to Enhanced Transformation Efficiency. *Biotechnology Journal*, *19*(12), e202400457. DOI: 10.1002/biot.202400457

Zhang, D., Hussain, A., Manghwar, H., Xie, K., Xie, S., Zhao, S., Larkin, R. M., Qing, P., Jin, S., & Ding, F. (2020). Genome editing with the CRISPR-Cas system: An art, ethics and global regulatory perspective. *Plant Biotechnology Journal*, *18*(8), 1651–1669. DOI: 10.1111/pbi.13383

Zhang, L., Zhu, M., Ren, L., Li, A., Chen, G., & Hu, Z. (2018). The SlFSR gene controls fruit shelf-life in tomato. *Journal of Experimental Botany*, *69*(12), 2897–2909. Advance online publication. DOI: 10.1093/jxb/ery116

Zhong, Q., Xu, Y., & Rao, Y. (2024). Mechanism of rice resistance to bacterial leaf blight via phytohormones. *Plants*, *13*(18), 2541. DOI: 10.3390/plants13182541

Chapter 4
Improving Nutritional Quality of Crops With Gene Editing

Xiaopei Qin
Shanghai Zhongqiao Vocational and Technical University, Shanghai, China

Iram Fatima
PCSIR Labs, Islamabad, Pakistan

Mamoona Munir
Rawalpindi Women University, Pakistan

Jazbia Shirin
Peking University, Shenzhen, China

Muhammad Waseem
http://orcid.org/0000-0001-7947-7722
Hainan University, Haikou, China

ABSTRACT

Crop nutritional quality is shaped by environment, variety, management, and post-harvest handling, with genetic variation driving differences in nutrient composition and adaptability. Gene editing is a targeted biotechnology that inserts, deletes, or replaces DNA sequences to redesign crop traits, advancing breeding toward genome-informed design. Compared with conventional gene transfer, gene editing is generally more precise and efficient, enabling site-specific changes without broadly altering gene copy number or relying on random DNA integration. As a leading genome-targeted tool, it is reshaping crop breeding and improving selection efficiency, with

DOI: 10.4018/979-8-3373-9534-0.ch004

major implications for agricultural development and food security. This chapter reviews key genome-editing platforms used in crop improvement and outlines three strategies to enhance nutritional quality: redirecting metabolic pathways, activating stress-response genes that influence nutrient profiles, and systematically optimizing multi-gene networks.

INTRODUCTION

With global population growth and the intensification of climate change, modern agriculture is confronted with numerous dilemmas and challenges, such as "high yield and supply security" and "nutritional health." Existing crop cultivation varieties urgently need improvement and optimization to cope with deteriorating environmental issues and a continuously growing world population. Although traditional crop breeding techniques have achieved remarkable results in increasing yield, they have inherent limitations in improving complex nutritional traits, such as the difficulty in removing anti-nutritional factors (such as phytic acid and oxalic acid), low efficiency of enriching micronutrients (such as vitamin D, iron, and zinc), and frequent fluctuations in crop nutritional quality due to environmental impacts. According to the United Nations Food and Agriculture Organization (FAO), over 2 billion people worldwide are in a state of "hidden hunger," and it is urgent to solve the problems of the accuracy, stability, and adaptability of nutritional fortification through technological innovation.

In the early 2000, scientists developed the first-generation gene-editing technology, ZFN technology, and systematically expounded the design principles of ZFN technology and its application in human cells (Durai et al., 2005,; Weinthal et al. 2010). In 2009, the second-generation gene-editing technology, transcription activator-like effector nuclease (TALEN) technology, emerged and has been widely applied (Boch et al., 2009). In 2012, two female scientists, Emmanuelle Charpentier and Jennifer A. Doudna, published a pioneering paper on the CRISPR-Cas9 gene editing system, proposing a gene editing technology that can rapidly, efficiently, and precisely modify the genome of living organisms(Chylinski et al., 2012), which was awarded the Nobel Prize in Chemistry in 2020, having a revolutionary impact on life sciences and giving rise to the third-generation gene editing technology, CRISPR/Cas9. In 2013, George Church from Harvard Medical School and Zhang Feng from the Broad Institute of MIT and Harvard, along with others, successfully applied the CRISPR-Cas system to gene editing in mammalian cells(Mali et al., 2013), and obtained the CRISPR-Cas9 technology patent in the following year.

Over the past decade, CRISPR-based gene-editing technology has developed at an astonishing speed and has been continuously updated (Table 1). Gene editing is different from transgenic technology; it involves guiding biological enzymes through probes to the positions that need editing, cutting the DNA, and separating the biological enzymes after cutting is completed. During the cell repair process, mutations occur, affecting traits such as crop yield, disease resistance, and nutritional quality. New varieties are bred, bringing unprecedented historical opportunities to increase grain production and improve people's health (Li et al., 2024).

Table 1. Features of precise editing technologies.

Editing technology		**Main effectors**	**Sequence changes**	**Product puritya**	**Editing window**	**Editing efficiency**
Single base editing system	Cytosine base editor	Rat rAPOBEC1	C:G>T:A	High	3-9	30%
	Adenine base editor	ecTadA-ecTadA*	A:T>G:C	Very high	4-9	50%
	Guanine base editor	APOBEC1	C-to-U	High	6-14	7%-12%
	Prime editor	Prime Editing Guide RNA	Any change between the four bases	low	1-50	40%-75%
Two-base gene editor	Saturated tareted endogenousmutagenesis editor	APOBEC3A-ecTadA-ecTadA	Any change between the four bases	High	30	60%
Other Editing technology	APOBEC-Cas fusion-induced deletion system	APOBEC3A	Designed multinucleotide deletion	low	Not available	50%
		Carboxy-terminal catalytic domain of APOBEC3B		low	Not available	50%

Gene-editing technology enables scientists to precisely target key nodes of metabolic pathways and regulate the synthesis, storage, and distribution of target components, as well as reshape the molecular mechanisms of crop-environment interactions. Currently, gene-editing-driven nutritional quality enhancement has formed three major technical paths: blocking and redirecting metabolic pathways, activating stress response genes, and systematically optimizing multi-gene networks. These strategies not only significantly enhance the nutritional density of crops but also improve their adaptability in resource-poor areas, providing multidimensional solutions to the global challenges intertwined with food security and nutritional health. This chapter systematically elaborates on the innovative logic of gene editing technology in crop nutritional improvement from aspects such as precise genomic editing tools and nutritional quality enhancement technical paths (metabolic regulation, environmental interaction, multi-gene network design, etc.) and explores its industrial application prospects and safety guarantees.

Precise genome editing tools

Since the early 21st century, gene-editing products have developed rapidly worldwide, and corresponding gene-editing tools have also been constantly innovated. This section elaborates on the core principles and application scenarios of precise genome-editing tools, reflecting the leap from "coarse cutting"to "molecular surgery" in gene editing.

ZFN and TALEN

ZFN and TALEN are first- and second-generation tools for gene editing, which emerged earlier than CRISPR technology. Their basic principle is to achieve gene positioning and cleavage by relying on the protein-DNA recognition mechanism (ZFN is based on zinc finger structure, and TALEN is based on transcription activator-like effector)(Moscou et al. 2009). The biggest difference between TALEN and ZFN is that it uses amino acid sequences as guides to recognize target gene sequences, which is more specific in binding to the target gene than zinc fingers, and has higher stability and lower cost. It is suitable for early crop gene knockout (Figure 1), but its design is relatively complex, and its efficiency is lower.

Figure 1. Gene knockout enhances crop stress resistance

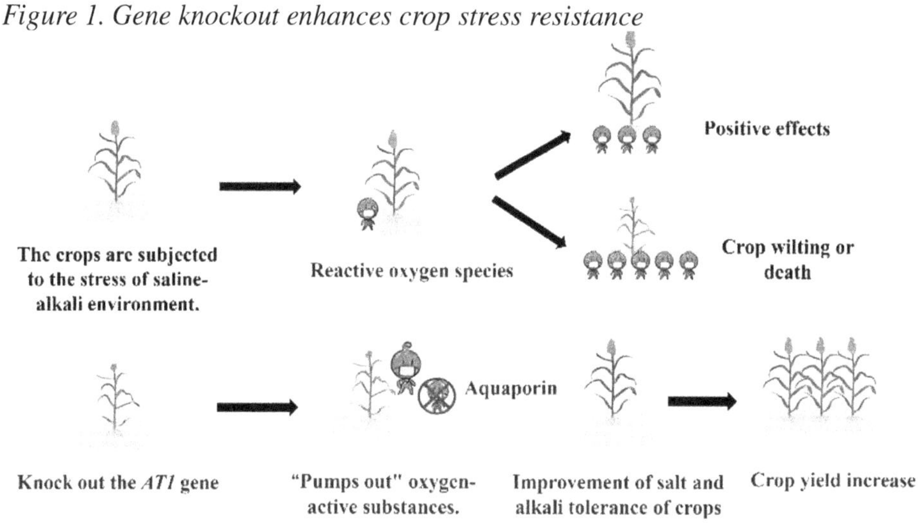

CRISPR-Cas9 system

The CRISPR-Cas9 system is a third-generation tool for genome editing, a biotechnology mediated by RNA, and is a pioneering genome editing tool. Clustered Regularly Interspaced Short Palindromic Repeats is natural defense mechanisms in single-celled microorganisms (Christian et al., 2010). During evolution, bacteria have developed the ability to absorb viral gene fragments and transcribe the corresponding sequences. By using these sequences to guide the Cas9 protein to bind with viral genes and cut them, they can affect the expression of the genes and achieve immune defense (Ishino et al. 1987). CRISPR-associated protein 9 is a key protein in the CRISPR system that is responsible for recognizing and cutting DNA molecules (Doudna et al., 2014). Its basic principle is to guide the Cas9 protein to cut DNA double strands through single guide RNA, triggering the repair mechanism for editing, fine-tuning the amino acid sequence of a certain protein, and changing the sequence that determines the expression level of a certain gene. The process of gene editing using the CRISPR-Cas9 system includes two main steps: the design of single guide RNA and the cutting of the Cas9 protein (Gasiunas et al., 2012). Single-guide RNA is composed of a CRISPR sequence and a specific target sequence, which can guide the Cas9 protein to the specific position of the target gene (Clarke et al., 2018). When the Cas9 protein binds to the target DNA sequence, it forms a double-strand break in the target DNA sequence, causing the DNA repair mechanism to intervene. The application of the CRISPR-Cas9 system in gene editing is extensive. It can be used for various crops, including wheat, rice, corn, and soybeans, and for the improvement of fruit trees, vegetables, and other crops. Through the CRISPR-Cas9 system, scientists have successfully improved traits such as yield, resistance, quality, and nutritional value of crops(Sander et al. 2014; Chen et al. 2013).

The Class 2 Type V-A CRISPR system

The Class 2 Type V-A CRISPR system is a modified gene-editing technology developed after the maturation of CRISPR-Cas9 technology. Its basic principle is to recognize PAM sequences rich in T, cut DNA to generate sticky ends, and reduce mismatches at target sites (Zetsche et al., 2015). Cas12a system is complementary to Cas9, and when it comes to the off-target effects in editing human cells, it is found that the off-target effect of Cas12a is lower, indicating that Cas12a is more suitable for precise operation at specific gene loci (Kleinstiver et al., 2016). Currently, the Cas12a system is mainly used to regulate stress-resistant genes in plants and for multiple gene editing. For example, it can perform multi-site editing on the soybean genome and optimize gene synthesis and metabolic pathways.

Base Editor (BE)

Base Editor (BE) is a tool developed in 2016 that can achieve single-base-level substitutions without generating DNA double-strand breaks. It has shown great potential in the fields of basic research and gene therapy (Komor et al. 2016; Gaudelli et al. 2017; Zhao et al. 2021). There are three main types of Base Editors: cytosine base editors (CBE), adenine base editors (ABE), and glycosylase base editors (GBE). They can achieve base editing of C-to-T, A-to-G, and C-to-G without the need for DNA double-strand breaks and editing templates(Chen et al. 2021, Komor, et al. 2016). Base Editors have high accuracy and are suitable for the targeted optimization of single nucleotide polymorphisms (SNPs). They are often used to precisely correct point mutations in genes related to herbicide resistance in maize (Anzalone et al. 2019; Molla, et al. 2020).

Prime Editor (PE)

Prime Editor (PE) is a novel editing technology proposed by Liu Ruicai, a core member of the Broad Institute of the United States and a Chinese biologist, and his team in a 2019 research paper. Its basic principle is to combine reverse transcriptase with Cas9 endonuclease and write new sequences through peg RNA templates, supporting complex targeted point mutations, precise insertions, precise deletions, and various combinations thereof without causing DNA double-strand breaks(Liu et al. 2019; Chen et al., 2023). Compared with Cas9 editing technology, PE technology is more efficient, has a wider editing window, fewer by-products, a lower off-target rate, and can handle complex editing that traditional HDR cannot achieve. Compared with base editing, prime editing is more flexible, as it can effectively perform all 12 types of single-base-free conversions without relying on DNA templates. Currently, it is mainly used to simultaneously repair multiple stress-resistant genes (such as OsSOS1 and OsDEP1) in rice, enhancing salt tolerance and alkali tolerance.

Orthogonal Cas protein systems and epigenetic editing

Orthogonal Cas protein systems and epigenetic editing are two derivative technologies of gene editing that have rapidly developed after 2020. Orthogonal Cas protein systems are based on the CRISPR-Cas9 system and integrate multiple non-competitive Cas proteins (such as Cas9 and Cas12a) to achieve multiple gene editing or parallel regulation, avoiding interference among tools (Feng et al. 2021; Tang et al., 2019). Epigenetic editing is a technology that precisely controls gene expression by targeting and regulating epigenetic marks (such as DNA methylation and histone modification) without changing the DNA sequence. It is a derivative tool of

CRISPR and an extension of base editing (Haldeman et al., 2019). The advantage of this technology is that it can precisely regulate genes and is safer.

After 2021, a research group led by Lai Liangxue from the Chinese Academy of Sciences fused the adenine base editor (ABE) and glycosylase base editor (CGBE) systems to develop a novel fusion double base editor, AGBE. This double base editor can induce single or simultaneous base conversions of the target sequence of the same allele using a single sgRNA, including four types of base conversions: A-to-G, C-to-G, C-to-T, and C-to-A(Liang et al., 2022). Unlike early STEME double-base editors, which simply concatenate the cytosine base editor with the adenine base editor, AGBE has a wider editing window and higher efficiency and is suitable for multi-fragment editing of complex genomes of organisms. However, the current research is aimed at mammals and has a relatively high editing efficiency for these animals. However, its applicability to crops requires further investigation.

These gene-editing tools have significantly enhanced the stress resistance, yield, and nutritional quality of crops by precisely regulating their genomes. They have become the core technology for modern agricultural biotechnology.

Precise regulation of nutrient metabolism pathways

Gene-editing technology significantly enhances the stress resistance and nutritional quality of crops by precisely regulating the key genes of their metabolic pathways. Its mechanism of action can be classified into four categories:

The key nodes of metabolic pathways

The first category involves optimizing the key nodes of metabolic pathways. By precisely modifying the genes of metabolic rate-limiting enzymes through base editing technology, the natural feedback inhibition mechanism can be relieved, and the synthesis efficiency of nutrients can be enhanced. For instance, modifying the MTHFR gene in the methylene tetrahydrofolate reductase of folic acid synthesis and the PSY1 gene in the synthesis of carotenoids can increase the folic acid content in rice endosperm by three times and the β-carotene content in tomatoes by six times(-Li et al. 2024; Zhang et al. 2021; Ahmar et al. 2020). By knocking out the caffeic acid-O-methyltransferase (COMT) gene in the lignin synthesis pathway of wheat, carbon metabolism can be redirected from fiber synthesis to starch accumulation, increasing the starch content in wheat grains by 12% and enhancing the lodging resistance of wheat stems(Liu et al., 2024).

The activation of metabolic network reconstruction systems

The second category is the activation of metabolic network reconstruction. By activating or regulating the transcription factor genes and nutrient component synthesis pathways of crops, crop resistance can be enhanced. For instance, editing OsNAC5 transcription factor genes related to aging, salt tolerance, drought resistance, and yield in rice can activate the proline synthesis pathway under drought conditions, thereby improving cell osmotic regulation and increasing water-use efficiency (Takasaki et al., 2010). Additionally, in various plants (such as rice), the expression level of the rate-limiting enzyme (PAL) in the phenylpropanoid metabolic pathway is positively correlated with the accumulation of lignin. By activating the PAL gene, the synthesis of lignin precursors (such as coumaric acid and ferulic acid) is directly affected, thereby regulating the degree of cell wall lignification and enhancing the crop's resistance to Fusarium head blight (Liu et al., 2025).

Precise enhancement of metabolic pathways

The third category is the precise enhancement of metabolic pathways. Through multiple gene-editing techniques, the crop's nutrient component synthesis pathways have been reconstructed and optimized, significantly enhancing their content. For instance, editing the HGGT and MPBQ-MT genes in corn enables the α-tocopherol content in the endosperm to reach 30% of the daily recommended intake (per 100 g) (Lu et al., 2024). Methionine, a basic unit of proteins, plays an important role in the growth of organisms. Studies have shown that salt stress regulates the molecular mechanism of plant salt tolerance and growth balance by inducing methionine synthesis and signal transduction. By strengthening the CGS and CBL genes in the methionine synthesis pathway, the methionine content can be increased (Shi et al. 2025).Chinese researchers have successfully created germplasm with high-yield seeds and high biomass using CRISPR-Cas9 to edit the floretin genes GmFT2a and GmFT5a in soybeans. The soluble protein and soluble sugar contents of the double mutant remained unchanged, whereas the phenylalanine content increased and the tannin content decreased, making it suitable for continuous development and utilization as a high-quality stem and leaf feed (Dong et al., 2025).

Dynamic regulation of metabolic networks

The fourth category is the dynamic regulation of metabolic networks. The dynamic regulation of nutrient enrichment sites and content is achieved by using promoters to drive animal genes. For instance, using an endosperm-specific promoter to drive the wheat iron transporter gene (TaVIT2), iron elements in the grains are enriched in the endosperm rather than the aleurone layer, thereby enhancing the nutritional value of the flour (Yang et al, 2023). In tomatoes, integrating photosensitive promoters to regulate anthocyanin synthesis genes (ANT1) enables the automatic activation of antioxidant substance synthesis under enhanced ultraviolet conditions, allowing the anthocyanin content in the fruits to be dynamically regulated by the intensity of light (Tan et al., 2025).

Tomatoes with high vitamin D3 content. On May 23, 2022, a research team from the Research Park of Norwich, UK, developed a tomato using CRISPR-Cas9 gene-editing technology mediated by Agrobacterium tumefaciens. This tomato can accumulate vitamin D3 and convert it into vitamin D3 upon sunlight exposure. The experimental results showed that in a medium-sized tomato (dry weight 8-10g) after gene editing, the vitamin D3 content reached 20% of the recommended daily intake, equivalent to the vitamin D3 content of two eggs; moreover, this study also indicated that the vitamin D3 content in tomato leaves was higher than that in tomato fruits, and tomato leaves could be used as a plant source for extracting vitamin D3 supplements (Li et al., 2022; Ledford, 2022).

Synergistic Optimization of Nutrient Absorption and Stress Response

In the field of gene-editing technology, the core of synergistic optimization of nutrient absorption and stress response lies in precisely regulating crop genes to achieve efficient generation and expression of nutrient components while simultaneously enhancing stress resistance. Currently, a considerable number of studies indicate that gene editing can enhance the nutrient absorption efficiency of crops under non-biological stress, indirectly improving their nutritional quality. The main mechanisms of action include the following two aspects:

Key gene targets should be identified

Gene editing technology can simultaneously target and regulate key genes involved in both nutrient absorption and stress response in plants. Gene-editing tools can recognize target DNA sequences through specific molecular structures. In the CRISPR-Cas9 system, single-stranded guide RNA binds to the target DNA through base complementary pairing, guiding the Cas9 nuclease to the precise genomic site, thereby achieving sequence recognition and obtaining key gene targets. Using the CRISPR/Cas9 system to identify key gene targets, such as ScVPP1 that regulates ion homeostasis, the nutrient absorption efficiency of crops under salt stress can be enhanced, and the nutrient accumulation ability in adverse conditions can be indirectly improved.

The genes of symbiotic systems should be regulated

Editing genes related to plant mycorrhizal symbiosis (such as the ammonium transporter gene ZmAMT3, the phosphorus homeostasis regulatory gene SPX, CCaMK, and SYMRK) can strengthen the symbiotic efficiency of arbuscular mycorrhizal fungi. These genes are involved in biological processes such as the invasion and symbiotic structure development of arbuscular mycorrhizal fungi and nutrient exchange. The research team of Wang Ertao was the first to draw the transcriptional regulatory network of rice-mycorrhizal symbiosis and found that both the direct phosphorus absorption pathway of plants (root pathway) and the phosphorus absorption pathway of mycorrhizal symbiosis (symbiotic pathway) are uniformly regulated by the phosphorus signal network of plants, which increased the phosphorus absorption efficiency of plants by more than 70% in low-phosphorus soil and simultaneously enhanced the adaptability of roots to salt-alkali stress (Shi et al., 2021).

Mineral-enriched crops under salt stress conditions. The research results of Wang Meng, Wu Jianhui, et al. indicate that in hexaploid wheat, the gene cluster TaCYP81Ds, composed of five tandemly repeated genes, significantly enhances salt tolerance and thereby protects wheat growth and yield potential under salt stress. This gene cluster regulates ion homeostasis and antioxidant systems, thereby reducing oxidative damage to cells caused by salt stress. The gene TaSPL6-D activates the sodium transporter gene TaHKT1;5-D, reducing sodium ion accumulation in cells and maintaining the absorption efficiency of essential minerals, such as potassium, thereby promoting mineral absorption and balance (Wang et al., 2020). Wang Mengcheng's research group used the overexpression wheat lines of the salt-tolerant gene TaCHP identified in the previous stage to conduct field yield experiments in saline-alkali land for 8 years and 4 years. The results showed that compared with the yield of the receptor wheat JN17, the yield of the TaCHP overexpression lines in saline-alkali land was significantly increased in each year and at each point, by 3.92–13.91% and 5.73–11.23%, respectively. At the same time, it significantly improved the salt tolerance of wheat in saline-alkali land for potassium, calcium, and other minerals, increasing the mineral content in grains by more than 20% while maintaining a stable yield (Chao et al., 2023). Chinese researchers Wang Qinnan and Gao Sanji collaborated on the overexpression of the ScVPP1 gene, which significantly enhanced plant salt tolerance. This gene can serve as a key site for salt tolerance genes in sugarcane and provide important genetic resources for breeding new high-salt-tolerance sugarcane varieties using marker-assisted breeding (Sun et al., 2025).

Multigene Synergistic Regulation and Network Design

Multigene synergistic regulation and network design are core development directions of gene editing technology, aiming to achieve targeted optimization of complex traits by precisely manipulating the interaction network of multiple genes. Combined with the multi-gene editing strategy, it is possible to simultaneously enhance the yield, stress resistance, and nutritional quality. The key technical paths in this field include two aspects: the development of multiple gene-editing tools using the compatibility of different nucleases, such as Cas9 and Cas12a (Cpf1), to form an orthogonal Cas protein system and simultaneously edit multiple gene loci through multiple single-stranded guide RNAs (sgRNA) arrays (Pacesa et al., 2024). It is also possible to form a gene circuit coupling by designing a "promoter-regulatory element-effect gene" chain reaction through synthetic biology, connecting stress resistance genes with nutrient absorption genes, and forming a regulatory network that links nutrient absorption and stress response (Figure 2). The other is to reconstruct gene regulatory networks (GRNs) by collaboratively editing target core transcription factors (such as NAC and MYB families) to change the downstream gene expression network. Alternatively, by editing long non-coding RNAs (such as miR399) through the CRISPR-dCas9 system, the cross-network of phosphorus starvation response and antioxidant pathways can be regulated to solve the collaborative challenges of low phosphorus and oxidative stress (Badia-i-Mompel et al., 2023).

Figure 2. Multigene regulation of functional traits in the interaction between microorganisms and nutrients.

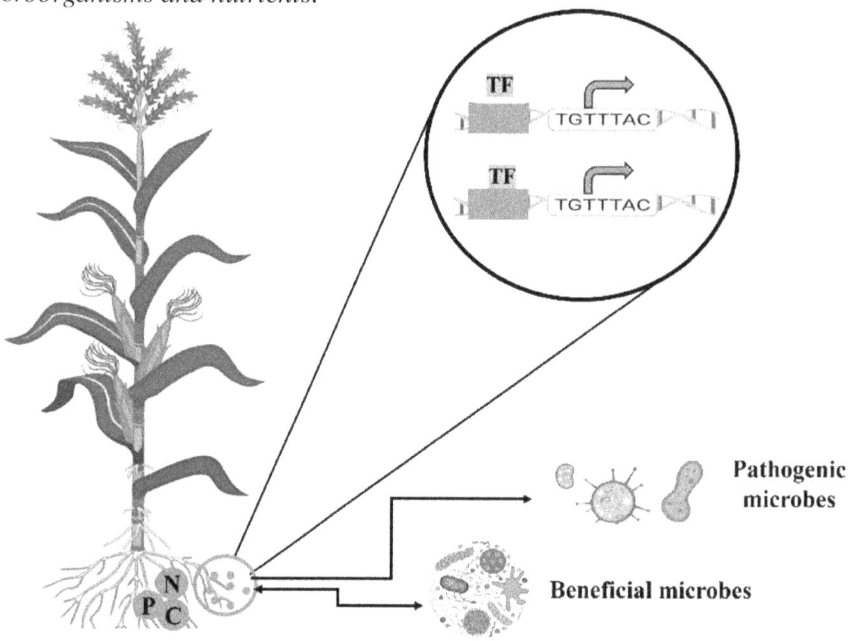

Wild tomato rapid domestication. In 2018, researchers utilized gene editing technology to edit the genes related to growth habits, fruit shape, fruit number, and nutritional quality in wild tomatoes, achieving rapid re-domestication of wild tomatoes and obtaining new materials with increased fruit size, increased fruit number, and significantly enhanced lycopene content. This resulted in a threefold increase in the yield per plant of wild tomatoes and a fivefold increase in GABA content, while retaining the disease resistance of the wild species (Zsögön et al., 2018). Chinese Academy of Sciences academician Li Jiayang and other researchers proposed four steps for the "re-domestication" of new crops using gene editing technology: the first step is to screen the best wild germplasm resources for comprehensive traits; the second step is to establish a re-domestication technology system for wild species, including the drawing of reference genomes, efficient genetic transformation systems, and genome editing technology systems; the third step is molecular design and re-domestication of varieties, including gene function verification, variety molecular design, gene editing, and field trait assessment; and the fourth step is the promotion and application of new crops(Yu et al., 2021). Their team successfully created the world's first rapid domesticated tetraploid rice material, which has significantly better yield potential and environmental adaptability than traditionally cultivated rice. In 2022, Xu Cao from the Chinese Academy of Sciences and other researchers developed a "two-in-one" rapid breeding strategy, which integrates the re-domestication breeding method and conventional hybrid breeding technology. This strategy has achieved the rapid introduction of stress-resistant traits and the "one-click" creation of diversified planting patterns in tomatoes, providing new ideas for the domestication breeding of efficient and rapid utilization of excellent wild resources and cultivated germplasm resources to cultivate excellent crop varieties (Yue et al., 2022).

These studies demonstrate that the renaissance of crop breeding through the integration of genome editing and precision breeding, known as "de novo domestication," will usher in a new wave of crop breeding. More wild crops will be domesticated into superior cultivated varieties, leading to the "third green revolution" in agriculture. Through the aforementioned multidimensional paths, gene editing technology is driving the transformation of crops from "high-yield and supply security" to "nutrition customization," providing a core driving force for the development of new quality productive forces in agriculture.

Improvement of nutritional quality and industrial application

Crop improvement through genome editing involves the targeted modification of genes to enhance crop traits, such as stress resistance, disease resistance, and nutrient content. It mainly includes two aspects: targeted enhancement of functional nutrients and content increase, and improvement of the structure and function of nutrients.

Targeted enhancement of functional nutrients

To meet human health needs, crop varieties with specific nutritional functions must be developed. By regulating the expression of key genes (such as PSY and LCYε) in the carotenoid metabolism pathway in banana fruits through gene editing, the β-carotene content in banana flesh was increased by 3 times and the lutein was increased by 2 times(Liu et al., 2021). This type of banana not only improves the color of the flesh but also enhances its antioxidant capacity, which has potential value for improving vitamin A deficiency. With the rapid development of gene-editing technology, many studies have been able to rapidly changed the genome structure of tomatoes, thereby changing their color (Figure 3). Currently, tomatoes can be red, yellow, purple, or green, and these colors are formed due to the combination and superposition of pigments such as carotenoids (such as lycopene and lutein), flavonoids (such as hesperetin chalcone), and chlorophyll. The research group of Li Chuanyou from the Institute of Genetics and Developmental Biology of the Chinese Academy of Sciences proposed a strategy for rapidly and directionally creating seven different fruit colors of tomato materials using red fruit tomato materials as the chassis materials through multiple gene editing technologies (Yang et al., 2023). These color changes are the result of the non-linkage of various loci according to Mendelian inheritance laws. Through gene-editing technology, we can create tomatoes of various colors, providing more choices for consumers.

Figure 3. Multifactorial gene editing enables the rapid customization of tomato fruit color.

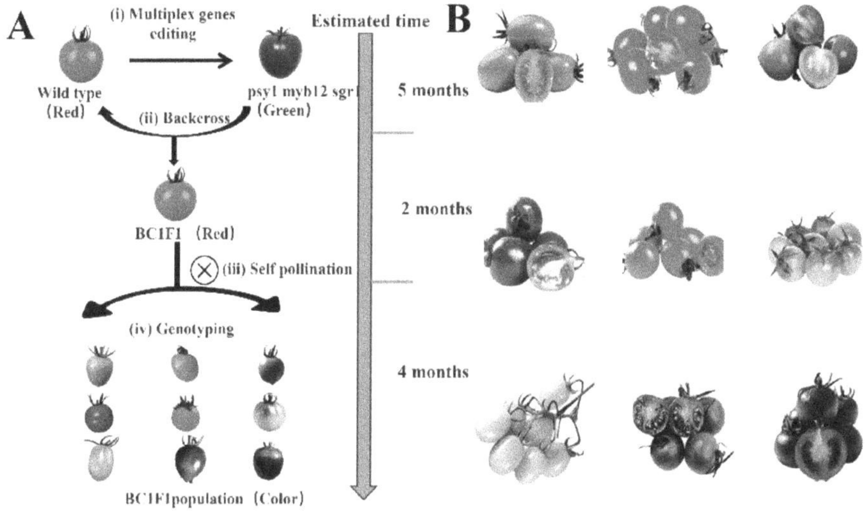

A:The process of multiple gene editing for tomato fruit coloration; B:Tomatoes of different fruit colors.

γ-Aminobutyric acid (GABA) regulates the human nervous system, alleviating mood, improving sleep quality, and regulating blood pressure. By knocking out the genes that limit GABA levels in tomatoes using gene-editing technology, the GABA content can be increased. Genetically edited tomatoes were launched for sale in Japan in 2022 and are quite popular among consumers owing to their potential health benefits. High oleic acid soybeans and rapeseed can be created by modifying the genes that control fatty acid synthesis. The high oleic acid soybeans led by Chinese scientist Zhu Jianhe obtained China's first agricultural use gene-edited biological production application safety certificate. They can increase the oleic acid content of soybeans from less than 20% to over 80%, exceeding the oleic acid content of olive oil. This variety has entered the industrialization phase. Through gene editing, rapeseeds with reduced phytic acid content can be created, providing higher nutritional value. By removing allergenic protein genes, low-allergenic peanuts and soybeans can be obtained, benefiting patients with allergies. By adjusting the amylose-to-amylopectin ratio through gene editing, potato varieties rich in anti-digestive starch can be developed, which not only reduces calorie absorption but also serves as a functional food raw material suitable for dietary development for patients with diabetes. CRISPR/Cas9 genome editing technology has been applied to potatoes to adjust the starch composition of tubers, enhance biotic stress resistance, and solve the problem of self-incompatibility, demonstrating its great application potential (Qu et al. 2024; Singh et al., 2024). Specific nutritional components can be enriched by editing and regulating key genes that control metabolic pathways. The vitamin C content of genetically edited lettuce was significantly higher than that of conventional varieties.

Improve the nutritional composition structure and function

Superior varieties of grain crops, such as wheat, rice, and corn, must exhibit high yield, disease resistance, stress tolerance, and excellent nutritional quality. Genetic editing of grain crops has also focused on these traits. An increasing number of research teams are adopting base editing and prime editing technologies to increase the yield, disease resistance, pest resistance, and attractiveness to consumers of commercial grain and commercial fruit and vegetable crops. In staple food crops, such as rice and wheat, starch can be transformed into resistant starch to limit the glycemic index. Using gene-editing technology to knock out the four MLO genes in wheat, wheat has obtained persistent resistance to powdery mildew. By using CRISPR/Cas9 to knockout the SS3a and SS3b genes in rice, a new variety of nutritious and healthy rice with high resistant starch content (up to 9.54% - 9.73%) and slow digestion rate was created, which is helpful for controlling blood sugar and promoting intestinal health(Rasheed et al., 2021; Huang et al., 2023; Huang et al., 2023). Chinese scientists used gene editing technology to knockout the Br2 gene controlling the height of corn plants and the ZmNL4 gene controlling the leaf width of corn, obtaining dwarf, robust, lodging-resistant corn, and corn with narrower leaves (Zhao et al. 2025). Both corn types are more suitable for high-density planting and naturally increase the per mu yield. Using gene editing technology to knock out the genes related to phytic acid synthesis in corn kernels, the phytic acid content in the kernels was reduced by more than 70%. Low-phytic-acid corn can increase the absorption rate of iron, zinc, and other trace elements by the human body, and is especially suitable for regions with malnutrition where corn is the main food (Zhang et al., 2024).

Safety of Genetically Engineered Crops

Genetically engineered crops, such as dwarf corn and low-cadmium rice, have entered the industrialization stage. Their safety and nutritional advantages have been certified by biosafety standards, providing technical support for the development of healthy foods in the future. Gene-editing technology is evolving from the improvement of a single trait to the coordinated regulation of multiple genes, providing an efficient tool for customized high-nutrition crops and promoting the upgrading of agriculture towards precision and functionality. Some countries, such as the United States, Japan, Australia, and South America, starting from the principle of gene editing technology, generally believe that genetically edited crops do not contain additional genes and are not subject to regulation. The United States has the fastest industrial development of genetically edited crops. Genetically edited high oleic acid soybeans, sweet corn, citrus rich in lycopene, herbicide-resistant rapeseed, anti-browning mushrooms, and non-sour mustard have been approved for commercialization. In January 2022, China released the "Trial Guidelines for Safety Evaluation of Agricultural Genetically Engineered Plants," clearly distinguishing "gene editing" from "genetic modification" and managing them separately. A simplified safety evaluation process was adopted for genetically edited plants that do not increase environmental or food safety risks. In April 2023, the "Trial Detailed Rules for Review of Agricultural Genetically Engineered Plants" was released to guide the safety management of genetically edited plants. In April 2023, Shandong Shunfeng Biotechnology Co., Ltd. in China obtained the first genetically edited biosafety certificate for staple grain crops in China for dwarf corn cultivation. This milestone event marked the rapid development of the industrialization of genetically edited crops in China.

CONCLUSIONS AND FUTURE PROSPECTS

An increasing number of scientific studies have shown that gene editing technology can be used to remove or weaken genes that are detrimental to the traits of crops and to enhance genes that are beneficial to crops, thereby helping to cultivate crops that are more tolerant to extreme weather, have higher yields, are more nutritious, and require less fertilizer and pesticides.

Gene editing technology is reconstructing the scientific path for improving crop nutritional quality with unprecedented precision and flexibility. This has also ushered in a new era of crop nutritional enhancement. By targeting key pathways, such as vitamin synthesis (e.g., beta-carotene and vitamin D), mineral absorption, and protein synthesis metabolism, scientists have successfully bred breakthrough varieties, such as vitamin A-enriched "golden rice" and iron-zinc dual-enriched wheat. This provides a molecular-level solution to the problem of global hidden hunger. Through base editing and guide editing technologies, the optimization of crop nutritional quality is no longer limited to the improvement of a single component but is moving towards the systematic design of multi-nutrient synergy enhancement and simultaneous reduction of anti-nutritional factors. The combination of synthetic biology and metabolic pathway network design enables crops to heterologously synthesize high-value nutritional components, such as DHA and astaxanthin, thereby expanding the scope of traditional agriculture.

With the development of orthogonal editing systems and dynamic regulatory networks, as well as the development of artificial intelligence technology, crop nutritional enhancement is breaking through the limitations of "unidirectional expression" and has achieved significant breakthroughs in achieving environmentally responsive nutritional enrichment. For example, proline content can automatically increase under drought conditions, and folic acid synthesis efficiency can be enhanced under low-light conditions. Although gene-editing technology is constantly trending towards innovation and development, it still needs to overcome technical bottlenecks, such as the complexity of gene interactions and metabolic resource competition. The artificial intelligence-driven genome prediction model and cross-species metabolic network transplantation technology have injected new energy into the precise design of " super-nutritional crops". From the laboratory to the field, gene editing technology not only redefines the connotation of "nutritional agriculture" but also elevates human ability to transform the food system to a new dimension at the molecular level, opening up new paths for ensuring global food security and sustainable healthy food.

REFERENCES

Ahmar, S., Gill, R. A., Jung, K.-H., Faheem, A., Qasim, M. U., Mubeen, M., & Zhou, W. (2020). Conventional and molecular techniques from simple breeding to speed breeding in crop plants: Recent advances and future outlook. *International Journal of Molecular Sciences*, *21*(7), 2590. DOI: 10.3390/ijms21072590

Anzalone, A. V., Randolph, P. B., Davis, J. R., Sousa, A. A., Koblan, L. W., Levy, J. M., Chen, P. J., Wilson, C., Newby, G. A., Raguram, A., & Liu, D. R. (2019). Search-and-replace genome editing without double-strand breaks or donor DNA. *Nature*, *576*(7785), 149–157. DOI: 10.1038/s41586-019-1711-4

Badia-i-Mompel, P., Wessels, L., Müller-Dott, S., Trimbour, R., Ramirez Flores, R. O., Argelaguet, R., & Saez-Rodriguez, J. (2023). Gene regulatory network inference in the era of single-cell multi-omics. *Nature Reviews. Genetics*, *24*(11), 739–754. DOI: 10.1038/s41576-023-00618-5

Boch, J., Scholze, H., Schornack, S., Landgraf, A., Hahn, S., Kay, S., & Bonas, U. (2009). Breaking the code of DNA binding specificity of TAL-type III effectors. *Science*, *326*(5959), 1509–1512. DOI: 10.1126/science.1178811

Chao, G., Wang, M., Li, X., Jiang, Z., Zhang, H., & Gao, D. (2023). TaCHP encoding C1-domain protein stably enhances wheat yield in saline-alkaline field. Journal of Integrative Plant Biology.

Chen, B., Gilbert, L. A., Cimini, B. A., Schnitzbauer, J., Zhang, W., Li, G.-W., Park, J., Blackburn, E. H., Weissman, J. S., Qi, L. S., & Huang, B. (2013). Dynamic imaging of genomic loci in living human cells by an optimized CRISPR/Cas system. *Cell*, *155*(7), 1479–1491. DOI: 10.1016/j.cell.2013.12.001

Chen, L., Park, J. E., Paa, P., Rajakumar, P. D., Prekop, H. T., Chew, Y. T., Mannebach, S. M., & Chew, W. L. (2021). Programmable C:G to G:C genome editing with CRISPR-Cas9-directed base excision repair proteins. *Nature Communications*, *12*(1), 1384. DOI: 10.1038/s41467-021-21559-9

Chen, L., Zhang, S., Xue, N., Hong, M., Zhang, X., Zhang, D., Yang, J., Bai, S., Huang, Y., Meng, H., Wu, H., Gao, C., & Li, D. (2023). Engineering a precise adenine base editor with minimal bystander editing. *Nature Chemical Biology*, *19*(1), 101–110. DOI: 10.1038/s41589-022-01163-8

Christian, M., Cermak, T., Doyle, E. L., Schmidt, C., Zhang, F., Hummel, A., Bogdanove, A. J., & Voytas, D. F. (2010). Targeting DNA double-strand breaks with TAL effector nucleases. *Genetics*, *186*(2), 757–761. DOI: 10.1534/genetics.110.120717

Clarke, R., Heler, R., MacDougall, M. S., Yeo, N. C., Chavez, A., Regan, M., Zhang, F., Doudna, J. A., Cate, J. H. D., & Merrill, B. J. (2018). Enhanced bacterial immunity and mammalian genome editing via RNA-polymerase-mediated dislodging of Cas9 from double-strand DNA breaks. *Molecular Cell*, *71*(1), 42–55. DOI: 10.1016/j.molcel.2018.06.005

Dong, S. S., Wang, X. R., Han, K. K., Tang, J. Q., Lu, W. T., Kong, X. Y., & Zhai, H. (2025). Creation of dual-purpose soybean germplasm for grain and forage by CRISPR/Cas9-mediated targeting mutation of GmFT2a and GmFT5a. *International Journal of Biological Macromolecules*, •••, 142234.

Doudna, J. A., & Charpentier, E. (2014). The new frontier of genome engineering with CRISPR-Cas9. *Science*, *346*(6213), 1258096. DOI: 10.1126/science.1258096

Durai, S., Mani, M., Kandavelou, K., Wu, J., Porteus, M. H., & Chandrasegaran, S. (2005). Zinc finger nucleases: Custom-designed molecular scissors for genome engineering of plant and mammalian cells. *Nucleic Acids Research*, *33*(18), 5978–5990. DOI: 10.1093/nar/gki912

Feng, W., Newbigging, A. M., Tao, J., Cao, Y., Peng, H., Le, C., Wu, J., Pang, B., Li, J., Qiu, D. L., Huang, J., Zhang, X., Zhang, X. C., & Le, X. C. (2021). CRISPR technology incorporating amplification strategies: Molecular assays for nucleic acids, proteins, and small molecules. *Chemical Science (Cambridge)*, *12*(13), 4683–4698. DOI: 10.1039/D0SC06973F

Gasiunas, G., Barrangou, R., Horvath, P., & Siksnys, V. (2012). Cas9–crRNA ribonucleoprotein complex mediates specific DNA cleavage for adaptive immunity in bacteria. *Proceedings of the National Academy of Sciences of the United States of America*, *109*(39), E2579–E2586. DOI: 10.1073/pnas.1208507109

Gaudelli, N. M., Komor, A. C., Rees, H. A., Packer, M. S., Badran, A. H., Bryson, D. I., & Liu, D. R. (2017). Programmable base editing of A•T to G•C in genomic DNA without DNA cleavage. *Nature*, *551*(7681), 464–471. DOI: 10.1038/nature24644

Haldeman, J. M., Conway, A. E., Arlotto, M. E., Slentz, D. H., Muoio, D. M., Becker, T. C., & Newgard, C. B. (2019). Creation of versatile cloning platforms for transgene expression and dCas9-based epigenome editing. *Nucleic Acids Research*, *47*(4), e23. DOI: 10.1093/nar/gky1286

Huang, L., Chao, Y., Zhao, W., Rao, Y., Shen, H., Gu, Z., & Liu, Q. (2023). Creating high-resistant starch rice by simultaneous editing of SS3a and SS3b. *Plant Biotechnology Journal*, *22*(4), 787–789. DOI: 10.1111/pbi.14053

Ishino, Y., Shinagawa, H., Makino, K., Amemura, M., & Nakata, A. (1987). Nucleotide sequence of the iap gene, responsible for alkaline phosphatase isozyme conversion in Escherichia coli, and identification of the gene product. *Journal of Bacteriology*, *169*(12), 5429–5433. DOI: 10.1128/jb.169.12.5429-5433.1987

Kleinstiver, B. P., Tsai, S. Q., Prew, M. S., Nguyen, N. T., Welch, M. M., Lopez, J. M., McCawley, T. R., Joung, J. K., & Joung, J. K. (2016). Genome-wide specificities of CRISPR-Cas Cpf1 nucleases in human cells. *Nature Biotechnology*, *34*(8), 869–874. DOI: 10.1038/nbt.3620

Komor, A. C., Kim, Y. B., Packer, M. S., Zuris, J. A., & Liu, D. R. (2016). Programmable editing of a target base in genomic DNA without double-stranded DNA cleavage. *Nature*, *533*(7603), 420–424. DOI: 10.1038/nature17946

Komor, A. C., Kim, Y. B., Packer, M. S., Zuris, J. A., & Liu, D. R. (2016). Programmable editing of a target base in genomic DNA without double-stranded DNA cleavage. *Nature*, *533*(7603), 420–424. DOI: 10.1038/nature17946

Ledford, H. (2022). Gene-edited tomatoes could provide new source of vitamin D. *Nature*. Advance online publication. DOI: 10.1038/d41586-022-01443-2

Li, B., Sun, C., Li, J., & Gao, C. (2024). Targeted genome-modification tools and their advanced applications in crop breeding. *Nature Reviews. Genetics*, *25*(9), 603–622. DOI: 10.1038/s41576-024-00720-2

Li, B., Sun, C., Li, J., & Gao, C. (2024). Targeted genome-modification tools and their advanced applications in crop breeding. *Nature Reviews. Genetics*, *25*(9), 603–622. DOI: 10.1038/s41576-024-00720-2

Li, J., Scarano, A., Gonzalez, N. M., D'Orso, F., Yue, Y., Nemeth, K., Saalbach, G., Hill, L., de Oliveira Martins, C., Moran, R., Santino, A., & Martin, C. (2022). Biofortified tomatoes provide a new route to vitamin D sufficiency. *Nature Plants*, *8*(6), 611–616. DOI: 10.1038/s41477-022-01154-6

Liang, Y., Zhang, Q., Wang, X., Gou, S., Lin, L., & Lai, L. (2022). AGBE: A dual deaminase-mediated base editor by fusing CGBE with ABE for creating a saturated mutant population with multiple editing patterns. *Nucleic Acids Research*, *50*(9), 5384–5399. DOI: 10.1093/nar/gkac353

Liu, D., He, J., Li, Q., Zhang, X., Wang, Y., Sun, Q., & Wan, J. (2025). A WRKY transcription factor confers broad-spectrum resistance to biotic stresses and yield stability in rice. *Proceedings of the National Academy of Sciences of the United States of America*, *122*(10), e2411164122. DOI: 10.1073/pnas.2411164122

Liu, D. R., Komor, A. C., & Kim, Y. B. (2019). Development of a broadly applicable gene-editing tool by fusion of Cas9 with high-fidelity cytidine deaminase. *Nature*.

Liu, J., Liu, M., Wang, J., Zhang, J., Miao, H., Wang, Z., & Jin, Z. (2021). Transcription factor MaMADS36 plays a central role in regulating banana fruit ripening. *Journal of Experimental Botany*, *72*(20), 7078–7091. DOI: 10.1093/jxb/erab341

Liu, Q. Cloning and characterizing candidate genes underlying major QTLs for wheat stem pith thickness and stem diameter (Doctoral dissertation, Murdoch University).

Lu, Q., Zhao, H., Zhang, Z., Bai, Y., Zhao, H., Liu, G., & Song, W. (2024). Genomic variation in weedy and cultivated broomcorn millet accessions uncovers the genetic architecture of agronomic traits. *Nature Genetics*, *56*(5), 1006–1017. DOI: 10.1038/s41588-024-01718-6

Mali, P., Yang, L., Esvelt, K. M., Aach, J., Guell, M., DiCarlo, J. E., Norville, J. E., & Church, G. M. (2013). RNA-guided human genome engineering via Cas9. *Science*, *339*(6121), 823–826. DOI: 10.1126/science.1232033

Molla, K. A., Qi, Y., Karmakar, S., & Baig, M. J. (2020). Base editing landscape extends to perform transversion mutation. *Trends in Genetics : TIG*, *36*(12), 899–901. DOI: 10.1016/j.tig.2020.09.001

Moscou, M. J., & Bogdanove, A. J. (2009). A simple cipher governs DNA recognition by TAL effectors. *Science*, *326*(5959), 1501. DOI: 10.1126/science.1178817

Pacesa, M., Pelea, O., & Jinek, M. (2024). Past, present, and future of CRISPR genome editing technologies. *Cell*, *187*(5), 1076–1100. DOI: 10.1016/j.cell.2024.01.042

Qu, L., Huang, X., Su, X., Zhu, G., Zheng, L., Lin, J., & Xue, H. (2024). Potato: From functional genomics to genetic improvement. *Molecular Horticulture*, *4*(1), 34. DOI: 10.1186/s43897-024-00105-3

Rasheed, A., Gill, R. A., Hassan, M. U., Mahmood, A., Qari, S., Zaman, Q. U., & Wu, Z. (2021). A critical review: Recent advancements in the use of CRISPR/Cas9 technology to enhance crops and alleviate global food crises. *Current Issues in Molecular Biology*, *43*(3), 1950–1976. DOI: 10.3390/cimb43030135

Sander, J. D., & Joung, J. K. (2014). CRISPR-Cas systems for editing, regulating and targeting genomes. *Nature Biotechnology*, *32*(4), 347–355. DOI: 10.1038/nbt.2842

Shi, B., Li, K., Xu, R., Zhang, F., Yu, Z., Ding, Z., & Tian, H. (2025). Methionine-mediated trade-off between plant growth and salt tolerance. *Plant Physiology*, *197*(3), kiaf074. DOI: 10.1093/plphys/kiaf074

Shi, J., Zhao, B., Zheng, S., Zhang, X., Wang, X., Dong, W., Qi, J., Wang, E., & Wang, E. (2021). A phosphate starvation response-centered network regulates mycorrhizal symbiosis. *Cell, 184*(22), 5527–5540. DOI: 10.1016/j.cell.2021.09.030

Singh, B., Lokya, V., Kaundal, P., & Tiwari, S. (2024). Genetic engineering for potato improvement: Current challenges and future opportunities. In *Genetic engineering of crop plants for food and health security* (Vol. 1, pp. 213–232). Springer Nature Singapore.

Sun, S. R., Wang, Z. Q., Lian, M., Chen, J. L., Qin, Y. X., Chang, H. L., & Wang, Q. N. (2025). Systematic identification of sugarcane vacuolar H+-translocating pyrophosphatase (VPP) gene family and the role of ScVPP1 in salt resistance. *Plant Cell Reports*, *44*(1), 11. DOI: 10.1007/s00299-024-03401-w

Takasaki, H., Maruyama, K., Kidokoro, S., Ito, Y., Fujita, Y., Shinozaki, K., Yamaguchi-Shinozaki, K., & Nakashima, K. (2010). The abiotic stress-responsive NAC-type transcription factor OsNAC5 regulates stress-inducible genes and stress tolerance in rice. *Molecular Genetics and Genomics*, *284*(3), 173–183. DOI: 10.1007/s00438-010-0557-0

Tan, X., Zeng, W., Yang, Y., Lin, Z., Li, F., Liu, J., Chen, S., Liu, Y.-G., Xie, W., & Xie, X. (2025). Genome-wide profiling of polymorphic short tandem repeats and their influence on gene expression and trait variation in diverse rice populations. *Journal of Genetics and Genomics*, *52*(6), 733–746. DOI: 10.1016/j.jgg.2025.03.005

Tang, Z., Chen, S., Chen, A., He, B., Zhou, Y., Chai, G., Guo, F., & Huang, J. (2019). CasPDB: An integrated and annotated database for Cas proteins from bacteria and archaea. *Database : The Journal of Biological Databases and Curation, 2019*, baz093. DOI: 10.1093/database/baz093

Wang, M., Yuan, J., Qin, L., Shi, W., & Liu, S. (2020). TaCYP81D5, one member in a wheat cytochrome P450 gene cluster, confers salinity tolerance via reactive oxygen species scavenging. *Plant Biotechnology Journal*, *18*(3), 791–804. DOI: 10.1111/pbi.13247

Weinthal, D., Tovkach, A., Zeevi, V., & Tzfira, T. (2010). Genome editing in plant cells by zinc finger nucleases. *Trends in Plant Science*, *15*(6), 308–321. DOI: 10.1016/j.tplants.2010.03.001

Yang, Q., Tan, Y., Ye, Y., Zhao, D., & Liu, Q. (2023). Serotonin enrichment of rice endosperm by metabolic engineering. *The Crop Journal*, *11*(6), 1943–1948. DOI: 10.1016/j.cj.2023.07.003

Yang, T., Ali, M., Lin, L., Li, P., He, H., Zhu, Q., & Deng, L. (2023). Recoloring tomato fruit by CRISPR/Cas9-mediated multiplex gene editing. *Horticulture Research*, *10*(1), uhac214. DOI: 10.1093/hr/uhac214

Yu, H., Lin, T., Meng, X., Du, H., Zhang, J., Liu, G., Chen, M., Jing, Y., Kou, L., Li, X., Gao, Q., Zhou, Y., Chen, Q., Liu, J., Huang, X., & Li, J. (2021). A route to de novo domestication of wild allotetraploid rice. *Cell*, *184*(5), 1156–1170. DOI: 10.1016/j.cell.2021.01.013

Zetsche, B., Gootenberg, J. S., Abudayyeh, O. O., Slaymaker, I. M., Makarova, K. S., Essletzbichler, P., Volz, S. E., Joung, J., van der Oost, J., Regev, A., Koonin, E. V., & Zhang, F. (2015). Cpf1 is a single RNA-guided endonuclease of a class 2 CRISPR-Cas system. *Cell*, *163*(3), 759–771. DOI: 10.1016/j.cell.2015.09.038

Zhang, D., Zhang, Z., Unver, T., & Zhang, B. (2021). CRISPR/Cas: A powerful tool for gene function study and crop improvement. *Journal of Advanced Research*, *29*, 207–221. DOI: 10.1016/j.jare.2020.10.003

Zhao, B., Sun, C., Yang, D., Zhao, Y., Cao, J., & Wang, B. (2025). CRISPR/Cas9-mediated genomic editing of Brachytic2 creates semi-dwarf mutant alleles for tailored maize breeding. Plant Biotechnology Journal.Zhang, Q., Zhao, Z., & Li, P. (2024). Research and progress of gene editing technology in maize. *Zhiwu Xuebao*, *59*(6), 978.

Zhao, D., Li, J., Li, S., Xin, X., Hu, M., Price, M. A., Rosser, S. J., Bi, C., & Zhang, X. (2021). Glycosylase base editors enable C-to-A and C-to-G base changes. *Nature Biotechnology*, *39*(1), 35–40. DOI: 10.1038/s41587-020-0592-2

Zhu, H., Li, C., & Gao, C. (2020). Applications of CRISPR–Cas in agriculture and plant biotechnology. *Nature Reviews. Molecular Cell Biology*, *21*(11), 661–677. DOI: 10.1038/s41580-020-00288-9

Zsögön, A., Čermák, T., Naves, E. R., Notini, M. M., Edel, K. H., Weinl, S., Freschi, L., Voytas, D. F., Kudla, J., & Peres, L. E. P. (2018). De novo domestication of wild tomato using genome editing. *Nature Biotechnology*, *36*(12), 1211–1216. DOI: 10.1038/nbt.4272

Chapter 5

Gene Editing and Crop Improvement Using CRISPR-Cas9 Technology

A. Mubarak Ali
http://orcid.org/0009-0009-6367-7864
Department of Microbiology, Sacred Heart College, Tirupattur, India

P. Saranraj
http://orcid.org/0000-0002-5052-6532
Department of Microbiology, Sacred Heart College, Tirupattur, India

S. Kanshiya
Department of Microbiology, Sacred Heart College, Tirupattur, India

P. S. Sruthi
Department of Microbiology, Sacred Heart College, Tirupattur, India

K. Gayathri
http://orcid.org/0009-0007-3786-2009
Department of Microbiology, Sacred Heart College, Tirupattur, India

M. Manigandan
http://orcid.org/0000-0002-8115-2713
Department of Microbiology, Sacred Heart College, Tirupattur, India

ABSTRACT

The CRISPR-Cas9 (Clustered Regularly Interspaced Short Palindromic Repeats paired with Cas9), genome editing techniques have significantly advanced. These technologies have changed functional genomics and agricultural breeding. This RNA-guided method makes use of a synthetic guide RNA (gRNA) that corresponds to the target DNA sequences. It instructs the Cas9 endonuclease to make double strand breaks at sites adjacent to a Protospacer Adjacent Motif (PAM). These breaks initiate cellular healing processes. Non-Homologous End Joining (NHEJ) results in insertion and deletion gene mutations. The targeted insertion or replacements accomplish the same goal as HDR, or homology-directed repair. This allows several modifications to be made without the need to add extra DNA. Compared to earlier

DOI: 10.4018/979-8-3373-9534-0.ch005

techniques like TALENs and Zinc Finger Nucleases (ZFNs), CRISPR-Cas9 is more user-friendly, flexible, and specific. Because of improved gRNAs and Cas variants, it has fewer off-target effects.

1. INTRODUCTION

Breaking the double-strands with a brief RNA sequence, the Cas9 protein, an endonuclease guided by RNA, is used to determine RNA targets in both plants and mammals. The development of Genetically Edited (GE) crops, such as those that are created using the conventional or mutation breeding method using this promising technique, is a promising and highly versatile approach to the delivery of sustainable productive agriculture to improve the nourishment of burgeoning populations in a changing climate. There have also been descriptions of the regulatory uncertainties and public acceptance of plant breeding using Cas9 genome editing. By overcoming biotic and abiotic challenges, the designer GE genetically modified plants would be able to increase production and improve weather adaptable and environmentally friendly agriculture in the future (Bhatta and Malla, 2020). Targeted editing of the plant genome using sequence-specific nucleases has enormous potential for improving crops to satisfy the world's expanding food demand and developing a successful, sustainable food production system. Improvement of crops in the past was done through conventional and mutant breeding of plants but currently it is fallen behind due to the loss of genetic diversity of plants that render it more challenging to produce food in the future. Enhancing crop quality, nutrition, yield, and resilience to stress from abiotic and biotic is the goal of crop improvement. New agricultural technologies have been developed during the past few decades, greatly increasing crop output. Because humans immediately absorb nutrients including protein molecules, fibre, vitamin, mineral, and bioactive compounds through the intake of crops, human health is associated more with agricultural quality. Additionally, scientists and breeders are focusing more on improving quality rather than boosting output. Nevertheless, particularly in the case of the breeding of polyploid crops, the conventional mutagenesis-based breeding techniques are tedious and challenging. Due to the precision and predictability of the plant genome modifications that are currently possible with the aid of genome editing technologies, Genome Editing (GE) technologies have recently achieved enormous success in crop breeding (Limbalkar et al., 2025).

One of the major applications for biochar in soil and environmental remediation is the detoxification of pesticides. Herbicides, insecticides, and fungicides often remain in soil and can have negative impacts on non-target species, water resources and human health. The use of biochar, which is produced by pyrolysis of biomass, offers a potential solution to these issues. Adsorption is one mechanism by which biochar reduces the toxicity of pesticides. Biochar has a large specific surface area and pore volume, and it adsorbs pesticides. The carboxyl and hydroxyl functional groups on biochar interact with pesticide molecules via non-polar interactions, hydrogen bonding and electrostatic attraction. This reduces the availability of pesticides to plants and microorganisms and therefore reduces their toxicity. Biochar also promotes microbial degradation of pesticides (Uranga et al., 2024). This makes the soil more hospitable to helpful microbes by improving soil structure, water retention and nutrient availability. These can biodegrade pesticide residues to non-toxic or less-toxic compounds. Biochar increases microbial diversity and enzyme activity, further speeding up the detoxification process. Moreover, biochar can change soil pH and other chemical properties that affect the stability of pesticides. In fact, a higher pH can enhance the rate of hydrolysis of certain pesticides, promoting their breakdown. Biochar may also play a role as a catalyst for oxidation and reduction reactions of pesticides. Finally, biochar can help to reduce leaching and surface runoff of pesticides. Thus, biochar is an environmentally friendly amendment that reduces the toxicity of pesticides by adsorption, microbial activation and soil pH improvement, leading to sustainable farming and environmental protection (Nadeem et al., 2017).

2. CRIPR TECHNOLOGY

The Clustered Regularly Between Shorter Palindromic Cas/Repeat system is a form of adaptive immunity that bacteria have been shown to have. This system aids the bacteria in defending themselves from foreign DNA infiltration, such as DNA from bacteriophage infections. The Cas9 protein is and CRISPR loci found in DNA make up this system. The genomic locus known as CRISPR is composed of tandem repeating patterns that arise from invasive elements and include protospacers and direct repeats. The CRISPR locus contains Cas9 genes, short transgene CRISPR RNA sequences identified as trans- activating RNA or crRNA (tracrRNA), and non-coding transcript patterns known as CRISPR genetic RNA (crRNA). The ends of the protospacer region cleave the target DNA, which is in the form of a double-strand DNA (Sharma et al., 2025). Once a guide RNA is bound to the protein, Cas9, which is an endonuclease associated with CRISPR loci, it will trigger the formation of a break at the junction (Double strands). The categorization of the Cas protein determines the five types and two categories of CRISPR and Cas systems. The type II CRISPR/SpCas9 pathway in *Streptococcus pyogenes* has been modified and converted into versatile GE tools for a range of applications. Its two principal components are guideRNA (gRNA or sgRNA) and the Cas9 protein. CRISPR RNA (crRNA) and trans-activating crRNA (tracrRNA) are made of gRNA (Tuncel et al., 2025). The former includes the Protospacer Adjacent Motif (PAM) in the target genome of interest following a complementary 20 nt region to a specific target gene position. Based on the gRNA, Cas9 nuclease produces DSBs approximately 3 bp above the PAM motif. The most common results of cleavage repair in NHEJ are gene deletion or lack of protein function. Alternatively, HDR can result from the introduction of an exogenous repair template in the optimal genomic region. CRISPR (Clustered Regularly Interspaced Short Palindromic Repeats), a genome editing technique that allows for extremely accurate DNA editing of an organism, is a revolutionary development in the field of genetics. The fundamental principle of the CRISPR technique is that a DNA endonuclease, like Cas9, is guided to a specific location in the DNA sequence by a guide RNA (Hasan et al., 2021).

This method allows for the selective cutting of the DNA strand, allowing specific genetic sequences to be added, removed, or modified. It was extracted from a defence mechanism in bacteria. The first step in the process is to create a complementary gRNA sequence to the DNA target sequence. The Cas9 enzyme can cut a target sequencing of DNA in a double-strand cut when it is within the cell, precisely where the Cas9-gRNA combination has attached itself to the target DNA sequence's base pairs. The cell's natural healing process can then be used to incorporate changes to the genetic code. CRISPR technology was a significant step as compared to the previous methods of genetic alterations. Over the generations, traditional techniques like selective breeding and mutagenesis caused by chemicals or radiation have been used to improve crop qualities. To direct the Cas enzyme to a specific region of the DNA and cause a double-strand break, CRISPR uses a guide RNA (Movahedi and Yang, 2025). The plant's repair mechanisms seal this fracture and result in specific genetic alterations. CRISPR precision has made it possible to produce crops with better characteristics, such as increased growth, disease resistance, nutritional composition, and resilience to abiotic stress. The possibilities of CRISPR have been significantly increased by more recent advancements like base editing and prime editing, which can make single nucleotide changes without producing DSBs. Such advancements have made CRISPR an important revolution in biological agricultural technology and have made it a powerful tool towards sustainable crop enhancement (Wan et al., 2021).

3. FUNCTIONAL MECHANISM OF CRISPR CAS9 TECHNOLOGY

The crystal structural investigations have shed information on the underlying process of the enzyme Cas9 action in a complicated structure with slightly twisted targeted DNA that contains the sgRNA and the PAM motif. This demonstrates how

Cas9 can be altered to create variants with distinct PAM specificities. However, for high throughput modification of genes, silencing, and DNA transcription control in a variety of systems, such as yeast, Drosophila, and other bacteria, living organisms, monkeys, zebrafish, human beings and plants, a range of commercially accessible Cas9 versions with improved specificities and fewer off-target consequences are available. One of the many potential applications of the CRISPR/Cas system made possible by the 2012 discovery of its molecular mechanism was RNA-Guided Engineered Nucleases (RGENs), which were used as sequence-specific nucleases to produce precise genetic alterations (Mešić et al., 2024). The two components of RGENs are the Cas9 nuclease and designed single guide RNA (sgRNA), which are genetically modified nucleases that cells must produce to perform genome editing. The 20 nucleotides at five ends of the sgRNA direct Cas9 to the proper target site. The first 20 nucleotide of the gRNA can be changed to any sequence of DNA in the form N20-NGG, in the case of innovative genome editing applications. The three stages of the CRISPR/Cas9 system's adaptive immunity are interference, expression, and adaptation. The process of adaptation entails the penetration of viral and plasmid genes pieces into the CRISPR locus. The transcription and processing of CRISPR loci produces small RNA (crRNA), which instructs effector endonucleases to use base complementarity to destroy the viral material (Ansori et al., 2023).

One Cas9 protein is needed for the Type IICRISPR/Cas system to perform DNA interference. The large protein Cas9 is made up of two inadequate RNAs, tracrRNA and crRNA, as well as additional domains like the Ruv C dominating amino terminus and the HNH protease domain in the core. Cas9 facilitates adaptability, helps transform pre-crRNA into crRNA, and produces targeted Double-Strand Breaks (DSBs) that are specific to tracrRNA and double-stranded RNA and directed toward RNase III. Furthermore, type III CRISPR is linked to the breakage protein Cas10 and is distinct from type II CRISPR in that it cleaves both DNA and RNA.A transcriptionally active promoter is present in the cleavage, a transcription-dependent alteration of the DNA sequence (Sharma et al., 2019). In certain cases, the Cas10 mechanism allows the bacterium to acquire spacer viral elements that result in some kind of resistance to foreign DNA. The host cell suffers from a resistance to foreign as well as viral genetic material because it prevents the activation of the lytic pathway. Additionally, these sequences may change the cell's physical properties, giving the host cell a chance to survive. To accomplish the deletion, insertion, or modification of genetic sequences, this technique also makes it possible to cut DNA several times at a particular location. It was brought in from a bacterial defense system (Kafle, 2022). Creating a gRNA nucleotide that is complimentary to the desired DNA sequence is the first step in the procedure. At the exact spot where the Cas9 enzyme and gRNA connect to a desired DNA sequence inside the cell, the Cas9 enzyme splits the target genetic material sequence into two separate strands. Then, by utilizing the cell's natural healing process, changes can be made to the genetic code. For many generations, conventional breeding techniques including natural selection and chemical or radiation-induced mutagenesis have been employed to enhance crop quality. Nevertheless, these techniques can result in unwanted genetic alterations and are frequently laborious and imprecise. These days, CRISPR-Cas systems are crucial tools for accurately altering the genomes of many different species, including plants (Atia et al., 2024).

CRISPR system Originally identified as a defensive mechanism for both archaea and bacteria against viral infection, the CRISPR system has recently been modified to modify the genomes of eukaryotic cells. Because it can produce specific DNA breakage in the form of two separate strands that are easily modified to modify genes, the CRISPR-Cas9 system—more especially, those obtained using Streptococcus pyogenes—is very popular. The CRISPR/Cas9 pathway includes multiple crucial steps in plant genome engineering, as shown in Figure 1. Plant tissue is converted, target genomic sequences are chosen, single-guide RNAs (sgRNA) are made and synthesized, the Cas9-sgRNA complex or its vectors are expressed into plant cells, and genetically modified lines with the necessary modifications are screened. While Cas12a (Cpf1) detects T-rich PAMs, Cas9 frequently detects 5'-NGG-3' PAM, suggesting that a PAM sequence adjacent to target DNA is necessary for accurate target identification. The specificity must now be high to ensure accurate genetic changes and minimise off-target effects. Together with the production of the related Cas genes into the precursor molecule known as pre-crRNA, the CRISPR locus has been expressed during the gene expression stage of the CRISPR genome mechanism (El-Mounadi et al., 2020). Cas proteins as well as auxiliary components subsequently transform this precursor RNA into smaller and more functional crRNAs. Enzymes like Cas6e or Cas6f cause site-specific fragmentation at junctions created by the combination of double and just one strand in Type I CRISPR-like systems. In Type I systems, hairpin loop topologies frequently produce junctions. A little RNA molecule called tracrRNA binds with the pre-crRNA in Type II CRISPR systems to create a double-stranded RNA complex in nature which is then cleaved by the enzymes Cas9 and RNase III working together (Arora and Narula, 2017).

4. CLASSIFICATION OF CRISPR TECHNOLOGY

CRISPR (Clustered Regularly Interspaced Short Palindromic Repeats) technology is a potent gene-editing technology, which enables scientists to edit the DNA with high precision. It is primarily categorized according to the construction and role of the CRISPR-associated (Cas) proteins and how they act to target genetic material. CRISPR systems can be classified into two major classes: Class 1 and Class 2, which are further subdivided into different types and subtypes. Class 1 CRISPR Systems feature multiple Cas proteins that co-exist as a complex to recognize and cleave target DNA. Such systems are more prevalent in nature and are present in several bacteria and archaea. There is Class 1 (Type I, Type III, and Type IV systems). In Type I systems, the target DNA is bound by a protein complex known as Cascade which recruits another protein known as Cas3 to cut the DNA (Mansi and Danai, 2026). Type III systems are versatile since they can be used to attack both DNA and RNA. Class 1 systems, however, are not as commonly used in genetic engineering since they are complex systems that consist of several proteins. Class 2 CRISPR Systems are less complex and more commonly applied in the biotechnology industry since they use a single and large Cas protein as a means of gene editing. There are Type II, Type V and Type VI systems under this classification. The most popular one is the Type II one that involves the Cas9 protein. Cas9 is directed to a target DNA sequence by a small RNA molecule, where it induces a double strand cut. This system is simple and efficient with wide use in research, medicine and agriculture. The Type V system that utilizes proteins like Cas12 is another key classification of Class 2. In contrast to Cas9, Cas12 produces staggered DNA cuts and even degrades single-stranded DNA (Chen et al., 2024).

This allows its application in gene editing as well as diagnostics. Likewise, the Type VI system that employs Cas13 targets RNA, rather than DNA. Cas13 is especially practical in the investigation of gene expression and the creation of tools to identify viral invasions. The technologies of CRISPR can also be categorized by their functional use (Song et al., 2025). To give an example, gene knockout and gene insertion are performed with the help of CRISPR-Cas9, gene expression suppression without the cutting of the DNA is performed with the help of CRISPR interference and gene expression stimulation is performed with the help of CRISPR activation. More recently, more advanced systems such as base editing and prime editing have enabled DNA to be edited accurately without introducing a double strand cut and therefore eliminates the possibility of unintended mutations (Ahmad, 2023).

5. THE CRISPR TECHNOLOGY FOR CROP IMPROVEMENT

The world's food needs cannot be met through enhancing crops through genetically modified crops or conventional breeding, nor can the losses brought on by climate change be made up for. As of right moment, additional sophisticated methods are needed in addition to these strategies. By effectively addressing the single nucleotide alterations that impact numerous important agronomic characteristics, base editing offers an opportunity to improve the crop type in this regard. Because of its precise targeting, mobility, and multiplexing capabilities, base editing is used to improve crops (Kumari et al., 2022). There are many uses for base editing in crop enhancement, which have garnered a lot of interest lately. In the same study, we also discuss how base modification can be used to close the gaps left by conventional genome editing and enhance crops. Many cell cycle events also activate the NHEJ repair pathway, which does not require a homologous repaired template. Because of this, interfering with genes by inserting small indels at the precise locations of target genes is now a regular practice. Because it may be used to insert donor sequences of genetic material without giving homology much care, NHEJ is an efficient gene stacking technique for crop development. Initiating Crossing Overs (COs) across homologous chromosomes, which aims to modify meiotic recombination genomic regions, appears to be a strategy for using specific structural alterations in plant genetics. This strategy aims to regulate CO density and rates. Homologous recombination (HR) fuses homologous chromosomes together in subsequent COs, preventing them from splitting apart during meiosis (Tuncel et al., 2023).

The Crop development depends heavily on meiotic recombination because it gives crops the ability to have specific allelic variants. Prime editing is a sophisticated type of genome editing that can be carried out in human cells and introduces all twelve base-to-base con variations, including transitions and transversions, in addition to indels. As of right now, it has been used to produce the intended mutations in plant cells, such as those from rice, wheat, tomatoes, potatoes, etc (Dong and Fan, 2024). Despite being a novel method, prime modification is a quick and effective way to precisely alter the plant genome. This powerful technique has been utilized by numerous researchers to cause desirable mutations in specific genes in agricultural crops. Because gene integration is random, the traditional random mutagenesis techniques were unable to target every gene to study its inactivation. To speed up plant breeding without adding a transgene, the CRISPR-Cas9 method would be very helpful in producing mutations in inaccessible genes, modifying several loci, and creating huge deletions.

However, GM crops might have been a way to improve agriculture if the debates surrounding the potential negative consequences of GM crops on the health of people and the environment had been avoided. The plants created by NPBT are identical, and their worth should be determined by the finished result rather than the production process. Therefore, CRISPR-edited plants pose additional hurdles to the management as well as acceptance from society of GE crops, even though they may not be covered by the current GMO legislation. Advisory and regulatory bodies are currently debating NPBTs to categorize genome editing GE crop varieties under non-GM to make them considered socially acceptable considering the GMO regulation. The discovery of innovative bioenergy crops that could maximize yield in wasteland and combat climate change will be greatly aided by genome editing. This method could enable any novel plant genome-editing concept, improving crop nutrition and food security. the potential ideas that could be used in crop development and plant biotechnology (Wani et al., 2022). The market values of crops have been significantly influenced by crop quality. Both internal and exterior characteristics have an impact on crop quality. Size, colour, texture, fragrance, and other physical and aesthetic characteristics are examples of external quality features. On the other hand, bioactive compounds (flavonoids, lycopene, carotenoids, and g-aminobutyric acid) and nutrients (protein, carbs, fats, etc.) are examples of internal quality factors. CRISPR/Cas9-mediated crop quality improvement focused primarily on fruit flavour, texture, nutritional value, edible quality, and physical attributes. Agricultural biotechnology has changed because of CRISPR technology's capacity to precisely and readily alter plant genomes. Its numerous and varied uses in agriculture address important issues like boosting crop yields, improving disease resistance, improving nutritional quality, and expanding resilience to the negative effects of the environment. CRISPR is being used to improve crop resilience to pests and diseases, adapt crops to new environmental factors, and lessen the environmental impact of agricultural practices (Adhikari and Poudel, 2020).

6. EFFECTS OF CRISPR TECHNOLOGY ON PLANT PRODUCTIVITY

The CRISPR/Cas system's off-target effects, which may arise from inadvertent Cas enzyme incisions at non-target locations, are one of the primary problems with genome editing. Inadvertent crop mutations can result in undesirable traits, reduced yields, or changed traits. Therefore, to ensure the safety and effectiveness of CRISPR-based technology and avoid off-target consequences, DNA cutting precision must be crucial. High-fidelity Cas versions and gRNA composition optimisation have made great strides in overcoming these obstacles. By decreasing nonspecific interactions with DNA, high-fidelity Cas9 variants such as eSpCas9, SpCas9-HF1, or FrCas9 have been produced to improve precision. The variations are suitable for usage because of their excellent on-target efficacy and modest off-target activity (Haider et al., 2026). Furthermore, research has shown that Cas12b is very promising for rice genome editing due to its low variant tolerance, high PAM needs, and negligible sgRNA-dependent off-target effects. The CRISPR/Cas system's capacity to precisely target genomic regions with little off-target consequences clarifies the significance of improving genetic RNA structure and Cas enzyme activity to improve. This will make CRISPR technological advancements safe and reliable in agricultural applications by increasing genome editing accuracy and reducing unintended consequences. The most recent developments in CRISPR technology must be incorporated into regulatory frameworks. Future research should primarily address safety issues, such as the long-term ecological effects of CRISPR-modified crops and off-target effects. This would require close collaboration between legislators, regulatory bodies, and scientists in addition to the development of appropriate risk assessment models. Reduced plant height, wilting leaves, and interference with buds and blooms—the latter signifying that a plant has already experienced dryness during its earliest vegetative stages—are the obvious signs of this. Severe drought frequently prevents plants from absorbing nutrients. The regulatory cascade that SlMAPK3 uses to support drought tolerance was explained by using the CRISPR-Cas9 technique to silence mitogen-activated tomato protein kinase 3 in tomatoes (Dormatey et al., 2025). In dry conditions, the stem curving and leaf wilting were worse in the slmapk3 mutant's lines than in the wild type lines. Additionally, compared to the wildtype plants, the mutant plants had much greater quantities of H 2 O 2, proline, and MDA, suggesting that the mutation lines were more significantly harmed by oxidative stressors and membrane damage under drought. These results in tomatoes shed light on the function of SlMAPK3 in drought resistance and reveal details about the SlMAPK3-controlled mechanism of drought tolerance. Other studies that have used genetic engineering to increase tomato SIMAPK3 transcription can improve drought tolerance and yield (Baloch et al., 2023).

Table 1. Effects of CRISPR Technology on Plant Productivity

Aspect	CRISPR-Based Modification	Effect on Plant Productivity	Examples
Yield Improvement	Editing yield-related genes	Increase grain size, number, and biomass	Rice (GW2, GS3 genes), Wheat
Disease Resistance	Knockout of susceptibility genes or insertion of resistance genes	Reduces crop losses due to pathogens	Tomato (resistance to bacterial speck), Rice (blast resistance)
Abiotic Stress Tolerance	Modification of stress-responsive genes	Enhances tolerance to drought, salinity, heat, and cold	Maize (drought tolerance), Rice (salt tolerance)
Nutrient Use Efficiency	Editing genes related to nutrient uptake and metabolism	Improves efficiency of nitrogen, phosphorus use	Rice and Wheat with enhanced nitrogen efficiency
Photosynthesis Efficiency	Engineering photosynthesis-related genes	Increases carbon fixation and biomass production	Tobacco and rice experimental models
Crop Quality Improvement	Modification of genes controlling nutritional traits	Enhance protein, oil, vitamin, and mineral content	Soybean (improved oil quality), Rice (biofortification)
Growth Rate and Development	Editing growth-regulating genes	Faster growth and early maturation	Tomato (early flowering), Arabidopsis

continued on following page

Table 1. Continued

Aspect	CRISPR-Based Modification	Effect on Plant Productivity	Examples
Weed Control Compatibility	Engineering herbicide resistance	Enables efficient weed management, improving yield	Maize, Soybean (herbicide-resistant varieties)
Reduction of Post-Harvest Loss	Editing genes affecting shelf life and ripening	Extends storage life and reduces spoilage	Tomato (delayed ripening)
Hybrid Seed Production	Editing fertility-related genes	Facilitate production of hybrid seeds with higher yield	Rice (male sterility systems)

7. CRISPR Cas9 TOOLS FOR CROP IMPROVEMENT

Important agricultural features including longer plant shelf life and resilience for heating, cold, and herbicides have been made possible by genome editing. India is the second largest of potato, a staple crop, and Indiana is the fourth-biggest growing commodity. However, growers and farmers are interested in enzymatic browning, which lowers the productivity of production and the level of quality of processed items (Bharti et al., 2025).

7.1. CRISPR/Cas9 for Abiotic Stress Tolerance

Artificial environmental conditions, such as high concentrations of elements like heat, water, cold, or drought, can lead to abiotic stress. Plants growth and development are greatly impacted by stress, which frequently causes growth retardation. With the CRISPR/Cas9 genome editing method, stress resistance can be readily increased. It offers a means of making plants more resilient to environmental stressors so they can withstand and thrive in abiotic environmental stresses (Kanth et al., 2025).

7.1.1. Drought Stress Tolerance

Plant drought tolerance is increased by transcription factors and the overexpression of numerous drought-sensitive genes, which promote the accumulation of signalling molecules. Drought-sensitive genes maintain drought vulnerability at a specific level while increasing the production of Reactive Oxygen Species (ROS) and decreasing antioxidant activity. AREB1 overexpression has been demonstrated to increase resistance to drought stress. enhanced drought resistance in the plants. Under drought stress, these plant mutations were found to be withering, to have lower antioxidant levels, and to have higher membrane damage. SIMAPK3 shields tomato cell membranes from drought stress, according to their research. The expression of CRISPR/Cas9 using a histone acetyl transferase gene increased Arabidopsis drought stress. Furthermore, drought sensitivity was found by modifying tomato dimensions, malondialdehyde, and H2O2 through base editing of the parthenogenesis tomato gene. Drought stress is largely determined by the SINPR1 gene in most crops (Sathiyabalan et al., 2024).

7.1.2. Salinity Stress Tolerance

Over half of the land used for agriculture is predicted to be severely salinized by 2050. Through changes in gene expression, salt stress causes a variety of changes in morphology and physio in plants. The primary symptoms of salinity stress include necrosis, the death of older leaves, and the accumulation of toxic ions in the cells. Many additional genes associated with plant stress tolerance have been found and described through genome editing using the CRISPR/Cas9 technique. Genes vulnerable to salt stress are controlled (Nirmal et al., 2025).

7.1.3. Heat Stress Tolerance

One of the main abiotic environmental constraints on plant growth is heat stress, which is mostly caused by global temperature increases linked to increased air pollution. Burning, yellowing, and wilting are signs of heat stress in plants. The plants complex molecular systems react to heat stress at the same time. Another important role is the activation of heat shock proteins, which are necessary to protect plants from heat stress. By controlling ROS like H2O2, these proteins alter intracellular transmission channels. As a result, CRISPR/Cas9 is a technique that helps plants improve crop resilience and get rid of heat stress (Patel et al., 2023).

7.1.4. Metal Stress Tolerance

Plant growth, yield, and productivity are all greatly impacted by metal stress. Plant cells are harmed by oxidative stress brought on by heavy metals. Plants have developed detoxifying mechanisms as a defense towards heavy metal stress because metal buildup interferes with biological processes. To withstand heavy metal ions, plants have developed to remove the ions from their roots. Tolerance to stress of metal is conferred by several genes. The oxp1/CRISPR mutant plant was discovered to be resistance to Cadmium (cd) in a study concerning the detoxifying of heavy metals in Arabidopsis using thexp1/CRISPR, suggesting that the mutation plants were more successful at detoxifying heavy metals than the Wild-Type (WT) oncs (Ray et al., 2023).

7.2. CRISPR/Cas9 for Biotic Stress Tolerance

7.2.1. Insect-Pest Resistance

The FAO estimates that insect infestations cost crops between 20 and 40% of their output. The usage of insecticides to control these pests and protect crops for human consumption is becoming more troublesome. However, it has been shown that the increasing use of chemical control techniques hurts consumers through soil toxicity and unsustainable environmental circumstances. The grower would surely benefit from any genetic improvement that could reduce the use of chemicals and, as a result, reduce insect losses. *Bacillus dinginess*, which was discovered in soil bacteria, has proven to be extremely potent, active, and successful in managing a wide range of Lepidopteran insects in several crops. In 1987, tomato insect resistance was reported using the Bt gene (Hamdan et al., 2023). Transgenic Bt the tomatoes offer defence against *Spodoptera litura* and *Heliothis virescens*. Transgenic persimmons with the Cry gene appear to be unaffected by Momena flavescens and Plodia inter punctata. One of the most widely consumed vegetables in Asia, particularly on the Indian subcontinent, is brinjal. However, *Leucinodesorbonali*, a lepidopteran bug, is a significant problem that gives insects an opportunity in Brinjal. The synthetic cry1AC gene, which produces an insecticidal tal protein was altered in 1998. These genes were examined via ELISA analysis. Using fruit bioassays, recombinant line sex showed significant differences in insect lethality. The potatoes are now resilient to the Rockies Potato Beetle. The Benee (Cry1Ac), which confers resistance to both the fruit and the shoot borer—the most devastating insect pest in Asia—has been effectively translated and expressed by okra (Ambrin and Eram, 2024).

7.2.2. Disease Resistance

There are many potentials to use the major modifying mechanism to speed up crop enhancement through genetic alteration, even if it is still in its early stages of research. In precision agricultural breeding, prime editing may make it possible to develop cultivars that are specifically resistant to a variety of biotic and abiotic risks. Due to single-nucleotide polymorphisms in various alleles, missense mutations can occasionally produce pseudogenes that expose plants to a variety of illnesses. Similarly, phytopathogens can multiply by taking advantage of plants' susceptibility (S) genes. Primary editing may be responsible for certain disease resistance by reactivating the pseudogenes and inactivating the S genes (Kumari et al., 2024). Prime editing can also correct these receptor genes, making the plants more resilient to disease. Furthermore, another crucial element that affects a plant's overall health is the interaction between bacteria and the plant. In addition to biotic stress tolerance, prime editing can result in abiotic stress tolerance. However, there is still need for improvement in the application of primary gene editing to treat cis-regulating factors, such as transcription components and other regulatory proteins, to counteract stresses like salt, drought, severe temperatures (Zhu et al., 2020).

7.2.3. Virus Resistance

Effective introduction of CRISPR elements is necessary for plant genome editing to be successful. By boosting cell absorption and shielding CRISPR components from degradation, one of the more modern techniques (mediated by nanoparticles) greatly improves trait improvement in maize. Viral vectors have been shown to be safer and more efficient in transient expression efforts, such as the development of virus resistance traits in tobacco and tomatoes using natural infection by viral routes. Based on the delivery of CRISPR component as proteins and RNA in a single complex, the Ribonucleoprotein (RNP) complexes delivery approach is intended to decrease off target (Kaur et al., 2025). This technique has proven successful in increasing agricultural yields and preventing illnesses in crops like wheat. By using sophisticated computational algorithms to create precise guide RNA and high-fidelity Cas variants, off-target activity is decreased. Using high-fidelity Cas9 variations, heat has been employed to increase drought resistance and decrease unwanted mutations. These developments expand the use of CRISPR, whether for the safety for genetic editing or the development of crops resistant to climate change (Liu et al., 2022).

Figure 1. CRISPR/Cas9 for Abiotic and Biotic Stress Tolerance

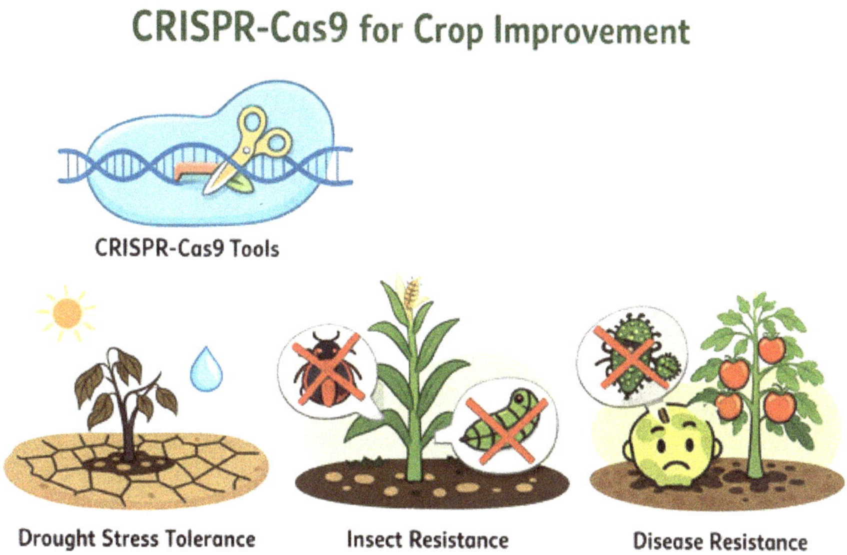

8. THE CRISPR TECHNOLOGY SPECIFICITY IN PLANT

The appropriate target sites, transformation techniques, and a suitable carrier network (codon-adapted Cas9 gene and Cas9 and sgRNA regulators) are necessary for efficient CRISPR/Cas9 genome alteration in plants. Single guide RNA (sgRNA) and the cas9 protein must be introduced and expressed in the targeted cell to perform CRISPR editing. A few vectors with expressions will be used to do this. Tissue-specific normal RNA polymerase type III inducers, including AtU6 and TaU6, which control the synthesis of short RNA in their specific species, typically regulate sgRNA (Angon and Habiba, 2023). The ubiquitin promoters, which regulate the presentation of longer RNAs, Cas9 is found beneath RNA polymerase II promoters. To transfer the Cas9 to the nuclear DNA, it is frequently functionalised with the Nuclear Localisation Sequence (NLS). The choice of vectors will be greatly influenced by the kind of Cas9 system, the expression technique that will be used, and the location of restrictions that will be present to insert the sgRNA. Additionally, independently sgRNA plasmids are made such that there is no co-expressed of Cas9, although thematic transcription can be combined with other forms of Cas9 to be beneficial. The best promoters to activate sgRNA or Cas9 expression and codon-optimized Cas9 must be selected to accomplish successful genome editing. Codon-optimized SpCas9 versions are used for most of the work in eukaryotic cells. It has been shown that modified Cas9 with other people or plant codons is effective (Das et al., 2025).

9. GENOME EDITING

A particular genome can be precisely and consistently changed using genome editing. Numerous genes in a variety of cell types and species may be altered by activity nucleases that provide modifications. DNA and non-specific cleaved domains of Fok1 endonuclease are bound by ZFNs, which are predicted restriction enzymes with specific sequence zinc finger motifs. ZFNs have successfully changed the genomes of numerous plants, including tobacco, maize, and soybeans (Dong, 2024).

9.1. Monocot

The CRISPR RGEN tools website was used to design sgRNA that targets specific regions with a higher outside-the-frame score to achieve optimal knockout efficacy using the freely available recombinant Cas9 protein (160 kDa). There have also been reports of successful targeted mutation with 0.1% and 0.5% - 0.69% indels (insertion or deletion) in the designated locus of MLO-7 and DIPM-1, 2, and 4, respectively, following direct delivery of CRISPR RNPs to plant protoplasts. However, protoplast formation is a difficult source of regeneration of plants in many cereal crops, particularly in monocots. Maize and other large monocots can now be transformed. The primary factor that increases transformation efficiency in maize (Zea mays) and other monocots is the overexpression of the Wuschel2 (Wus2) and Baby Boom genes. It was discovered that over-expression of these genes related to development causes many maize non-transforming inbred lines to have a high transforming rate. Additionally, Wuschel2 and Baby Boom, two morphogenic regulators, help change sorghum cultivars. Due to their need for callus cultivation and regeneration, most other cultivars are unable to maintain the beneficial wheat transformation process. Numerous issues related to cultures of tissues and regeneration can be avoided by using the Planta transformation method. Recent research indicates that sperm, zygotes, and egg cells are among the most dependable targets for genetic manipulation. The muscular tissues of mature embryos may be utilized by the planta wheat transforming system, which envisions the use of biolistic particle delivery (Mattiello et al., 2022).

Since these cells ultimately produce gametes, the transport could also occur through a shoot apical meristem. Furthermore, the most challenging aspect of using the plant gene-editing technology would be the transfer of CRISPR/Cas9 payload Efficiency is significantly impacted by the recipient genotype, particularly in monocots, biolistic attack, and Agrobacterium-mediated transformation. Some elite rice types are typically hard to adapt because they can't elicit the attributes needed for an environment and regeneration. Furthermore, it is impossible to eliminate T-DNA inclusion, and the ensuing plant regeneration procedures are typically difficult and time-consuming. Without the aid of an outside force, nanomaterials like Carbon Nanotubes (CNTs) and Nanoparticles (NPs) enable gene or the DNA of a plasmid to enter plant cells with walls. This is advantageous for the CRISPR/Cas9 system. In 2017, a novel technique known as "pollen magnet fiction" was employed to transfer DNA within pollen grains of different model crops to add exogenous genes. Approximately 1% of transgenic organisms were created after pollen was exposed to a magnet. Other researchers, however, were dubious about the reproducibility of pollen magnets. If CRISPR/Cas9 payload could be delivered to reproductive cells and continuously expressed by the pollen magnetoception approach, this will expedite the crucial step toward producing heritable gene change in transgenic seeds without utilizing tissue culture. HDR has several interesting applications in both basic and applied research. The CRISPR/Cas controlled resistance gene insertion in monocot plants by HDR has been established by Vu and associates. ABA-based ecTadA variants have been developed for rapeseed, wheat, rice. They don't work for all targets, though. Using a simpler ecTadA monomer form, improving the gRNA scaffold to produce better gRNAs, and inserting three SV40 nuclear distribution gene sequences within the C terminal region of nCas9 are some of the methods used to improve their editing of monocots (Ntsefong et al., 2023).

9.2. Dicot

Although Premier Editor appears to be effective and works best with monocot crops, there is need for improvement in its application to dicotyledons plants. It appeared that codon and promoter tweaking have been successfully applied to prime editors in tomatoes. GAI, ALS2, and PDS1 are three tomatoes naturally expressed genes that were edited using the customized primer editor pCXPE03. Only two genes, however, were accurately identified by the modifications discovered in the sequencing data: inserting the tip of a cg in PDS1 and multi-nucleotide changes in ALS2. According to their editing results, pCXPE03 could be utilised for tomato prime editing. They simultaneously caused transversions and nucleotide changes in the StALS1 gene using a main editor tuned to dicot codons. Sanger analysis revealed three expected substitutions at the StALS1 target locus using the PPE2 strategy, even though both the PPE2 and PPE3 techniques were used to induce mutation. There is a great deal of potential for precise breeding in polyploid and vegetatively propagated crops if the current primary modifying technology is enhanced and the planned base replacement is more varied. Additional sequencing verified the high mutation efficiency (26–84%). Another study discovered that the genes associated with albinism in Arabidopsis, magnesium-chelatase subunit I (CHLI1) and CHLI2, may be altered by CRISPR/Cas9 (Bhavanee et al., 2024). Fragment Length Polymorphism was used to select mutant plants. They demonstrated the importance of removing large sections of a plant's genome and making the required gene changes using the newly developed genetic editing tool. The effectiveness, heritability, specificity, and pattern of modified genes using CRISPR/Cas9 genome editing were investigated by looking at a set of seven genes that targeted 12 loci in Arabidopsis in successive generations. Few mutations were significantly changed in the T1 to T3 generations, and they were more likely to contain 1-bp insertions and base deletions. No off targets were found, and the homozygous mutants were passed on to the following generation in the absence of changes. This study showed that CRISPR/Cas9 genome editing in plants produced the heritable changes (Dormatey et al., 2020).

10. TARGETED GENOME MODIFICATION IN CROP PLANTS

10.1. Genome Modification for Nutrition Improvement

The CRISPR/Cas9 system can generate stable and hereditary changes without modifying the advantageous properties that already exist. The consequence of this is that the homozygous modified transgene free plants are transferred to successive generations in a single generation. Compared to ZFNs and TALENs, Cas9 was an improved tool with a rather high cleavage efficiency. The conventional ways of producing potatoes, apples, and mushrooms without acryl amide in them, without any bruising, and without browning is through low phytic acid in maize and mutation in the Polyphenol Oxidase (PPO) genes (Singh et al., 2025).

10.2. Biotic and Abiotic Stress Resistance through CRISPR/Cas9.

The primary editing system has enormous potential for accelerating crop improvement through genetic modification, even though it is still in its early stages of development. Future cultivars that are resistant to a range of biotic and abiotic stresses may be produced with the aid of precision crop breeding and prime editing. Single nucleotide variations in alleles can sometimes cause missense mutations through pseudogenes, leaving plants susceptible to a range of diseases. Similarly, to multiply, phytopathogens may look for plant susceptibility (S) genes. Primary editing, which aims to deactivate the S genes and restore the function of the pseudogenes, may cause some diseases to become resistant. To defend against invasive infections, Effector-Triggered Immunity (ETI) makes use of Nucleotide-Binding Leucine-Rich Repeat (NLR) immunity receptors found within cells. Prime editing can be used to change these receptor genes, giving the resulting plants more resistance to disease. Additionally, a significant component influencing the general health of plants is the interaction between microorganisms and plants. By using primary editing, it is feasible to target the DNA sequences that control the signalling systems that are advantageous to the plant in its interactions with the microbe. In addition to biotic stress resistance, prime editing can provide abiotic stress tolerance. However, there is still much to learn about using primary editing to target cis-regulatory factors, such as transcription variables and other regulatory proteins, to combat stressors like salt, drought, severe temperatures (Atak et al. 2022). Future crops of genomic edited sustainable productive agriculture will be more nutritious, more pest-resistant, and able to adapt to changing climate circumstances. The future of crop development will involve climate resilient agriculture that uses genome editing to study transcriptional regulation and analyse transcription cross talk between physiological and gene regulation cross talk in combination stress, coupled with manipulations mediated by mutagenesis. By focusing on the Coat Protein (CP), Intergenic Region (IR), and Replication Related Protein (RRP), single-guided RNAs (sgRNAs) cause the two sequence breaks in the viral genome. Drought is an abiotic stressor that contributes significantly to food poverty. According to recent research, CRISPR/Cas9 may effectively modify the trehalose gene, reduce its activity, and raise the amount of trehalose (a-D-glucopyranosyl-1, 1-a-D glucopyranoside) in plants. This technique produces genetically modified crops that are more drought resistant. Indica rice cultivars with the Drought Salt Tolerance (DST) gene (MTU1010) can be altered using the CRISPR/Cas9 genome editing approach to produce indica mega rice that is more resistant to salt and drought. Another gene that has been shown to have decreased ethanol sensitive in maize is ARGOS8, which functions as negative regulator. Additionally, when subjected to drought resistance, the ARGOS8 gene may be over expressed, leading to an increase in crop production. According to the analysis's findings, the endogenous ARGOS8 gene has extremely low mRNA expression. The most effective method for increasing the expression of the ARGOS8 gene in maize has been shown to be CRISPR/Cas9. Modern tomato hybridization cultivars have been produced by breeders through rigorous breeding cycles; however, these new cultivars are frequently the result of numerous biotic and abiotic stresses. It also results in loss of genetic variety and fitness consequences. One of the most dangerous abiotic stressors is cold stress, which causes plant cell tissues to freeze before being implanted. Low germination rates, injured seedlings, and poor development are all signs of cold stress (Faizal et al., 2024).

11. RECENT ADVANCEMENT IN CRISPR CAS9 IN CROP IMPROVEMENT

CRISPR-Cas9 technology has developed very fast over the past few years and has been made one of the most essential instruments in crop enhancement. Its capacity to precisely edit genes has allowed scientists to come up with crops that have a higher yield, greater nutritional value as well as resistance to environmental stress. In agriculture, this technology has also been made more efficient, accurate, and applicable with the recent advancements. Among the significant improvements, the high-fidelity Cas9 variants must also be mentioned. These are superior forms of the Cas9 protein and are created to minimize off-target effects which are unintended alterations in the genome. This has rendered gene editing safer and more dependable. As an example, high-fidelity Cas9 has been employed successfully in crops such as wheat to increase drought tolerance with minimal undesirable mutations. This has been especially significant in the development of climatically resistant crops. The other significant trend is the introduction of prime editing and base editing technologies. These new methods can make specific-base changes in the DNA rather than cutting the entire strand of DNA which is what traditional CRISPR-Cas9 does. This minimizes the possibility of making mistakes and makes editing more efficient. This is being done to alter the metabolic pathways in crops to enhance the nutritional value of crops by enhancing the levels of vitamins and bioactive compounds (Wang et al., 2022).

New developments have also been aimed at enhancing the production and quality characteristics of crops. Gene editing in crops such as rice and maize has been carried out using CRISPR-Cas9 to modify the genes that determine grain size, weight and composition. As an example, specific gene editing of genes like GS3 and GW2 has led to higher grain size and enhanced nutrition in rice. Likewise, it has been applied through gene editing to produce tomatoes that have a longer shelf life and better color, and potatoes that have fewer harmful compounds that are produced during storage and processing. The other major development is the application of CRISPR-Cas9 to increase biotic and abiotic stress resistance. Resistance to diseases, pests, and adverse environmental factors like drought and salinity can now be introduced into crops by scientists as they can knock out susceptibility genes or introduce resistance in the crops. This plays a vital role in making sure that food security exists in the climate change scenario. Another recent development is the combination of AI and computational tools with CRISPR technology. AI can be applied to more effective guide RNAs, predicting off-target effects, and enhancing precision in editing. This saves time on experiments and makes genome editing in crops more successful. Also, CRISPR combined with speed breeding methods has made it much faster to create new varieties of crops. Innovations have also been achieved regarding ways of delivering components of CRISPR to plant cells. Better methods like nanoparticle-based delivery and DNA-free editing methods minimize regulatory issues and make the technology more acceptable in commercial applications. These methods assist to generate gene-edited crops without the implantation of alien DNA that resemble the spontaneous mutations. Also, CRISPR-Cas9 is being applied to make crops with consumer-desired characteristics, including fruit without seeds, non-browning, and better taste and texture. An example is the creation of seedless berries and non-browning bananas and avocados with the help of CRISPR and has increased the market value of these foods and decreased food waste (Farooq et al., 2026).

12. POTENTIAL APPLICATION OF CRISPR TECHNOLOGY IN CROP IMPROVEMENT

The world's food needs cannot be met by using transgenic technology or traditional breeding to improve crops, nor can the losses brought on by climate change be compensated for. These days, more sophisticated procedures are needed, therefore such strategies are insufficient. By accurately rectifying single nucleotide mutations that impact different agronomic features of crops, base editing might be used to enhance a particular kind of crops. Because base editing is focused, non-invasive, and multiplexing-adaptable, it is used to improve crops. The numerous applications of base modification to enhance crops have garnered significant interest in the past few years. In this paper, we also address the potential of base editing to enhance crops by completing the gaps left by conventional genome editing.

13. REGULATION AND SOCIAL ACCEPTANCE OF CRISPR EDIT CROP

Plant biology research will benefit from the directed mutagenesis that is frequently carried out on plants utilizing the CRISPR/Cas9 system. Because the genes were included at random, the traditional methods of random mutation were unable to inactivate every gene. To speed up plant breeding procedures without introducing a transgene, CRISPR-Cas9 technology might be very helpful in generating mutants in the genes that are difficult to access, large-scale mutation, and enormous deletions. However, if the arguments about GM crops' possible detrimental effects on the environment and human health could have been avoided, GM crops would have been seen as a means of improving agriculture. Genetic engineering allows for the random insertion of a DNA construct into one or more chromosomal to regulate the genome. However, when significant, abrupt changes—like adding an entire metabolism to a plant—are needed and can be unpredictable, random gene insertion is not the ideal option. Using random mutagenesis and the inherent genetic variation already present, plant breeding techniques have also been widely utilized to bring novel features into domesticated species (Kumar et al. 2023). To produce transgene-free plants for the improvement of agricultural types, the CRISPR/Cas system presents a precise concept of modifications to a plant's genome that is robust and enables the removal of transgenic regions after target gene editing. CRISPR-Cas9 technology challenges. The main issue with CRISPR/Cas9 technology is unexpected modifications brought on by off-target impacts. Off-target cleavage may result from an excess or inadequate amount of Cas9 to sgRNA (single-guide RNA). The impact of cleavage off target increases with the ratio of improper concentration of Cas9: sgRNA. PAM sites may be promiscuous, which could result in undesired DNA cleavage. Guide-RNA: Cas9 complexes cut off-target locations inside or close to the protospacer adjacent motif (PAM) at greater concentrations. However, with low Cas9: guide-RNA concentrations, DNA cleavage might not happen. Compared to current alignment methods, the Cas9/sgRNA target area is quite short, which is crucial for effectively identifying any off-target area. Because the format of a sgRNA varies among sgRNAs and depends on several factors, it is seen to be a significant issue with this approach. The resistant sgRNA/target is another major problem with CRISPR/Cas9 technology. Several factors, such as the hairpin form of the sgRNA and the chromatin structure of the target locus, may render some sgRNAs ineffective (Mohamed et al., 2024).

14. CONCLUSION

CRISPR-Cas9 technology, which delivers levels that were unheard of accuracy and efficacy in crop development, has revolutionized gene editing. It tackles important issues in agriculture, like increasing yield, nutritional value, and resistance to both biotic and abiotic stress, by permitting modifications to plant genomes. CRISPR-Cas9 allows multiplexed edits, rapid trait stacking, and the development of distinct alleles without including foreign DNA, all of which are critical advantages for producing non-GMO crops that meet consumer and regulatory requirements, in contrast to traditional breeding methods, which are time-consuming and often inaccurate. In practice, CRISPR has yielded groundbreaking outcomes. For instance, disease-resistant rice cultivars modified for bacterial blight genes that are susceptible have boosted yields by up to 30% in field tests, while drought-tolerant wheat and maize lines construct with stress- responsive promoters promise nutrient availability in climate-sensitive areas. Nutrient-enhanced crops, such biofortified tomatoes with higher GABA levels or golden bananas with lots of provitamin A, best illustrate its capacity to combat malnutrition. By optimizing resource use and reducing dependency on pesticides, these advancements promote sustainable agri-food systems, which is crucial for regions like India that are coping with rising populations and erratic monsoons. CRISPR's integration with state-of-the-art methods including base modification, prime editing, and artificial intelligence-driven predictive modelling will speed up discovery and implementation in the future. There are still issues with off-target effects, delivery optimisation for numerous crops, and ethical and legal barriers, particularly in developing nations. International frameworks like those from the FAO as well as national health and safety committees need to be modified to achieve a balance between innovations and biosafety. forth the end, CRISPR-Cas9 helps scientists develop resilient, nutrient-dense crops that sustain global food systems, ushering forth a new era in precision agriculture. As use increases, it may help promote equitable agri-food chains by preventing yield losses due to climate change, which are expected to reach 20–40% by 2050.

REFERENCES

Adhikari, P., & Poudel, M. (2020). CRISPR-Cas9 in agriculture: Approaches, applications, future perspectives, and associated challenges. *Malaysian Journal of Halal Research*, *3*(1), 6–12. DOI: 10.2478/mjhr-2020-0002

Ahmad, M. (2023). Plant breeding advancements with "CRISPR-Cas" genome editing technologies will assist future food security. *Frontiers in Plant Science*, *14*, 1133036. DOI: 10.3389/fpls.2023.1133036 PMID: 36993865

Ambrin, G., & Eram, R. (2024). Perspective chapter: Major insights into CRISPR-Cas9 in edible oilseeds research. In *IntechOpen eBooks*. IntechOpen., DOI: 10.5772/intechopen.114967

Angon, P. B., & Habiba, U. (2023). Application of the CRISPR/Cas9 gene-editing system and its participation in plant and medical science. *Current Applied Science and Technology*, *23*(3). Advance online publication. DOI: 10.55003/cast.2022.03.23.003

Ansori, A. N. M., Antonius, Y., Susilo, R. J. K., Hayaza, S., Kharisma, V. D., Parikesit, A. A., Zainul, R., Jakhmola, V., Rebezov, M., & Ullah, M. E. (2023). Application of CRISPR-Cas9 genome editing technology in various fields: A review. *Narra J*, *3*(2), e184. DOI: 10.52225/narra.v3i2.184 PMID: 38450259

Arora, L., & Narula, A. (2017). Gene editing and crop improvement using CRISPR-Cas9 system. *Frontiers in Plant Science*, *8*, 1932. DOI: 10.3389/fpls.2017.01932 PMID: 29167680

Atak, A., Gizachew, H., & Gidamo, A.. (2022). Case studies of breeding strategies in major plant species. In *IntechOpen eBooks*. IntechOpen., DOI: 10.5772/intechopen.100882

Atia, M., Jiang, W., Sedeek, K., Butt, H., & Mahfouz, M. (2024). Crop bioengineering via gene editing: Reshaping the future of agriculture. *Plant Cell Reports*, *43*(4), 98. DOI: 10.1007/s00299-024-03183-1 PMID: 38494539

Baloch, F. S., Altaf, M. T., Liaqat, W., Bedir, M., Nadeem, M. A., Cömertpay, G., Çoban, N., Habyarimana, E., Barutçular, C., Cerit, I., Ludidi, N., Karaköy, T., Aasim, M., Chung, Y. S., Nawaz, M. A., Hatipoğlu, R., Kökten, K., & Sun, H.-J. (2023). Recent advancements in the breeding of sorghum crop: Current status and future strategies for marker-assisted breeding. *Frontiers in Genetics*, *14*, 1150616. DOI: 10.3389/fgene.2023.1150616 PMID: 37252661

Bharti, D. R., Pramanik, B., Paul, A., & Debnath, S. (2025). CRISPR-based precision breeding for crop improvement: current achievements and future directions. In *Plant Stress Tolerance* (pp. 118–138). CRC Press. DOI: 10.1201/9781003457237-6

Bhatta, B. P., & Malla, S. (2020). Improving horticultural crops via CRISPR/Cas9: Current successes and prospects. *Plants*, *9*(10), 1360. DOI: 10.3390/plants9101360 PMID: 33066510

Bhavanee, K. N. D., Krishnamoorthi, A., Rathva, H. M., Mareguddikar, S. C., Singh, A., Singh, B. P., Nageshwar, , & Chittibomma, K. (2024). Advancements in genetic engineering for enhanced traits in horticulture crops: A comprehensive review. *Journal of Advances in Biology & Biotechnology*, *27*(4), 90–105. DOI: 10.9734/jabb/2024/v27i2702

Das, P., Dutta, A. K., & Das, A. (2025). CRISPR/Cas9 applications for nutritional enhancement in major oilseed, vegetable, and cereal crops: A comprehensive review. *Plant Molecular Biology Reporter*, *43*(4), 1732–1755. DOI: 10.1007/s11105-025-01592-4

Dong, G., & Fan, Z. (2024). CRISPR/Cas-mediated germplasm improvement and new strategies for crop protection. *Crop Health*, *2*(1), 2. DOI: 10.1007/s44297-023-00020-x PMID: 41649714

. Dong, H. (2024). Application of genome editing techniques to regulate gene expression in crops. BMC Plant Biology, 24, Article number. DOI: 10.1186/s12870-024-04786-2

Dormatey, R., Antwi-Boasiako, A., Chimsah, F. A., Kojo, K. Y. K., Mensah, J., Keteku, A. K., & Dapaah, H. K. (2025). Applications of CRISPR in Crop Improvement and Global Food Security. In *CRISPR Technologies-Advances in Genome Editing*. Applications, and Ethical Implications. IntechOpen. DOI: 10.5772/intechopen.1011717

Dormatey, R., Sun, C., Ali, K., Coulter, J. A., Bi, Z., & Bai, J. (2020). Gene pyramiding for sustainable crop improvement against biotic and abiotic stresses. *Agronomy (Basel)*, *10*(9), 1255. DOI: 10.3390/agronomy10091255

El-Mounadi, K., Morales-Floriano, M. L., & García-Ruiz, H. (2020). Principles, applications, and biosafety of plant genome editing using CRISPR-Cas9. *Frontiers in Plant Science*, *11*, 56. DOI: 10.3389/fpls.2020.00056 PMID: 32117392

Faizal, A., Nugroho, S., Sembada, A. A., Theda, Y., Komariyah, T., & Esyanti, R. R. (2024). Genome editing in future crop protection: Utilizing CRISPR/Cas9 to improve crop resistance against diseases, pests, and weeds. *Discover Agriculture*, *2*(1), 104. DOI: 10.1007/s44279-024-00124-0

Farooq, M., Khan, A., Hassan, A., & Shah, M. M. (2026). Advances in CRISPR/Cas systems for engineering abiotic stress tolerance in plants: Mechanisms and future prospects. *Planta*, *263*(3), 72. DOI: 10.1007/s00425-026-04942-y PMID: 41665762

Haider, S., Singh, A. P., Panthi, B., Sindhu, S. R., Safa, N. T., Malik, S., & Rahimi, M. (2026). Advances in CRISPR/Cas9 genome editing for crop improvement and global food security. *Current Plant Biology*, *100593*, 100593. Advance online publication. DOI: 10.1016/j.cpb.2026.100593

Hamdan, M. F., Hensel, G., Alok, A., & Tan, B. C. (2023). Editorial: Genome editing and biotechnological advances for crop improvement and future agriculture. *Frontiers in Plant Science*, *14*, 1132821. DOI: 10.3389/fpls.2023.1132821 PMID: 36798708

Hasan, N., Choudhary, S., Naaz, N., Sharma, N., & Laskar, R. A. (2021). Recent advancements in molecular marker-assisted selection and applications in plant breeding programmes. *Journal of Genetic Engineering and Biotechnology*, *19*(1), 128. DOI: 10.1186/s43141-021-00231-1 PMID: 34448979

Kafle, S. (2022). CRISPR/CAS9: A new paradigm for crop improvement revolutionizing agriculture. *Journal of Agriculture and Food Research*, *11*, 100484. DOI: 10.1016/j.jafr.2022.100484

. Kanth, K., Mane, R. S., & Prasad, B. D. (2025). Editing the future: CRISPR/Cas9 for climate-resilient crops. Genome Editing-Recent Advances and Applications: Recent Advances and Applications, 95.

Kaur, N., Qadir, M., Francis, D. V., Alok, A., Tiwari, S., & Ahmed, Z. F. R. (2025). CRISPR/Cas9: A sustainable technology to enhance climate resilience in major staple crops. *Frontiers in Genome Editing*, *7*, 1533197. DOI: 10.3389/fgeed.2025.1533197 PMID: 40171546

Kumar, M., Prusty, M. R., Pandey, M. K., Singh, P. K., Bohra, A., Guo, B., & Varshney, R. K. (2023). Application of CRISPR/Cas9-mediated gene editing for abiotic stress management in crop plants. *Frontiers in Plant Science*, *14*, 1157678. DOI: 10.3389/fpls.2023.1157678 PMID: 37143874

Kumari, P., Jasrotia, P., Kumar, D., Kashyap, P. L., Kumar, S., Mishra, C. N., Kumar, S., & Singh, G. P. (2022). Biotechnological approaches for host plant resistance to insect pests. *Frontiers in Genetics*, *13*, 914029. DOI: 10.3389/fgene.2022.914029 PMID: 35719377

Kumari, V., Kumawat, P., Yeri, S., & Rajput, S. S. (2024). CRISPR-Cas9: A genome editing tool in crop plants: A review. *Agricultural Reviews (Karnal)*, *45*(2), 329–334.

Limbalkar, O. M., Srivastava, P., Reddy, K. R., Lali, S., Lal, S. K., Bishi, S. K., Soren, K. R., Bhadana, V. P., Rakshit, S., & Tribhuvan, K. U. (2025). Genome editing and its impact on crop improvement: Current approaches and future prospects. *Discover Plants*, *2*(1), 358. DOI: 10.1007/s44372-025-00410-1

Liu, H., Chen, W., Li, Y., Sun, L., Chai, Y., Chen, H., Nie, H., & Huang, C. (2022). CRISPR/Cas9 technology and its utility for crop improvement. *International Journal of Molecular Sciences*, *23*(18), 10442. DOI: 10.3390/ijms231810442 PMID: 36142353

Mansi, M., & Danai, P. (2026).. . *Frontiers in Plant Science*, *15*, 1478398. DOI: 10.3389/fpls.2024.1478398

Mattiello, L., Rütgers, M., Sua-Rojas, M. F., Tavares, R., Soares, J. S., Begcy, K., & Menossi, M. (2022). Molecular and computational strategies to increase the efficiency of CRISPR-based techniques. *Frontiers in Plant Science*, *13*, 868027. DOI: 10.3389/fpls.2022.868027 PMID: 35712599

Mešić, A., Jurić, M., Donsì, F., Maslov Bandić, L., & Jurić, S. (2024). Advancing climate resilience: Technological innovations in plant-based, alternative and sustainable food production systems. *Discover Sustainability*, *5*(1), 423. DOI: 10.1007/s43621-024-00581-z

Mohamed, H. I., Khan, A., & Basıt, A. (2024). CRISPR-Cas9 system mediated genome editing technology: An ultimate tool to enhance abiotic stress in crop plants. *Journal of Soil Science and Plant Nutrition*, *24*(2), 1799–1822. DOI: 10.1007/s42729-024-01778-x

Movahedi, A., & Yang, L. (2025). CRISPR-Cas genome editing in plants: Revolutionizing precision agriculture and plant biotechnology. *Plant Innovation Journal*, *1*(1), 1–7. DOI: 10.62762/PIJ.2025.263896

Nadeem, M. A., Nawaz, M. A., Shahid, M. Q., Doğan, Y., Comertpay, G., Yıldız, M., Hatipoğlu, R., Ahmad, F., Alsaleh, A., Labhane, N., Özkan, H., Chung, G., & Baloch, F. S. (2018). DNA molecular markers in plant breeding: Current status and recent advancements in genomic selection and genome editing. *Biotechnology, Biotechnological Equipment*, *32*(2), 261–285. DOI: 10.1080/13102818.2017.1400401

Nirmal, N. P., Anyimadu, C. F., Khanashyam, A. C., Bekhit, A. E. A., & Dhar, B. K. (2025). Alternative protein sources: Addressing global food security and environmental sustainability. *Sustainable Development (Bradford)*, *33*(3), 3958–3975. DOI: 10.1002/sd.3338

Ntsefong, G. N., Seyum, E. G., Kingsley, T. M., Kassie, F. C., Gabriel, M. S. T., Shariati, M. A., Karapetkovska-Hristova, V., Martin, B. J., & Youmbi, E. (2023). CRISPR-Cas-based genome editing for crop improvement: Progress, challenges and future prospects. *Global Journal of Botanical Science*, *11*, 28–40. DOI: 10.12974/2311-858X.2023.11.3

Patel, A., Miles, A. J., Strackhouse, T., Cook, L., Leng, S., Patel, S., Klinger, K., Rudrabhatla, S., & Potlakayala, S. (2023). Methods of crop improvement and applications towards fortifying food security. *Frontiers in Genome Editing*, *5*, 1171969. DOI: 10.3389/fgeed.2023.1171969 PMID: 37484652

Ray, S., Sneha, K., & Jangid, C. (2023). CRISPR-Cas9 for sustainable food production: Impacts, recent advancements and future perspectives. *Food and Humanity*, *1*, 1458–1469. DOI: 10.1016/j.foohum.2023.10.014

Sathiyabalan, A., Amir, M., Sandhu, R., & Bhattacharya, S. (2024). Evaluating CRISPR/Cas9's progressive role in crop enhancement and sustainable agriculture. *Journal of Advances in Biology & Biotechnology*, *27*(11), 855–870. DOI: 10.9734/jabb/2024/v27i6949

. Sharma, S., Lee, S., & Huo, H. (2019). Current status of research, regulations, and future challenges for CRISPR gene-editing in crop improvement. EDIS, 2019(4). DOI: 10.32473/edis-hs1333-2019

Sharma, U., Nisha, , & Ray, A. (2025). An insight into the journey of CRISPR-CAS9 and its application in crop improvement. *Discover Plants*, *2*(1), 266. DOI: 10.1007/s44372-025-00343-9

Singh, P., Markam, N. K., Shinde, S. B., Yadav, T., Mohanty, S., Prasad, M., Pagoch, S. S., Swain, S., Musmade, N. A., & Saikia, B. (2025). Applications of CRISPR/Cas gene editing in vegetable crop disease and stress resistance. *International Journal of Advanced Biochemistry Research*, *9*(1), 960–970. DOI: 10.33545/26174693.2025.v9.i8Sn.5290

Song, Y., Hu, Z., Liu, S., Luo, S., He, R., Yang, X., Li, S., Yang, X., An, Y., & Lu, Y. (2025). Utilization of microalgae and duckweed as sustainable protein sources for food and feed: Nutritional potential and functional applications. *Journal of Agricultural and Food Chemistry*, *73*(8), 4466–4485. DOI: 10.1021/acs.jafc.4c11610 PMID: 39879156

Tuncel, A., Pan, C., Clem, J. S., Liu, D., & Qi, Y. (2025). CRISPR–Cas applications in agriculture and plant research. *Nature Reviews. Molecular Cell Biology*, *26*(6), 419–441. DOI: 10.1038/s41580-025-00834-3 PMID: 40055491

Tuncel, A., Pan, C., Sprink, T., Wilhelm, R., Barrangou, R., Li, L., Shih, P. M., Varshney, R. K., Tripathi, L., Van Eck, J., Mandadi, K., & Qi, Y. (2023). Genome-edited foods. *Nature Reviews Bioengineering*, *1*(11), 799–816. DOI: 10.1038/s44222-023-00115-8

Uranga, M., Martín-Hernández, A. M., Storme, N. D., & Pasin, F. (2024). CRISPR–Cas systems and applications for crop bioengineering. *Frontiers in Bioengineering and Biotechnology*, *12*, 1483857. DOI: 10.3389/fbioe.2024.1483857 PMID: 39479297

Wan, L., Wang, Z., Tang, M., Hong, D., Sun, Y., Ren, J., Zhang, N., & Zeng, H. (2021). CRISPR-Cas9 gene editing for fruit and vegetable crops: Strategies and prospects. *Horticulturae*, *7*(7), 193. DOI: 10.3390/horticulturae7070193

Wang, Y., Zafar, N., Ali, Q., Manghwar, H., Wang, G., Yu, L., Ding, X., Ding, F., Hong, N., Wang, G., & Jin, S. (2022). CRISPR/Cas genome editing technologies for plant improvement against biotic and abiotic stresses: Advances, limitations, and future perspectives. *Cells*, *11*(23), 3928. DOI: 10.3390/cells11233928 PMID: 36497186

Wani, S. H., Choudhary, M., Barmukh, R., Bagaria, P. K., Samantara, K., Razzaq, A., Jaba, J., Ba, M. N., & Varshney, R. K. (2022). Molecular mechanisms, genetic mapping, and genome editing for insect pest resistance in field crops. *Theoretical and Applied Genetics*, *135*(11), 3875–3895. DOI: 10.1007/s00122-022-04060-9 PMID: 35267056

Zhu, H., Li, C., & Gao, C. (2020). Applications of CRISPR–Cas in agriculture and plant biotechnology. *Nature Reviews. Molecular Cell Biology*, *21*(11), 661–677. DOI: 10.1038/s41580-020-00288-9 PMID: 32973356

Chapter 6

Gene Editing for Abiotic Stress Tolerance

Tauqeer Ahmaed Qadri
http://orcid.org/0009-0000-9617-4512
COMSATS University, Islamabad, Pakistan

Muhammad Waseem
http://orcid.org/0000-0001-7947-7722
Hainan University, Haikou, China

Sana Basharat
http://orcid.org/0000-0002-6737-5611
Hainan University, Sanya, China

Pingwu Liu
Hainan University, Sanya, China

Amjid Khan
http://orcid.org/0009-0001-5475-5156
University of South Africa, South Africa

Madiha Imtiaz
Bacha Khan University, Charsadda, Pakistan

ABSTRACT

Gene editing is rapidly transforming plant biology and crop improvement by enabling precise, efficient genome modifications to address food security, climate change, and sustainability. CRISPR/Cas systems, alongside TALENs and base editors, target genes controlling yield, nutritional quality, and resistance to drought, salinity, pests, and diseases—delivering traits that often exceed traditional breeding speed and scope. Despite strong lab and greenhouse progress, field-scale deployment remains limited, with few varieties commercialized and performance still being tested across diverse environments. Key barriers include off-target effects, regulatory uncertainty, and public acceptance. Achieving durable gains will require integrating gene editing with sustainable agronomy, genomic selection, high-throughput phenotyping, and data-driven analytics, supported by responsible governance and societal engagement.

DOI: 10.4018/979-8-3373-9534-0.ch006

INTRODUCTION

Overview of Abiotic Stresses in Agriculture

One of the most formidable challenges to global agriculture today remains the abiotic stresses like non-living environmental factors which adversely affect plant growth, development and productivity (Chaffai et al., 2024). Whereas biotic stresses result from living organisms like pathogens, insects, or weeds, abiotic stresses are physical or chemical conditions of the environment like extreme temperature, drought, salinity, flooding, heavy metals, or nutritional imbalances, of the environment that interfere with the physiological and metabolic processes of crop (Anas et al., 2025; Umar et al., 2021). Abiotic stresses are ubiquitous at least 90% of the world's cultivated land is subjected to one or more of them, and they have been estimated to cause 50–70% yield losses in major crops. With the global increase in abiotic stresses due to global climate change, food and systems that sustain them are under threat. In general, abiotic stresses are classified into three main groups which are stresses with respect to weather, growth, and agronomic management. Weather-related stresses include freezing, drought, excessive or insufficient solar radiation, floods, and high or low temperatures (Mafakheri et al., 2021). Drought and heat stress often coincide and present a real threat to plant biomass, grain yield, and crop quality. For example, under drought stress, the available water is limited, stomata close, CO_2 uptake is inhibited, and photosynthesis is disrupted, resulting in carbohydrate, protein, lipid, and mineral metabolism (Ostmeyer et al., 2020). Under heat stress, proteins can denature, membranes can destabilize, and reproductive development can be impaired. Under cold stress, cellular structures can be damaged, and enzymatic activity can be inhibited. Waterlogging or flooding causes hypoxia and root decay due to the deprivation of oxygen (Tiwari et al., 2020).

Growth-related stresses are often internal, such as nutrient deficiencies or toxicities, or physiological challenges of growth and developmental transitions (e.g., from vegetative to reproductive stages). Nutrient imbalances can limit essential metabolic processes, decrease the photosynthetic capacity of plants, and impair their immune systems (V. V. Kumari et al., 2022). Industrial pollution or excessive fertilizer can disturb enzyme activity, damage DNA, and retard growth. Due to irrigation with poor-quality water and soil degradation, salinity, which is a major agronomic management-related stress, is becoming a more common problem. Saline soil disturbs the osmotic balance, causes ion toxicity, affects water absorption, and disturbs metabolic processes (Tessema, Yadeta, Kebede, & Ayele, 2022). Abiotic stress can be induced by agronomic management practices. Heavy machines, excessive tilling, and the sitting of environmental pollutants restrict root growth, reduce aeration, and limit the availability of nutrients (Choudhary et al., 2021).

Furthermore, these factors combine with the effects of natural abiotic stresses and are exacerbated by unsustainable farming practices to further threaten crop productivity. Crops are subjected to multifarious impacts due to abiotic stresses at the morphological, physiological, and molecular levels (Magdy et al., 2023). Stress conditions can often initiate a cascade of metabolic adjustments for survival if it is at the expense of growth and yield. For example, drought and salinity impose osmotic stress on plants, where osmoprotectants (compatible solutes) such as proline and glycine betaine accumulate to sustain cell turgor. The generation of these stresses also leads to the generation of reactive oxygen species (ROS), which can cause oxidative damage to proteins, lipids, and nucleic acids unless neutralized by antioxidant defense systems (Dikilitas et al., 2020).

Abiotic stress has a high sensitivity to photosynthesis. This means that drought, heat, or salinity in plants can decrease chlorophyll content, damage photosystems, and inhibit the Calvin cycle, resulting in lower carbohydrate synthesis and energy production (Sharma et al., 2020). It also disrupts protein synthesis, enzyme activity, and growth, development, and reproductive success. When added together, this has a substantial effect on both the yield and nutritional quality of agricultural crops, with cereals and grain legumes being particularly susceptible (Weih et al., 2021). Abiotic stress has immediate economic repercussions. Drought, heat, and salinity account for a major share of global crop failures, directly straining farmer incomes, the economy of nations, and food security. Areas in developing regions with limited resources for irrigation and soil management are particularly vulnerable to abiotic stress. In addition, with the limitations of arable land expansion in terms of environmental and socioeconomic aspects, there is a need to increase productivity on stress-prone land to meet the increasing demands of a growing global population (Rathod & Verma, 2023).

Climate change worsens abiotic stress. The increase in global temperature, modified precipitation patterns, intensified occurrence of extraordinary climate events, and raised carbon dioxide levels in the environment are changing the scene of farming risk. In addition, these changes exacerbate existing plant stresses, such as drought, heat, and salinity, and introduce new combinations and sequences of stress that plants must deal with. For example, the simultaneous occurrence of drought and heat stress can have synergistically negative effects on yield that are more significant than those of each stress alone (Malhi et al., 2021). Climate-induced abiotic stresses pose unpredictable weather conditions for crops, and thus their management is quite complicated; therefore, the development of resilient crop varieties and the use of adaptive agronomic practices have become a necessity (Hossain et al., 2021). Plants have evolved a variety of abiotic stress coping strategies involving morphological, physiological, and molecular mechanisms. These include modifications in root and shoot architecture, changes in stomatal conductance, synthesis of stress-protective proteins and metabolites, and triggering of complex signaling pathways that involve phytohormones and transcription factors (Dixit et al., 2024). There is a huge variation among species, cultivars, and even developmental stages in the extent of tolerance, as would be expected from the leeway in genetic diversity in breeding for stress tolerance (Marone et al., 2021).

Recent advances in molecular genetics, genomics, and biotechnology have begun to reveal the complex networks that direct plant stress responses. These mechanisms provide critical insights for the development of crops that can maintain yield and quality under adverse conditions (Morales et al., 2020). Precise modification of stress-responsive genes presents a powerful application for gene-editing technologies, such as CRISPR/Cas, and engineering of crops with stress tolerance can be achieved through such gene editing (Zafar et al., 2020). Addressing abiotic stress in agriculture requires an integrated approach that combines genetic improvement, advanced crop management, and sustainable agriculture. The effects of stress can be mitigated by conservation tillage, efficient irrigation, soil amendments, and judicious use of fertilizers and micronutrients. Stress-tolerant crop varieties, developed by means of conventional breeding, marker-assisted selection, or gene editing, are necessary to maintain productivity in stress-prone areas (Gonzalez Guzman et al., 2022). Additionally, there is an increasing recognition of the necessity of adopting climate-smart agriculture as a means of achieving future food security, which focuses on enhancing resilience, resource efficiency, and environmental stewardship. This goes beyond technological innovations and requires policy support, farmer education, and investments in research and infrastructure (Kabato et al., 2025).

Traditional Breeding vs. Modern Gene Editing Approaches

Two supplementary methods for crop improvement exist: traditional breeding and modern gene editing technologies, which provide different benefits and challenges, together with agricultural consequences (Qaim, 2020). For thousands of years, traditional breeding has involved the selection of parent plants with positive attributes, followed by their controlled crossbreeding to develop new genetic variability in breeding populations. This process involves successive intersections between plants, followed by the selection of suitable traits to create breeding lines with multiple beneficial characteristics. Breeders worldwide can access this process because it is inexpensive, technically simple, and has few regulatory constraints. Traditional breeding faces inherent constraints because plants from the same gene pool show restricted genetic choices, and they need to have matching sexual characteristics to unite their genes during reproduction (Begna, 2021). Strain improvement through this method requires numerous breeding cycles to develop combined traits, as it works poorly with sterile or vegetative propagation systems. The necessity is large populations and the need for extensive selection to find plants with perfect traits because all traits must segregate simultaneously, which could lead to co-inheritance of undesired traits with desired traits (Lamichhane & Thapa, 2022). The plant breeding field has experienced a revolutionary breakthrough with the introduction of CRISPR/Cas9, which functions as a modern illustration of gene editing techniques. Gene editing grants breeders the ability to perform exact DNA modifications in plants by adding, moving, or eliminating chosen genes that contain preferred characteristics (Mushtaq et al., 2021). Traditional genetic engineering through foreign DNA introduction differs from gene editing because this method operates within the natural genomic framework of species, which enables evolutionary and conventional breeding changes. The exact nature of this technique enables breeders to overcome the requirements of sexual reproduction by editing several genes simultaneously to accelerate trait enhancement through a single generational cycle. The implementation of gene-editing technologies enables researchers to quickly implement drought protection and nutritional enhancement combined with disease resistance applications in a shorter timeline than traditional breeding methods (Martínez-Fortún et al., 2022). The genetic modifications in gene-edited plants are undetectable from conventionally produced plants since they lack foreign DNA, and many nations, including the USA, Japan, and Canada, do not classify them as transgenic organisms (Caradus, 2023).

The scientific dangers between traditional breeding and gene editing methods still match, although new crops from any development method require extensive examination ahead of market introduction. Gene editing provides fast and precise solutions, but the technical requirements exceed the average capabilities, and specific national regulations may control these approaches based on the genetic alteration type (Turnbull et al., 2021). Traditional breeding methods continue to be essential for complex traits that require many genetic components, together with crops that lack complete genome research. Traditional breeding supplies a wide genetic range and adaptability features, but gene editing achieves quick targeted improvements for traits when used together as an integrated approach. Traditional breeding alongside modern gene editing serves as an important method for enhancing crop improvement. Traditional breeding remains an established method for achieving multi-trait combinations because it is accessible to all and has proven its effectiveness; however, the restriction comes from the slow speeds of development and limited genetic diversity (Begna, 2021). The gene-editing approach provides both increased accuracy and augmented efficiency, allowing researchers to create specific crop varieties intended for contemporary agricultural needs, such as climate change and food safety requirements. Both methods require sound regulatory frameworks and equitable access to achieve sustainable agricultural requirements worldwide in the twenty-first century (Bohra et al., 2020).

Mechanisms of Abiotic Stress Response in Plants

Some of these abiotic stresses include drought, salinity, heat, cold, and toxicity or the presence of heavy metals, which also have a detrimental effect on growth, development, and yield in plants. This is because plants are aware that to survive, they have to develop hard-wired strategies that now work at the level of the whole body, biochemistry, as well as at the molecular level (Kopecká et al., 2023). These responses are regulated by a complex of leading genes and pathways, along with essential regulatory systems, where phytohormones are considered master conductors of stress signals and adaptation processes (Zhao et al., 2021).

Key Genes and Metabolic Pathways

Stress induces a sequence of processes associated with the perception of stress by signaling proteins at the cell membrane or within organelles. Such perception leads to the rapid generation of secondary messengers, including calcium ions, ROS, and lipid signaling molecules, such as sphingolipids and phosphatidic acids. (Medina-Puche & Lozano-Durán, 2022). These messengers then lead to signal transduction and alter gene expression, thereby promoting the production of proteins and metabolites involved in the stress response. One of the response patterns that occurs is the upregulation of stress-responsive genes that code for proteins that protect or repair cell damage (Jogawat et al., 2021). These include:

Osmoprotectant biosynthesis genes: In response to drought and salinity stress, plants produce osmoprotectants, including proline, glycine betaine, trehalose, and soluble sugars, to maintain cell turgor pressure and protect cellular structures. These molecules assist in osmotic regulation, thereby helping plants cope with water deficit, water stress, and ion homeostasis (Dikilitas et al., 2020).
Oxidative stress gene: A biotic stress regularly results in the overproduction of ROS, which in turn affects proteins, lipids, and DNA. To underpin this, plants increase the synthesis of antioxidant proteins, including superoxide, dismutase, catalase, and peroxidase. These enzymes neutralize ROS and prevent oxidative effects (Suman et al., 2021).
Heat shock protein (HSP) genes: When cells are exposed to high temperatures, HSPs are produced to act as molecular chaperones for folding-associated proteins to counteract heat stress and damage (Zeba Khan & Shahwar, 2020).
Ion transporter genes: In particular, under saline stress, plants control the transport, trapping, and localization of ions. Ion transporters and pumps (SOS1, NHX1, and HKT1) are some of the genes that assist in removing toxic ions,

such as sodium, and retaining useful ions, thereby maintaining homeostasis (Malakar & Chattopadhyay, 2021).

***AT genes*:** Autophagy is a catabolic process that helps plants remove damaged molecules, such as proteins and organelles, under stress conditions. It has been identified that ATG gene family plays the crucial role in stress resistance in variety of species (Su et al., 2020).

This is chiefly managed by transcription factors (TFs), which act as the main regulators of gene expression changes. Some of the chief classes reported to be involved in the regulation of stress-responsive genes are MYB, AP2/ ERF, bZIP, WRKY, NAC, and NF-Y. These TFs may work together with other TFs, although there is redundancy in the function of the various types of TFs depending on the nature and intensity of stress (Prusty et al., 2024). Physiologically, plants can alter their metabolism to sustain themselves. For instance, the production of compatible solutes and secondary metabolites, such as flavonoids and phenolics, is enhanced to deal with challenges such as the protection of cellular structures and the removal of free radicals (Anjali et al., 2023). Changes in more specific tissues, including the cell wall, have the potential to increase mechanical strength and mineral loss with stress, such as more lignin deposits (Fig 1).

Figure 1. Plants stress response mechanisms

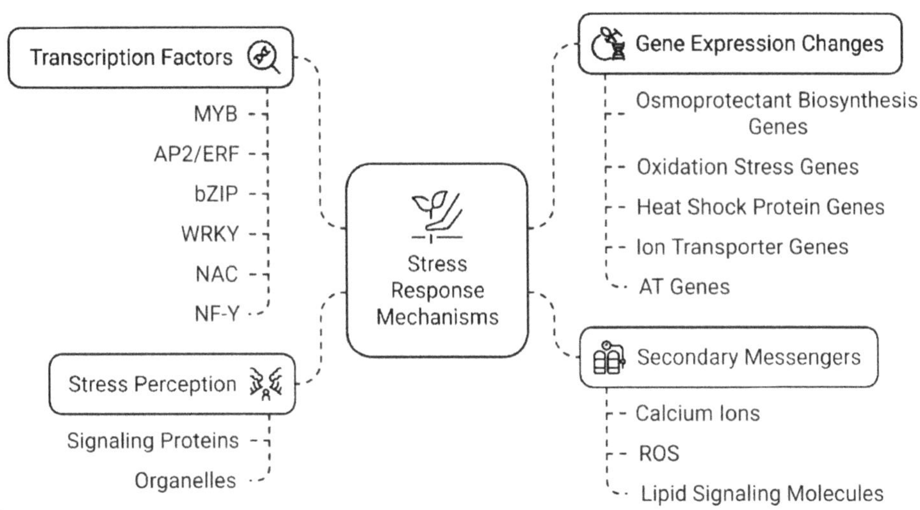

Role of Phytohormones and Regulatory Networks

Phytohormones are involved in the coordination and regulation of processes connected to abiotic stress, local and systemic signals, and physiological and molecular level changes. Among these, abscisic acid is considered one of the prominent examples of "stress hormones," the concentration of which is increased during stress related to drought, salinity, and cold. ABA also plays the role of closing stomata and reducing water leakage, stimulates the synthesis of genes against stress, and controls growth cessation for the conservation of resources (Y. Zhang, Berman, & Shani, 2023). The ABA signaling pathway includes well-characterized receptors such as PYR/PYL/RCAR, protein phosphatases (PP2Cs), and kinases (SnRK2s), which regulate the transcription of genes that are responsive to ABA and establish cellular cooperation (Fidler et al., 2022). Other phytohormones are also involved and participate in the following processes:

Salicylic acid (SA): SA plays a role in abiotic and biotic stress reactions, improving plant stress, salt stress, osmotic stress, and oxidative stress through the activation of antioxidants and stress-associated genes (Song, Shao, Zheng, Zhao, & Xu, 2023).
Ethylene: Ethylene production is upregulated by various stresses and is involved in ABA and SA-mediated regulation of stress-responsive gene expression, leaf senescence, and root development. Ethylene response factors (ERFs) are significant TFs involved in this network as significant TFs4 (Nazir et al., 2024).
Jasmonic acid (JA): JA is involved in stress signaling, the most common of which includes wounding, cold, and osmotic stress. This hormone regulates defense genes and affects other hormones to maintain growth-stress homeostasis (J. Wang, Song, Gong, Xu, & Li, 2020).

These hormones are also useful for growth but are forbidden during stress to enable the organism to handle stress or a certain stimulus. This suggests that their antagonistic compromise with ABA helps control developmental changes during stress. Each of these hormones works in harmony with the others to establish a complex interrelationship that includes interacting feedback and communication pathways (Choubey, Dey, Dey, & Boruah, 2025). For instance, ABA and ethylene can either positively or negatively interact in the processes of stomatal closure and root growth, depending on the stress conditions applied at different periods. They make the necessary hormonal communication so that plants can quickly and adequately react to constantly changing environmental conditions (Mearaji, Ansari, Igdelou, Lajayer, & Pessarakli, 2021). Stress on the organism triggers specific changes at the transcriptional level that lead to the reapportioning of the genome. Hormonal and environmental signals are perceived through transcription factors such as WRKY, NAC, and bZIP, propagating gene networks for osmoregulation, detoxification, and repair. Transcriptomics, proteomics, and metabolomics have all contributed significantly to the understanding of these signaling networks owing to the discovery of genes, proteins, and metabolites associated with stress tolerance (Thilakarathne, Liu, & Zou, 2025). This chapter provides information on some identified key genes, metabolic pathways, and regulatory networks controlling plant responses to abiotic stress, with phytohormones acting as master regulators. This knowledge forms the basis for producing crops that are genetically modified to withstand promoted stresses and marker-assisted selection and breeding techniques that are notably important in the current world where climate change has become evident.

GENOME EDITING TECHNOLOGIES FOR ABIOTIC STRESS TOLERANCE

CRISPR/Cas Systems: Principles and Variants

CRISPR/Cas System, also called CRISPR-associated proteins, is a breakthrough genome editing technique because of its simplicity and high efficiency (Afzal et al., 2020). A better-known part of this system was discovered to be a prokaryotic defense mechanism against the hands of parasitic genetic elements, in which it captures fragments of parasitic genes and uses them as spacers within the host genome. These spacers are then transcribed and processed into guide RNAs (gRNAs) that facilitate Cas nucleases to bind with complementary sites in the invading DNA or RNA and consequently cleave and inactivate it (Gerashchenkov et al., 2020). Among them, CRISPR/Cas9 is the most employed system; it consists of the Cas9 nuclease and a single-guide RNA (sgRNA) that combines the functions of both crRNA and tracrRNA. The specific targets are located on the DNA using the sgRNA complementary to the CAS9, and the CAS9 cleaves the DNA at that certain site. It then uses either NHEJ, which usually results in insertions or deletions (indels), or HDR, if a template is available, to fix this DSB (Burmistrz et al., 2020). CRISPR/Cas9 system is programmable in that it is possible to swap out the sgRNA sequence for any nucleotide sequence and target almost any genomic location if it is within a few bases of a PAM. The natural Cas9 from Streptococcus pyogenes, known as SpCas9, targets an NGG PAM, and cas9 can only be utilized in target sites that contain the PAM. To overcome this challenge, several Cas9 variants and orthologs have been developed or identified for use in scFv platforms (Gupta et al., 2021). For instance, xCas9, Cas9-NG, and SpRY can bind to broader or different PAM sequences, which increases the target range of DNA. eSpCas9(1.1), SpCas9-HF1, HypaCas9, and evoCas9 have been generated to avoid off-target issues due to changes in the amino acids that increase specificity. Other Cas enzymes, such as Cas12 or Cpf1 and Cas13, are different: Cas12 cuts DNA at T-rich PAMs and creates staggered double-strand breaks, whereas Cas13 edits RNA rather than DNA molecules and targets the transcriptome instead of the genome (Kovalev, Davletshin, & Karpov, 2024).

It has been ascertained that the CRISPR/Cas system is not only vaster than previously thought, but it also has more characteristics, classifications, classes, types, and subtypes. It differentiates two main classes: one is multi-protein, and the second class is single protein effectors, there are six types of them and more than thirty subtypes of them, currently Class 2 system of cas9, cas12 and cas13 is commonly used for genetic engineering because of its easy approach (Zulqurnain Khan et al., 2022). These systems are highly modulatory and can be multiplexed efficiently, allowing them to target multiple genes simultaneously in a single experiment. To improve CRISPR/Cas-based methods for genome editing, improvement strategies such as the design of sgRNA, optimization of Cas protein, development of the delivery system, and regulation of the repair pathway have emerged. However, some challenges that are associated with the use of the CRISPR/Cas technology include: Off-target effects; it is not very efficient and variable; the delivery of the Cas proteins to the target site is still a challenge; and the Cas proteins only function if they are accompanied by specific PAM sequences (Manghwar et al., 2020). Current efforts are thus focused on lighter sgRNA design, anti-CRISPR proteins, a new generation of Cas, and efficient delivery systems. Thus, the development of novel CRISPR/Cas systems, as well as improved methods of utilizing them, is set to progress on a constant basis beyond simple lab use and into agriculture, medicine, and other areas of biotechnology (Nidhi et al., 2021).

Other approaches include TALENs and ZFNs

ZFN and TALENs were other earlier approaches to genome editing prior to the invention of CRISPR/Cas. Both technologies use an individualized domain that can bind to DNA and a FokI nuclease, which needs to be dimerized to form site-specific double-strand breaks (Shamshirgaran, Liu, Sumer, Verma, & Taheri-Ghahfarokhi, 2022).

Zinc Finger Nucleases (ZFNs)

ZFNs are formed by fusing 3–6 zinc finger segments that target individual triplets with the FokI nuclease domain. When two ZFN architectures targeting opposite DNA strands at the site are designed, the FokI domains are brought together and cause a DSB (Singh et al., 2021). The cell repair equipment then proceeds to fix the breakthrough either through NHEJ or HDR, which allows gene targeting, repair, or insertion (Nambiar et al., 2022). Although ZFNs have provided a new approach to targeted genome modification, their construction is challenging and time-consuming because a different set of zinc finger domains is typically needed for each new target. Off-target effects may also occur because the individual domains of zinc fingers do not have high selectivity (Rafiq, 2024).

Transcription Activator-Like Effector Nucleases (TALENs)

TALENs are designed from transcription activator-like effectors (TALEs), which were originally isolated from Xanthomonas bacteria. Each TALE repeat recognizes a single DNA base, which means that it is relatively easy to link DNA-binding domains (Teper et al., 2023). Similar to ZFNs, TALENs are linked to FokI and act as dimers to generate DSBs at the targeted locus 35. Technically, TALENs are easier to design because they have a generally higher specificity with less cytotoxicity than ZFNs. Nevertheless, they are less efficient for delivery into cells due to their large size, and although their design is somewhat less rigid than that of ZFNs, assembling them is still a nontrivial task compared to CRISPR/Cas systems (Shamshirgaran et al., 2022).

Comparison and Limitations

ZFNs and TALENs have created a niche for precision genome editing and can be effectively used in plants, animals, and human cells. However, they are slow, expensive, and have a limited number of units that can be analyzed simultaneously (Singh et al., 2021). Originally, the design of new protein pairs was required for each target sequence, which created obstacles for large-scale or multiplexed editing. In contrast, CRISPR/Cas systems require simple RNA-DNA base pairing for targeting, which makes them more adaptable for large-scale, rapid, and high-throughput mutagenesis than previous methods, without requiring much modification of the proteins involved (de Dieu Habimana et al., 2022).

Advances in Base and Prime Editing

Unlike earlier methods, such as the CRISPR/Cas system or any of its precursors that work using DSBs and the regular repair machineries in the cell, there are improved methods that allow for single nucleotide or small DNA segment editing, which is safer and faster than DSBs (Ray & Raghavan, 2020).

***Base Editing*:** Base editors are derived from proteins that are designed by fusing the catalytically deactivated Cas nuclease with a DNA-modifying enzyme, such as cytidine deaminase or adenosine deaminase. These editors allow for direct, efficient, and accurate changes of one DNA base substitute to another at a defined locus without generating DSBs or using donor templates (Rallapalli & Komor, 2023). The two main classes are as follows:
Cytosine Base Editors (CBEs): Use deaminase activity to convert C-G to T-A base pairs by converting part of cytosine to uracil, which is then converted to thymine during replication or repair processes (Tong et al., 2024).
CRISPR- Cas9: A highly efficient gene-editing tool and targeted method for modifying base pairs in DNA sequences, which uses ribonucleoprotein components to directly cleave DNA sequences. Base editing has been used with high target efficiency and specificity compared to previously used techniques based on the formation of DSB, as it has a low formation of indels and low cytotoxicity. Li1 5 Moreover, it is most beneficial in gene-targeted mutations for genetic disorders and when attempting to enhance certain desired characteristics in plants eight; p25 (Lee et al., 2021).
***Prime Editing*:** Prime editing is a more recent advancement that extends the capability of achieving exact edits to the genome. Prime editors are Cas9 nickases combined with reverse transcriptase enzymes and are dependent on a prime editing guide RNA (pegRNA) that defines the target position and encodes the change (Mikhaylova et al., 2024). The Cas9 nickase makes a single-strand break in the target DNA, after which the reverse transcriptase starts synthesizing the edit from the pegRNA to the displaced strand. This permits the introduction of all twelve potential base-to-base conversions, along with a few insertions and deletions, without DSB or donor templates (Ochoa-Sanchez et al., 2021).

Thus, prime editing is promising and versatile, with less off-target activity and fewer by-products than the original genome editing. It has been further used in various model organisms, such as plants and animals, and has immense potential for use in the future for gene therapy and improvement of other characteristics. However, base and prime editing technologies are not without shortcomings, including delivery efficiency, editing window limitations, off-target effects, and the requirement of cellular repair pathways (Hosseini et al., 2024). Current studies are focused on increasing the number of targeted sites, increasing the rate of editing, creating new deaminases and reverse transcriptases, and choosing the methods of their delivery (Li et al., 2020). The development of genome editing techniques has progressed significantly, and the more common and advanced system with high efficiency is the CR1/RNACas system. However, the advancement of high-fidelity Cas variants, other Cas proteins, and delivery systems has made the CRISPR/Cas approach more versatile (Lee, 2023). Although still used, ZFNs and TALENs have been succeeded by CRISPR/Cas in most applications because of their complexity and cost. This will usher in a new era of precise and programmable chassis modification and the utilization of base and prime editing for correcting point mutations and inadvertent creation of complex edits without DSBs. These technologies are expected to become more established and useful in the future for basic research, agriculture, and medicine, aiding in gene function studies, crop enhancement, and genetic disease treatment (Rafiq, 2024).

Application of Gene Editing in Major Abiotic Stresses

The rapid development of gene technology has produced crops that better resist major environmental threats through genetic material editing. Through traditional breeding methods and transgenic methods, scientists have successfully improved plant defense systems, but CRISPR/Cas tools now allow them to work faster and control gene changes at the network level with high precision (Wang et al., 2022). The following sections explore how gene editing can increase crop tolerance to major environmental pressures that affect global farming.

Drought Tolerance

The lack of sufficient rainfall seriously reduces the crop growth potential worldwide. Drought-activated plants employ a coordinated network of responses from cellular activity changes to trigger new protective gene expression (Sun et al., 2021). Targeted editing of these pathways aids in the development of drought tolerance. Scientists have used CRISPR/Cas9 systems to disable or control drought response regulators, which increased the plant's ability to use less water and enhanced crop production under limited water conditions (Rai et al., 2023). Researchers enhanced drought tolerance and tested these changes in real maize field conditions using CRISPR to edit ARGOS (Rasheed et al., 2023). Researchers have edited SlLBD and OsDREB transcription factors in tomato and rice, which boosted drought tolerance by activating protective genes against stress (Krishna et al., 2022). The alteration of SAPK2 ABA-regulated genes in rice improves plant water resistance through effective root growth and controlled stomatal function (Zhang et al., 2022). Multiple CRISPR techniques allow scientists to edit several genes that regulate osmoprotectant production and root growth while adjusting stress communication (Chauhan et al., 2023). Editing the P5CS and code A genes involved in proline and glycine betaine production improves drought tolerance in soybean, wheat, and rice plants. Genome editing technology now produces fast modifications in plants to improve their tolerance to dry spells through their core functioning and molecular networks (Feng et al., 2022).

Salinity Tolerance

Millions of hectares of croplands worldwide are affected by salt problems that slow crop growth due to water pressure imbalances and harmful salts. Gene editing helps plants resist salty conditions by improving their ion transfer and stress reaction functions (Shelake et al., 2022). CRISPR/Cas9 eliminated ABA-induced transcription repressors from soybean cells, which led to enhanced sensitivity to ABA and improved root development, alongside better tolerance to saline conditions. Genetic changes in plants, such as in transport genes like NHX1, HKT1, and SOS1, help block sodium-ion entry or store them safely to maintain cell balance under saline conditions (Malakar & Chattopadhyay, 2021). Arabidopsis plants engineered using the CRISPR system acquired better salt tolerance through their enhanced capability to withstand osmotic stress at ACQOS mutational sites. Research has shown that using gene-editing techniques helps enhance the production of both proline and glycine betaine to protect plant cells from the effects of salt stress (Kim, Choi, Bae, & Kim, 2021). Researchers have raised P5CS and codA gene production levels in various plants, creating more osmolytes that enhance their ability to survive salt stress. Deletion of the negative regulators AtWRKY3 and AtWRKY4 in plants leads to increased antioxidant production and decreased salt stress caused by cell damage (Li et al., 2021). Researchers are now testing methods to edit multiple salt tolerance pathways in plants using simultaneous genetic engineering without foreign genes (Fig 2).

Figure 2. Unveiling genetics strategies for salt tolerance

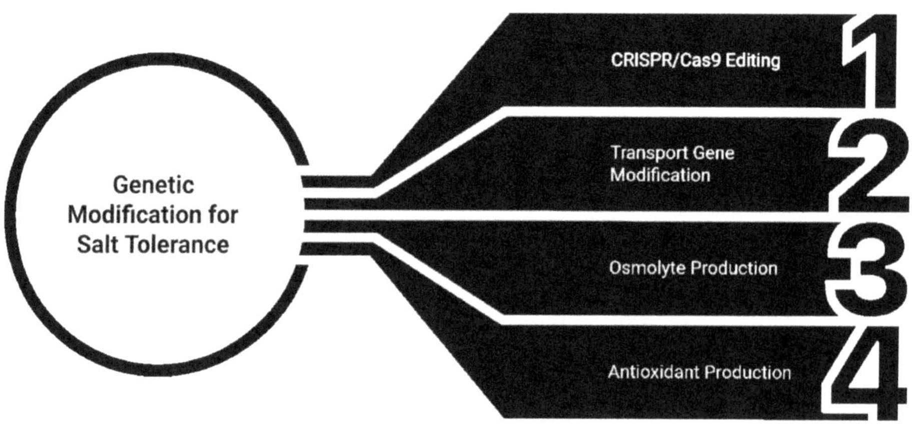

Heat and Cold Stress Tolerance

High temperatures and freezing conditions weaken crop development and affect production output. High temperatures damage proteins and tear cell membranes, whereas low temperatures slow enzymes and harm cell parts. Scientists have used CRISPR technology to improve plant tolerance to fluctuating temperatures. CRISPR/Cas9 technology has been used to modify heat shock protein (HSP) genes and genes such as HSFA1 and DREB2A, which trigger heat stress responses in rice and tomato cells to better resist high temperatures (Mondal et al., 2023). Plants with edited ROS-scavenging genes, SOD and CAT, develop stronger resistance against oxidative heat damage (Shah et al., 2020). Scientists use gene editing to change C-repeat binding factors (CBFs) and their regulatory elements, which activate cold-resistant genes (Kumari et al., 2025). Researchers have used CRISPR technology to change or improve CBF gene function in Arabidopsis, rice, and wheat, which raises their freezing and chilling tolerance by producing more protective substances. Changing the expression of fatty acid desaturase genes enhances the natural cold tolerance of plants (Kumari et al., 2025). Genetic modification techniques transform cells to produce epigenetic changes that help plants tolerate heat. Genome alterations through DNA remodeling increase ABA/AREB1-regulated protein activity, making Arabidopsis more drought- and cold-resistant. New genetic engineering technology can help plants handle heat and cold temperatures in different parts of the world (de Melo et al., 2020).

Heavy Metal and Oxidative Stress Tolerance

Crops suffer from growth reduction and safety problems when exposed to environmental pollutants and heavy metals. Research has shown that plants grow stronger when their metal stress-related genes are modified using gene-editing technology (Venegas-Rioseco, Ginocchio, & Ortiz-Calderón, 2021). Using CRISPR/Cas9 technology, scientists have altered OsHMA rice metal transporter genes, which decreased Cd content in grains, making food safer and helping plants resist stress better (Yan et al., 2024). Research has shown that phytochelatin synthase and metallothionein genes perform better at capturing and detoxifying metals when their genetic structures are modified through scientific approaches in different plant species (Anas et al., 2024; Liu et al., 2021). The transformation of plants through gene editing allows for increased SOD, CAT, and peroxidase activity in both WRKY and NAC transcription factors to prevent oxidative stress caused by heavy metals and environmental stress. Scientists use gene-editing methods to build non-enzymatic antioxidant pathways by modifying the genes that regulate glutathione and ascorbate production. Plant modifications increase the production of cellular defense tools that protect cells against oxidation under stressful conditions (Abbasi et al., 2025; Huang et al., 2024). Scientists have studied advanced gene-editing techniques, such as base editing and prime editing, to make plants tolerant to heavy metals and oxidative stress at the genetic level without including foreign DNA. This exact control benefits the creation of crops that follow government and customer requirements (Falak et al., 2025; Kumar et al., 2023).

GENE EDITING STRATEGIES FOR ENHANCED STRESS TOLERANCE

Editing Negative Regulators of Stress Response

Changing negative stress response regulators offers the best method for making plants more weather tolerant. Plants maintain or increase their stress defense capabilities when researchers edit or reduce the activity of key genes (Fujita & Hasanuzzaman, 2022). Scientists have widely applied the CRISPR/Cas9 system to disable stress response control genes in diverse agricultural species. Scientists in maize used CRISPR/Cas9 to edit the ARGOS8 gene and elevate its level under strong promoter control to create plants that resisted drought and maintained higher yields under low-water conditions (Rai et al., 2023). Studies in rice have confirmed that targeted disruption of the OsERA1 gene enhances binge tolerance by strengthening the ABA response and functioning in the stomatal system (Ma et al., 2024). SlARF4's negative control over plant responses to drought and high salt now stands weakened in tomato plants because SlARF4 deletion created better stress tolerance, more effective antioxidant protection, and stronger water control during stress periods (Chen et al., 2021). Plant editing can target more than one negative regulator simultaneously without restriction. The joint removal of multiple ABA receptor genes improved both resistance to cold temperatures and plant productivity in rice during research studies. When scientists aim their edits at stress-response regulators, these systems become stronger against environmental challenges (Ding et al., 2025).

Multiplex and Transgene-Free Editing Approaches

Stress tolerance to non-living factors depends on the activity of multiple related genetic traits. Changing a single gene does not result in major or ongoing improvements. Scientists use simultaneous gene editing to address the complex genetic traits that control plant stress response (Manna et al., 2021). CRISPR/Cas systems allow us to develop systems with multiple guide RNAs that target several locations that affect various aspects of stress tolerance, including osmoprotectant production, ion movement, and antioxidant protection. Scientists have shown that stacking multiple CRISPR systems helps plants become stronger against stress (Manghwar et al., 2020). Scientists have combined multiple stress tolerance gene edits to AVP1 vacuolar H+-pyrophosphatase PP2A-C5 protein phosphatase and AtCLCc chloride channel in Arabidopsis and cotton to improve plant performance under drought, heat, and salinity without harming growth and yield. These combination editing tools better match the stressful realities experienced by plants in their natural environments. Scientists have made progress in creating gene edits free of transgenic components (Balasubramaniam et al., 2022). People oppose the use of GMO technology because foreign DNA makes up the altered genes. Gene editing creates specific plant mutations without adding foreign DNA, making it indistinguishable from natural breeding outcomes (Martínez-Fortún et al., 2022). Researchers can produce regular plants beginning with CRISPR/Cas ribonucleoproteins (RNPs) or expression systems that require editing tools to disappear from future generations before any plant breeding. The faster and easier approval of gene-edited plants for sale to consumers improves their market prospects (Mushtaq et al., 2021).

Integration with Conventional Breeding

The merger of gene-editing technology with traditional plant breeding helps improve crops while maintaining the speed and precision of this method. The selection of traits from nature (traditional breeding) plus genetic editing technologies helps achieve fast progress in the development of specific traits and preserves plant diversity and versatility. Experts use marker-assisted selection to identify top-quality plant lines as targets before editing stress tolerance genes. Genomic prediction helps in selecting the best genetic material (Ahmar et al., 2020). Combining these methods helps improve crops by adding many good genes along with a wide natural genetic background, quickly developing plant strains that resist climate change. Scientists alter the genes of stress-resistant local cultivars to maintain their strong performance levels when developing new crop varieties. This technology helps integrate desired edits into both breeding products and destination genetic variations. Two types of crop breeding work together effectively to help plants survive in combinations of field conditions (Huang, 2024). By using traditional and modern breeding methods simultaneously, crop developers can produce plants that survive stress better and produce abundant high-quality yields in their area (Fig 3).

Figure 3. Comparison of gene editing and traditional plant breeding

Characteristic	Gene Editing	Traditional Plant Breeding
Trait Selection	Specific trait development	Selection from nature
Speed	Fast progress	Slower progress
Precision	High precision	Lower precision
Diversity	Preserves plant diversity	Preserves plant diversity
Integration	Integrates edits into breeding products	Integrates edits into breeding products
Stress Resistance	Improves stress tolerance	Improves stress tolerance

Challenges and Limitations

With the current advances in nanotechnology, plant science, and crop improvement have been boosted by the use of advanced gene-editing tools, particularly CRISPR/Cas systems. Nevertheless, several limitations and challenges must be acknowledged to ensure their safe, efficient, and ethical use in agriculture. These challenges include targeting side effects and genetic instability, legal and ethical aspects, and concerns related to field application and environmental effects (Piergentili et al., 2021).

Off-Target Effects and Genome Stability

There are several issues associated with gene editing, one of which is off-target effects, which is becoming technical in ensuring that the editing occurs where it is required and not anywhere else. Even though gRNA is complementary to the target DNA, off-target effects are observed with CRISPR/Cas systems because they cut other similar but non-related sequences. Some of these mutations may affect coding or regulatory areas, leading to negative phenotypes, decreased fitness, or undesired characteristics (Manghwar et al., 2020). Some investigations have pointed out that the delivery of Cas, choice of gRNA, and target site play critical roles in off-target effects. Off-target activity can be greatly reduced by optimizing Cas9 variants (e.g., SpCas9-HF1 and eSpCas9) and distinct orthologous Cas12a nucleases. Advanced computational approaches and artificial intelligence have also been used to develop gRNAs that are least likely to target unwanted sequences and regions, thus enhancing efficiency. Another principal challenge is the stability of the genome after gene editing (Manghwar et al., 2020). DSBs generated by nucleases initiate errors in DNA homologous recombination or non-homologous end joining, which could lead to chromosomal alterations and large deletions or insertions in the cell when the lesion repair process is erroneous (Gerelchuluun, 2020). In some cases, editing may trigger cellular stress responses such as p53 – mediated apoptosis that can decrease the efficiency or favor editing of cells with lower DNA fidelity. Mosaicism, where one or more cells undergo modifications in a non-uniform manner, presents another problem, especially for multicellular individuals, as it complicates phenotypic analysis and breeding (Smirnikhina et al., 2022). Considering that base editing and prime editing do not involve the formation of DSBs, these approaches are considered safer, with a lower risk of off-target mutagenesis and genome instability. Nevertheless, they come with challenges, including a limited editing window and the possibility of off-target deaminase activity, which requires enhancement (Kantor et al., 2020).

Regulatory, Ethical, and Biosafety Considerations

This is because the rules governing edited crops are still undecided, and each country has different policies. This is because many gene-edited plants, which may include those with a small number of modifications as well as those without transgenes, might be outside the definition of what is considered GMOs (Caradus, 2023). Almost all regulatory agencies have their own definitions and standards, which makes them different from each other, leaving developers and farmers confused. It also raises ethical issues, with attention paid to the possible negative consequences for the environment and the equal distribution of such gene technologies among people. Concerns have been raised over off-target mutations and gene flow to the related wild type, which raises biosafety concerns about possible effects on the existing bi- and tri-diversity and the balance of ecosystems. However, the present cautionary note is that the application of gene editing to germplasm collected from indigenous or traditional varieties raises issues of patents and benefit sharing. It is evident that perception and acceptance by the public is one of the key aspects in relation to crops that undergo gene editing (Trump et al., 2023). There is a need to inform consumers, policymakers, and other stakeholders about the positive aspects of using the technology, its dangers, and its legal permissiveness. There is a need to create an ethical arsenal that will allow the use of genetic technology to benefit agriculture and simultaneously exclude anything pathological. Pre-release also involves several biosafety assessments aimed at the prospects of allergenicity, toxicity, and environmental stability of the edited traits. It is also suggested that the subjects be tracked for any such impacts in the long run after the product is released (Harfouche et al., 2021).

Field Performance and Environmental Impact

Although genetic modifications offer a lot of potential when tested under contained laboratory and greenhouse conditions, transferring these successes to real field environments is often not easy. Drought stress in the field is usually accompanied by salinity, temperature, and other biotic stresses that are constantly changing. Such individual stress tolerance traits may not work efficiently when combined or fluctuating stresses are involved (Abdelaal et al., 2022). This means that field trials are of great significance in identifying the agronomic performance, yield reliability, and adaptability of gene-edited crops under experimental conditions. Management practices and variations in genotype by environment and soil must be factored into these trials. In addition, edited plants may display multiple phenotypic changes, such as growth and developmental changes, which affect competitiveness and adaptation (Metje-Sprink et al., 2020). Consequently, there are consequences related to the effect of gene-edited crop plants on organisms other than the target organism, soil microorganisms, and the spread of gene alterations to wild relatives. Admittedly, the use of gene editing in the production of crops has the advantage of reducing the application of chemical inputs by enhancing stress tolerance; however, such effects on the ecological environment should not be taken lightly (Sabat & Tripathy, 2024). Some risks can be managed using measures including gene containment, use of sterile lines for domestication genes, or even editing out the domestication genes. However, there is much evidence that indicates that strategies for the sustainable release of gene-edited crops involve using conventional breeding improvements concomitantly with IPM and abiotic stress tolerance. Therefore, management should follow the current results and modify courses of action to achieve a sustainable overall positive impact and minimize adverse effects (Wang et al., 2022). This chapter portrays gene-editing technologies as having massive potential for improving crop resilience and productivity under increasing adverse environmental conditions. However, issues such as off-target effects, genetic stability, governmental policies, ethical concerns, and field efficacy must be addressed to allow for safe use. Therefore, it is important to continue research, make regulations transparent, involve stakeholders, and conduct rigorous field tests to fully tap into the benefits of gene editing in sustainable agriculture without polluting the environment and undermining the welfare of society.

FUTURE PERSPECTIVES

Emerging Technologies and Innovations

The future applications of gene editing in agriculture will develop through technological advancements that solve key problems in food production involving sustainability. The CRISPR/Cas system is a genome modification technology, but new variant developments alongside additional technologies boost efficiency and effectiveness while decreasing operational costs. A significant technological advancement in gene editing is the creation of enhanced CRISPR/Cas versions, such as base editors and prime editors, which perform single-base edits and small modifications through mechanisms other than double-strand break events (Saber Sichani et al., 2023). These tools reduce both random genetic harm and unstable genomic sequences, making them particularly suitable for enhancing difficult-to-change traits, such as crop stress tolerance, disease immunity, and nutritional benefits. By editing regulatory elements beyond coding sequences, scientists can achieve better control of gene expression to maximize plant resistance to various environmental factors (Han et al., 2025). DNA sequencing laboratories utilize automated systems to digitize the genome-editing methods. Automated research and breeding solutions accelerate investigations by enabling high-volume genetic editing, phenotype screening, and data assessment of thousands of altered lines within shorter durations than traditional methods. Advanced technologies improve both reproduction capability and scalability, thus enabling practical field applications after laboratory discoveries (Penna & Jain, 2023). The combined use of gene editing technology and genomic selection, along with big data analytical approaches, shows great potential. Scientists use editing tools together with genome-wide association studies to find and modify essential genes that control complex traits by shortening breeding periods and producing more effective improvements (W. Huang, 2024). The use of synthetic biology methods to build new metabolic pathways, together with synthetic promoters, enables scientists to develop crops that possess new functional abilities, such as increased nutrients and reduced allergens with increased shelf stability. The safety and efficiency of gene editing have been significantly enhanced by newly developed gene delivery technologies that combine transient expression systems and nanoparticle-mediated transformation, especially for hard-to-edit crop varieties. These techniques enable the creation of gene-edited plants without transgenes, making them more acceptable to both regulatory bodies and consumers (Hamdan & Tan, 2024).

Prospects for Climate-Smart and Sustainable Agriculture

Gene editing will drive agricultural change toward sustainable and climate-resilient farming methods. The quick creation of crops to survive droughts, saline soil risks, heat waves, and new plant threats will help us fight the effects of climate change. Genome-edited crops are developed through updates that help plants use fewer resources without affecting their productivity (Sarma et al., 2024). Farmland communities that experience limited environments benefit when farmers use gene editing because it helps protect resources and creates job opportunities. Our farming resources benefit from new nutrients obtained through gene editing. Using gene-editing technology, scientists have created vitamin- and mineral-enhanced crops and reduced harmful elements for better health in people. Tomatoes with slow pectin breakdown help farmers decrease product waste, making storage and shipping easier while meeting sustainability targets. Genetic modification allows us to preserve diverse plant species and improve their quality. Introducing beneficial traits from specific local plant types with gene editing protects crop and animal genetic diversity as part of their adaptation to changing environmental trends (Baloglu et al., 2022). Food safety requires this protection to achieve stable and resistant food systems. To achieve the possible benefits of gene editing, existing rules, safety concerns, and public acceptance issues must be resolved (Vågsholm et al., 2020). Different regions have confusing laws for testing and releasing gene-edited crops, which slow down the process of new crop beneficial traits from reaching the market. Standards must align with science to distinguish transgenic from gene-edited organisms while allowing technology to improve. People need to understand and support our work through open discussions for consumers and company stakeholders to trust us (Rozas et al., 2022).

CONCLUSION

Gene editing technology has received a lot of attention over the years, being seen as a revolution in modern agriculture through the enhancement of crop improvement. These are CRISPR/Cas9, TALENs, and base editors, and have paved way to the modification of plant genomes that has I accelerated the development of crops with resistance to diseases, pests, and abiotic stresses, increased yield and nutrition. This specificity in relation to the modification of the native DNA in organisms makes the process of gene editing different from other forms of genetic modification, and the development of this technology has been fast-tracked in both research and breeding industries. In fact, there are many benefits of gene editing, as stated below. From an environmental perspective, GM crops enable the cutting of the use of chemical pesticides and fertilizers through natural built-in resistance against biotic and abiotic factors that hinder the growth of crops, thus easing the environmental burden of agricultural production. From an economic perspective, they may raise output, lower the cost of production, and reduce the cost of food, which are worthy goals, especially in developing countries, where the scarcity factor is a major issue. From the perspective of its impact on health, gene editing also makes it possible to create varieties that are richer in vitamins and contain fewer allergens. with the aim of contributing to health-related programs that combat malnutrition. However, with gene editing there are also certain challenges that come with being able to carry out gene editing procedures. Off-target effects and other unintended changes in the genome are technical issues that require strict verification and assessment to guarantee the genetic stability and safety of food crops. Regulatory and ethical issues feature prominently here, since many country guidelines concerning genetically designed organisms are not well-developed and differ extensively. Another factor arising from traditional thinking that also slows the rate at which gene editing is embraced is the perception that the public has towards emerging technology, the concern for the type of crops that are going to be developed, and other issues related to seeds, food safety, and the concern over the future control of seeds by large corporations. Some of these issues are being addressed with the recent advancement of transgene-free editing using ribonucleoprotein complexes, which may help reduce regulatory challenges and public rejection. Broad-spectrum resistance with gene-silencing technology and attention to susceptibility genes (S-genes) rather than resistance genes (R-genes) presents an opportunity for prominence and superiority over existing disease resistance technologies, simultaneously enhancing plant physiology. CRISPR technology is regarded as one of the most promising tools in modern science, as it contributes to the development of sustainable systems of agriculture and food provision in the context of climate shift and increased population pressure. Its future remains uncertain because it must weigh the positive changes enabled by technology with respect to the public, including governance and the proper regulation of new developments. Over time, with the discovery of more technologies, gene editing is expected to be at the center of defining future crops that are resilient, productive, and nutritious for the fast-growing world.

REFERENCES

Abbasi, A., Anas, M., Elahi, M., Khan, A., Khattak, W. A., Saleem, M. H., & Quraishi, U. M. (2025). Restoring Wheat Productivity and Nutrient Balance Under Cadmium Stress Through Reducing Toxicity, Metal Uptake, and Oxidative Damage Using Selenium Nanoparticles. *Journal of Trace Elements in Medicine and Biology : Organ of the Society for Minerals and Trace Elements (GMS)*, *127644*, 127644. Advance online publication. DOI: 10.1016/j.jtemb.2025.127644

Abdelaal, K., Alsubeie, M. S., Hafez, Y., Emeran, A., Moghanm, F., Okasha, S., & Ibrahim, M. F. (2022). Physiological and biochemical changes in vegetable and field crops under drought, salinity and weeds stresses: Control strategies and management. *Agriculture*, *12*(12), 2084. DOI: 10.3390/agriculture12122084

Afzal, S., Sirohi, P., & Singh, N. K. (2020). A review of CRISPR associated genome engineering: Application, advances and future prospects of genome targeting tool for crop improvement. *Biotechnology Letters*, *42*(9), 1611–1632. DOI: 10.1007/s10529-020-02950-w

Ahmar, S., Gill, R. A., Jung, K.-H., Faheem, A., Qasim, M. U., Mubeen, M., & Zhou, W. (2020). Conventional and molecular techniques from simple breeding to speed breeding in crop plants: Recent advances and future outlook. *International Journal of Molecular Sciences*, *21*(7), 2590. DOI: 10.3390/ijms21072590

Anas, M., Khan, A., Bibi, S., Farwa, U., & Quraishi, U. (2024). Manganese oxide nanoparticles elicit cadmium tolerance in wheat seedlings by ionomic and phenomic adjustment, regulation of AsA–GSH pathway, cellular thickness recovery, and antioxidant modulation. *Environmental Science. Nano*, *11*(11), 11. DOI: 10.1039/D4EN00490F

Anas, M., Khattak, W. A., Fahad, S., Alrawiq, N., Alrawiq, H. S., Abdelsalam, N. R., Jaremko, M., & Quraishi, U. M. (2025). Mitigating heavy metal pollution in agriculture: A multi-omics and nanotechnology approach to safeguard global wheat production. *Journal of Hazardous Materials Advances*, *17*, 100584. DOI: 10.1016/j.hazadv.2024.100584

Anjali, K., Kumar, S., Korra, T., Thakur, R., Arutselvan, R., Kashyap, A. S., Nehela, Y., Chaplygin, V., Minkina, T., & Keswani, C. (2023). Role of plant secondary metabolites in defence and transcriptional regulation in response to biotic stress. *Plant Stress*, *8*, 100154. DOI: 10.1016/j.stress.2023.100154

Balasubramaniam, T., Wijewardene, I., Hu, R., Shen, G., Zhang, J., & Zhang, H. (2022). Co-overexpression of AVP1, PP2A-C5, and AtCLCc in Arabidopsis thaliana greatly increases tolerance to salt and drought stresses. *Environmental and Experimental Botany*, *200*, 104934. DOI: 10.1016/j.envexpbot.2022.104934

Baloglu, M. C., Celik Altunoglu, Y., Baloglu, P., Yildiz, A. B., Türkölmez, N., & Özden Çiftçi, Y. (2022). Gene-editing technologies and applications in legumes: Progress, evolution, and future prospects. *Frontiers in Genetics*, *13*, 859437. DOI: 10.3389/fgene.2022.859437

Begna, T. (2021). Conventional breeding methods widely used to improve self-pollinated crops. *International Journal of Research*, *7*(1), 1–16.

Bohra, A., Chand Jha, U., Godwin, I. D., & Kumar Varshney, R. (2020). Genomic interventions for sustainable agriculture. *Plant Biotechnology Journal*, *18*(12), 2388–2405. DOI: 10.1111/pbi.13472

Burmistrz, M., Krakowski, K., & Krawczyk-Balska, A. (2020). RNA-targeting CRISPR–Cas systems and their applications. *International Journal of Molecular Sciences*, *21*(3), 1122. DOI: 10.3390/ijms21031122

Caradus, J. R. (2023). Processes for regulating genetically modified and gene edited plants. *GM Crops & Food*, *14*(1), 1–41. DOI: 10.1080/21645698.2023.2252947

Chaffai, R., Ganesan, M., & Cherif, A. (2024). Abiotic Stress in Plants: Challenges and Strategies for Enhancing Plant Growth and Development. In *Plant Adaptation to Abiotic Stress: From Signaling Pathways and Microbiomes to Molecular Mechanisms* (pp. 1-30): Springer.

Chauhan, P. K., Upadhyay, S. K., Tripathi, M., Singh, R., Krishna, D., Singh, S. K., & Dwivedi, P. (2023). Understanding the salinity stress on plant and developing sustainable management strategies mediated salt-tolerant plant growth-promoting rhizobacteria and CRISPR/Cas9. *Biotechnology & Genetic Engineering Reviews*, *39*(2), 311–347. DOI: 10.1080/02648725.2022.2131958

Chen, M., Zhu, X., Liu, X., Wu, C., Yu, C., Hu, G., Chen, L., Chen, R., Bouzayen, M., Zouine, M., & Hao, Y. (2021). Knockout of Auxin Response Factor SlARF4 Improves Tomato Resistance to Water Deficit. *International Journal of Molecular Sciences*, *22*(7), 3347. DOI: 10.3390/ijms22073347

Choubey, K., Dey, S. K., Dey, G., & Boruah, T. (2025). Understanding the role of phytohormones in mitigation of biotic and abiotic stress: Current status and future prospects. *Proceedings of the National Academy of Sciences. India. Section B, Biological Sciences*, *95*(1), 9–14. DOI: 10.1007/s40011-024-01566-8

Choudhary, V., Kumar, K. K., Sahu, B., Muthappa, S. K., & Das, A. D. A. A. (2021). Agronomic innovations in biotic stress management and its combined effect with abiotic stresses in crop production. *Indian Journal of Agronomy*, *66*(5), 237–257.

de Dieu Habimana, J., Huang, R., Muhoza, B., Kalisa, Y. N., Han, X., Deng, W., & Li, Z. (2022). Mechanistic insights of CRISPR/Cas nucleases for programmable targeting and early-stage diagnosis: A review. *Biosensors & Bioelectronics*, *203*, 114033. DOI: 10.1016/j.bios.2022.114033

de Melo, B. P., Lourenço-Tessutti, I. T., Paixão, J. F. R., Noriega, D. D., Silva, M. C. M., de Almeida-Engler, J., Fontes, E. P. B., & Grossi-de-Sa, M. F. (2020). Transcriptional modulation of AREB-1 by CRISPRa improves plant physiological performance under severe water deficit. *Scientific Reports*, *10*(1), 16231. DOI: 10.1038/s41598-020-72464-y

Dikilitas, M., Simsek, E., & Roychoudhury, A. (2020). Role of proline and glycine betaine in overcoming abiotic stresses. *Protective chemical agents in the amelioration of plant abiotic stress: biochemical and molecular perspectives*, 1-23.

Ding, G., Li, Z., Iqbal, Z., Zhao, M., Cui, Z., Cao, L., & Bai, L. (2025). Identifications of genes involved in ABA and MAPK signaling pathways positively regulating cold tolerance in rice. *Plants*, *14*(4), 498. DOI: 10.3390/plants14040498

Dixit, S., Sivalingam, P. N., Baskaran, R. M., Senthil-Kumar, M., & Ghosh, P. K. (2024). Plant responses to concurrent abiotic and biotic stress: Unravelling physiological and morphological mechanisms. *Plant Physiology Reports*, *29*(1), 6–17. DOI: 10.1007/s40502-023-00766-0

Falak, A., Anas, M., Khan, A., Hayat, A., Shaheen, Z., Saleem, M. H., Fahad, S., & Quraishi, U. M. (2025). Efficacy of ascorbic acid coated quantum dots in alleviating lead-induced oxidative damage and enhancing growth parameters in rice (Oryza sativa L.) for sustainable cultivation. *Journal of Trace Elements in Medicine and Biology : Organ of the Society for Minerals and Trace Elements (GMS)*, *88*, 127603. DOI: 10.1016/j.jtemb.2025.127603

Feng, X., Hu, Y., Zhang, W., Xie, R., Guan, H., Xiong, H., & Zheng, D. (2022). Revisiting the role of delta-1-pyrroline-5-carboxylate synthetase in drought-tolerant crop breeding. *The Crop Journal*, *10*(4), 1213–1218. DOI: 10.1016/j.cj.2022.04.002

Fidler, J., Graska, J., Gietler, M., Nykiel, M., Prabucka, B., Rybarczyk-Płońska, A., Muszyńska, E., Morkunas, I., & Labudda, M. (2022). PYR/PYL/RCAR receptors play a vital role in the abscisic-acid-dependent responses of plants to external or internal stimuli. *Cells*, *11*(8), 1352. DOI: 10.3390/cells11081352

Fujita, M., & Hasanuzzaman, M. (2022). *Approaches to enhancing antioxidant defense in plants* (Vol. 11). MDPI. DOI: 10.3390/books978-3-0365-4185-3

Gerashchenkov, G., Rozhnova, N., Kuluev, B., Kiryanova, O. Y., Gumerova, G., Knyazev, A., & Matniyazov, R. (2020). Design of guide RNA for CRISPR/Cas plant genome editing. *Molecular Biology*, *54*(1), 24–42. DOI: 10.1134/S0026893320010069

Gerelchuluun, A. (2020). DNA Damage, Repair Mechanisms, and Chromosomal Aberrations. *Proton Beam Radiotherapy: Physics and Biology*, 183-208.

Gonzalez Guzman, M., Cellini, F., Fotopoulos, V., Balestrini, R., & Arbona, V. (2022). New approaches to improve crop tolerance to biotic and abiotic stresses. *Physiologia Plantarum*, *174*(1), e13547. DOI: 10.1111/ppl.13547

Gupta, R., Gupta, D., Ahmed, K. T., Dey, D., Singh, R., Swarnakar, S., & Ghosh, D. (2021). Modification of Cas9, gRNA and PAM: Key to further regulate genome editing and its applications. *Progress in Molecular Biology and Translational Science*, *178*, 85–98. DOI: 10.1016/bs.pmbts.2020.12.001

Hamdan, M. F., & Tan, B. C. (2024). Genetic modification techniques in plant breeding: A comparative review of CRISPR/Cas and GM technologies. *Horticultural Plant Journal*. Advance online publication. DOI: 10.1016/j.hpj.2024.02.012

Han, X., Li, S., Zeng, Q., Sun, P., Wu, D., Wu, J., & Kang, Z. (2025). Genetic engineering, including genome editing, for enhancing broad-spectrum disease resistance in crops. *Plant Communications*, *6*(2), 101195. DOI: 10.1016/j.xplc.2024.101195

Harfouche, A. L., Petousi, V., Meilan, R., Sweet, J., Twardowski, T., & Altman, A. (2021). Promoting ethically responsible use of agricultural biotechnology. *Trends in Plant Science*, *26*(6), 546–559. DOI: 10.1016/j.tplants.2020.12.015

Hossain, A., Skalicky, M., Brestic, M., Maitra, S., Ashraful Alam, M., Syed, M. A., & Bhadra, P. (2021). Consequences and mitigation strategies of abiotic stresses in wheat (Triticum aestivum L.) under the changing climate. *Agronomy (Basel)*, *11*(2), 241. DOI: 10.3390/agronomy11020241

Hosseini, S. Y., Mallick, R., Mäkinen, P., & Ylä-Herttuala, S. (2024). Insights into Prime Editing Technology: A Deep Dive into Fundamentals, Potentials, and Challenges. *Human Gene Therapy*, *35*(17-18), 649–668. DOI: 10.1089/hum.2024.043

Huang, W. (2024). The current situation and future of using GWAS strategies to accelerate the improvement of crop stress resistance traits. *Molecular Plant Breeding*, •••, 15. DOI: 10.5376/mpb.2024.15.0007

Huang, Y., Sun, Z., & Zhou, X. (2024). WRKY Transcription Factors in Response to Metal Stress in Plants: A Review. *International Journal of Molecular Sciences*, *25*(20), 10952. DOI: 10.3390/ijms252010952

Jogawat, A., Yadav, B., Chhaya, , Lakra, N., Singh, A. K., & Narayan, O. P. (2021). Crosstalk between phytohormones and secondary metabolites in the drought stress tolerance of crop plants: A review. *Physiologia Plantarum*, *172*(2), 1106–1132. DOI: 10.1111/ppl.13328

Kabato, W., Getnet, G. T., Sinore, T., Nemeth, A., & Molnár, Z. (2025). Towards climate-smart agriculture: Strategies for sustainable agricultural production, food security, and greenhouse gas reduction. *Agronomy (Basel)*, *15*(3), 565. DOI: 10.3390/agronomy15030565

Kantor, A., McClements, M. E., & MacLaren, R. E. (2020). CRISPR-Cas9 DNA base-editing and prime-editing. *International Journal of Molecular Sciences*, *21*(17), 6240. DOI: 10.3390/ijms21176240

Khan, Z., Ali, Z., Khan, A. A., Sattar, T., Zeshan, A., Saboor, T., & Binyamin, B. (2022). History and classification of CRISPR/Cas system. *The CRISPR/Cas Tool Kit for Genome Editing*, 29-52.

Khan, Z., & Shahwar, D. (2020). Role of heat shock proteins (HSPs) and heat stress tolerance in crop plants. *Sustainable agriculture in the era of climate change*, 211-234.

Kim, S.-T., Choi, M., Bae, S.-J., & Kim, J.-S. (2021). The functional association of ACQOS/VICTR with salt stress resistance in Arabidopsis thaliana was confirmed by CRISPR-mediated mutagenesis. *International Journal of Molecular Sciences*, *22*(21), 11389. DOI: 10.3390/ijms222111389

Kopecká, R., Kameniarová, M., Černý, M., Brzobohatý, B., & Novák, J. (2023). Abiotic stress in crop production. *International Journal of Molecular Sciences*, *24*(7), 6603. DOI: 10.3390/ijms24076603

Kovalev, M. A., Davletshin, A. I., & Karpov, D. S. (2024). Engineering Cas9: Next generation of genomic editors. *Applied Microbiology and Biotechnology*, *108*(1), 209. DOI: 10.1007/s00253-024-13056-y

Krishna, R., Ansari, W. A., Soumia, P., Yadav, A., Jaiswal, D. K., Kumar, S., Singh, A. K., Singh, M., & Verma, J. P. (2022). Biotechnological interventions in tomato (Solanum lycopersicum) for drought stress tolerance: Achievements and future prospects. *Biotech*, *11*(4), 48. DOI: 10.3390/biotech11040048

Kumar, M., Prusty, M. R., Pandey, M. K., Singh, P. K., Bohra, A., Guo, B., & Varshney, R. K. (2023). Application of CRISPR/Cas9-mediated gene editing for abiotic stress management in crop plants. *Frontiers in Plant Science*, *14*, 1157678. DOI: 10.3389/fpls.2023.1157678

Kumari, K., Gusain, S., & Joshi, R. (2025). Engineering cold resilience: Implementing gene editing tools for plant cold stress tolerance. *Planta*, *261*(1), 2. DOI: 10.1007/s00425-024-04578-w

Kumari, V. V., Banerjee, P., Verma, V. C., Sukumaran, S., Chandran, M. A. S., Gopinath, K. A., Venkatesh, G., Yadav, S. K., Singh, V. K., & Awasthi, N. K. (2022). Plant nutrition: An effective way to alleviate abiotic stress in agricultural crops. *International Journal of Molecular Sciences*, *23*(15), 8519. DOI: 10.3390/ijms23158519

Lamichhane, S., & Thapa, S. (2022). Advances from conventional to modern plant breeding methodologies. *Plant Breeding and Biotechnology*, *10*(1), 1–4. DOI: 10.9787/PBB.2022.10.1.1

Lee, H. K., Oh, Y., Hong, J., Lee, S. H., & Hur, J. K. (2021). Development of CRISPR technology for precise single-base genome editing: A brief review. *BMB Reports*, *54*(2), 98–105. DOI: 10.5483/BMBRep.2021.54.2.217

Lee, M. (2023). Deep learning in CRISPR-Cas systems: A review of recent studies. *Frontiers in Bioengineering and Biotechnology*, *11*, 1226182. DOI: 10.3389/fbioe.2023.1226182

Li, H., Yang, Y., Hong, W., Huang, M., Wu, M., & Zhao, X. (2020). Applications of genome editing technology in the targeted therapy of human diseases: Mechanisms, advances and prospects. *Signal Transduction and Targeted Therapy*, *5*(1), 1. DOI: 10.1038/s41392-019-0089-y

Li, P., Li, X., & Jiang, M. (2021). CRISPR/Cas9-mediated mutagenesis of WRKY3 and WRKY4 function decreases salt and Me-JA stress tolerance in Arabidopsis thaliana. *Molecular Biology Reports*, *48*(8), 5821–5832. DOI: 10.1007/s11033-021-06541-4

Liu, J., Zhang, J., Kim, S. H., Lee, H.-S., Marinoia, E., & Song, W.-Y. (2021). Characterization of Brassica rapa metallothionein and phytochelatin synthase genes potentially involved in heavy metal detoxification. *PLoS One*, *16*(6), e0252899. DOI: 10.1371/journal.pone.0252899

Ma, Y., Tang, M., Wang, M., Yu, Y., & Ruan, B. (2024). Advances in Understanding Drought Stress Responses in Rice: Molecular Mechanisms of ABA Signaling and Breeding Prospects. *Genes*, *15*(12), 1529. DOI: 10.3390/genes15121529

Mafakheri, M., Kordrostami, M., & Al-Khayri, J. M. (2021). Abiotic stress in plants: socio-economic consequences and crops responses. *Nanobiotechnology: Mitigation of Abiotic Stress in Plants*, 1-28.

Magdy, M., Mostofa, M. G., Rahimi, M., & Abd El Moneim, D. (2023). Abiotic stress alleviation in plants: morpho-physiological and molecular aspects. In (Vol. 14, pp. 1295638): Frontiers Media SA.

Malakar, P., & Chattopadhyay, D. (2021). Adaptation of plants to salt stress: The role of the ion transporters. *Journal of Plant Biochemistry and Biotechnology*, *30*(4), 668–683. DOI: 10.1007/s13562-021-00741-6

Malhi, G. S., Kaur, M., & Kaushik, P. (2021). Impact of climate change on agriculture and its mitigation strategies: A review. *Sustainability (Basel)*, *13*(3), 1318. DOI: 10.3390/su13031318

Manghwar, H., Li, B., Ding, X., Hussain, A., Lindsey, K., Zhang, X., & Jin, S. (2020). CRISPR/Cas systems in genome editing: Methodologies and tools for sgRNA design, off-target evaluation, and strategies to mitigate off-target effects. *Advanced Science (Weinheim, Baden-Württemberg, Germany)*, *7*(6), 1902312. DOI: 10.1002/advs.201902312

Manna, M., Thakur, T., Chirom, O., Mandlik, R., Deshmukh, R., & Salvi, P. (2021). Transcription factors as key molecular target to strengthen the drought stress tolerance in plants. *Physiologia Plantarum*, *172*(2), 847–868. DOI: 10.1111/ppl.13268

Marone, D., Russo, M. A., Mores, A., Ficco, D. B., Laidò, G., Mastrangelo, A. M., & Borrelli, G. M. (2021). Importance of landraces in cereal breeding for stress tolerance. *Plants*, *10*(7), 1267. DOI: 10.3390/plants10071267

Martínez-Fortún, J., Phillips, D. W., & Jones, H. D. (2022). Natural and artificial sources of genetic variation used in crop breeding: A baseline comparator for genome editing. *Frontiers in Genome Editing*, *4*, 937853. DOI: 10.3389/fgeed.2022.937853

Mearaji, H. S., Ansari, A., Igdelou, N. K. M., Lajayer, B. A., & Pessarakli, M. (2021). Phytohormones and abiotic stresses: Roles of phytohormones in plants under abiotic stresses. In *Handbook of plant and crop physiology* (pp. 175–213). CRC Press. DOI: 10.1201/9781003093640-16

Medina-Puche, L., & Lozano-Durán, R. (2022). Plasma membrane-to-organelle communication in plant stress signaling. *Current Opinion in Plant Biology*, *69*, 102269. DOI: 10.1016/j.pbi.2022.102269

Metje-Sprink, J., Sprink, T., & Hartung, F. (2020). Genome-edited plants in the field. *Current Opinion in Biotechnology*, *61*, 1–6. DOI: 10.1016/j.copbio.2019.08.007

Mikhaylova, E., Kuluev, B., Gerashchenkov, G., Chemeris, D., Garafutdinov, R., Kuluev, A., Baymiev, A. K., Baymiev, A. K., & Chemeris, A. (2024). Prime-editing methods and pegRNA design programs. *Molecular Biology*, *58*(1), 17–32. DOI: 10.1134/S0026893324010084

Mondal, S., Karmakar, S., Panda, D., Pramanik, K., Bose, B., & Singhal, R. K. (2023). Crucial plant processes under heat stress and tolerance through heat shock proteins. *Plant Stress*, *10*, 100227. DOI: 10.1016/j.stress.2023.100227

Morales, F., Ancín, M., Fakhet, D., González-Torralba, J., Gámez, A. L., Seminario, A., Soba, D., Ben Mariem, S., Garriga, M., & Aranjuelo, I. (2020). Photosynthetic metabolism under stressful growth conditions as a bases for crop breeding and yield improvement. *Plants*, *9*(1), 88. DOI: 10.3390/plants9010088

Mushtaq, M., Ahmad Dar, A., Skalicky, M., Tyagi, A., Bhagat, N., Basu, U., & Dar, T.-U.-H. (2021). CRISPR-based genome editing tools: Insights into technological breakthroughs and future challenges. *Genes*, *12*(6), 797. DOI: 10.3390/genes12060797

Nambiar, T. S., Baudrier, L., Billon, P., & Ciccia, A. (2022). CRISPR-based genome editing through the lens of DNA repair. *Molecular Cell*, *82*(2), 348–388. DOI: 10.1016/j.molcel.2021.12.026

Nazir, F., Peter, P., Gupta, R., Kumari, S., Nawaz, K., & Khan, M. I. R. (2024). Plant hormone ethylene: A leading edge in conferring drought stress tolerance. *Physiologia Plantarum*, *176*(1), e14151. DOI: 10.1111/ppl.14151

Nidhi, S., Anand, U., Oleksak, P., Tripathi, P., Lal, J. A., Thomas, G., Kuca, K., & Tripathi, V. (2021). Novel CRISPR–Cas systems: An updated review of the current achievements, applications, and future research perspectives. *International Journal of Molecular Sciences*, *22*(7), 3327. DOI: 10.3390/ijms22073327

Ochoa-Sanchez, A., Perez-Sanchez, G., Torres-Ledesma, A. M., Valdez, J. P. R., Rinaldi, G., Moguel, B. B., & Molina-Aguilar, C. (2021). Prime Editing, a novel genome-editing tool that may surpass conventional CRISPR-Cas9. *Re:GEN Open*, *1*(1), 75–82. DOI: 10.1089/regen.2021.0016

Ostmeyer, T., Parker, N., Jaenisch, B., Alkotami, L., Bustamante, C., & Jagadish, S. K. (2020). Impacts of heat, drought, and their interaction with nutrients on physiology, grain yield, and quality in field crops. *Plant Physiology Reports*, *25*(4), 549–568. DOI: 10.1007/s40502-020-00538-0

Penna, S., & Jain, S. M. (2023). Fruit crop improvement with genome editing, in vitro and transgenic approaches. *Horticulturae*, *9*(1), 58. DOI: 10.3390/horticulturae9010058

Piergentili, R., Del Rio, A., Signore, F., Umani Ronchi, F., Marinelli, E., & Zaami, S. (2021). CRISPR-Cas and its wide-ranging applications: From human genome editing to environmental implications, technical limitations, hazards and bioethical issues. *Cells*, *10*(5), 969. DOI: 10.3390/cells10050969

Prusty, A., Panchal, A., Singh, R. K., & Prasad, M. (2024). Major transcription factor families at the nexus of regulating abiotic stress response in millets: A comprehensive review. *Planta*, *259*(5), 118. DOI: 10.1007/s00425-024-04394-2

Qaim, M. (2020). Role of new plant breeding technologies for food security and sustainable agricultural development. *Applied Economic Perspectives and Policy*, *42*(2), 129–150. DOI: 10.1002/aepp.13044

Rafiq, A. R. (2024). Genomic editing techniques for ensuring food security: CRISPR Cas, TALEN, ZFN, RNAi and mutagenesis. *Journal of Pure and Applied Agriculture*, *9*(2).

Rai, G. K., Khanday, D. M., Kumar, P., Magotra, I., Choudhary, S. M., Kosser, R., & Rouphael, Y. (2023). Enhancing crop resilience to drought stress through CRISPR-Cas9 genome editing. *Plants*, *12*(12), 2306. DOI: 10.3390/plants12122306

Rallapalli, K. L., & Komor, A. C. (2023). The design and application of DNA-editing enzymes as base editors. *Annual Review of Biochemistry*, *92*(1), 43–79. DOI: 10.1146/annurev-biochem-052521-013938

Rasheed, A., Jie, H., Ali, B., He, P., Zhao, L., Ma, Y., & Hamid, M. R. (2023). Breeding drought-tolerant maize (Zea mays) using molecular breeding tools: Recent advancements and future prospective. *Agronomy (Basel)*, *13*(6), 1459. DOI: 10.3390/agronomy13061459

Rathod, A., & Verma, N. S. (2023). *Impact of abiotic stress on agronomical crops. Frontiers of Agronomy*. Elite Publishing House.

Ray, U., & Raghavan, S. C. (2020). Modulation of DNA double-strand break repair as a strategy to improve precise genome editing. *Oncogene*, *39*(41), 6393–6405. DOI: 10.1038/s41388-020-01445-2

Rozas, P., Kessi-Pérez, E. I., & Martínez, C. (2022). Genetically modified organisms: Adapting regulatory frameworks for evolving genome editing technologies. *Biological Research*, *55*(1), 55. DOI: 10.1186/s40659-022-00399-x

Sabat, M., & Tripathy, A. (2024). Genetically Modified and Gene-Edited Food Crops: Recent Status and Future Prospects. *Food Production, Diversity, and Safety Under Climate Change*, 211-222.

Saber Sichani, A., Ranjbar, M., Baneshi, M., Torabi Zadeh, F., & Fallahi, J. (2023). A review on advanced CRISPR-based genome-editing tools: Base editing and prime editing. *Molecular Biotechnology*, *65*(6), 849–860. DOI: 10.1007/s12033-022-00639-1

Sarma, H. H., Borah, S. K., Dutta, N., Sultana, N., Nath, H., & Das, B. C. (2024). Innovative approaches for climate-resilient farming: Strategies against environmental shifts and climate change. *International Journal of Environment and Climate Change*, *14*(9), 217–241. DOI: 10.9734/ijecc/2024/v14i94407

Shah, Z., Iqbal, A., Khan, F. U., Khan, H. U., Durrani, F., & Ahmad, M. Z. (2020). Genetic manipulation of pea (Pisum sativum L.) with Arabidopsis's heat shock factor HsfA1d improves ROS scavenging system to confront thermal stress. *Genetic Resources and Crop Evolution*, *67*(8), 2119–2127. DOI: 10.1007/s10722-020-00966-9

Shamshirgaran, Y., Liu, J., Sumer, H., Verma, P. J., & Taheri-Ghahfarokhi, A. (2022). Tools for efficient genome editing; ZFN, TALEN, and CRISPR. *Applications of genome modulation and editing*, 29-46.

Sharma, A., Kumar, V., Shahzad, B., Ramakrishnan, M., Singh Sidhu, G. P., Bali, A. S., & Khanna, K. (2020). Photosynthetic response of plants under different abiotic stresses: A review. *Journal of Plant Growth Regulation*, *39*(2), 509–531. DOI: 10.1007/s00344-019-10018-x

Shelake, R. M., Kadam, U. S., Kumar, R., Pramanik, D., Singh, A. K., & Kim, J.-Y. (2022). Engineering drought and salinity tolerance traits in crops through CRISPR-mediated genome editing: Targets, tools, challenges, and perspectives. *Plant Communications*, *3*(6), 100417. DOI: 10.1016/j.xplc.2022.100417

Singh, A. K., Ramalingam, S., Rao, D. N., & Chandrasegaran, S. (2021). Genome editing revolution in life sciences. *Resonance*, *26*(7), 971–998. DOI: 10.1007/s12045-021-1195-z

Smirnikhina, S. A., Zaynitdinova, M. I., Sergeeva, V. A., & Lavrov, A. V. (2022). Improving homology-directed repair in genome editing experiments by influencing the cell cycle. *International Journal of Molecular Sciences, 23*(11), 5992. DOI: 10.3390/ijms23115992

Song, W., Shao, H., Zheng, A., Zhao, L., & Xu, Y. (2023). Advances in roles of salicylic acid in plant tolerance responses to biotic and abiotic stresses. *Plants, 12*(19), 3475. DOI: 10.3390/plants12193475

Su, T., Li, X., Yang, M., Shao, Q., Zhao, Y., Ma, C., & Wang, P. (2020). Autophagy: An intracellular degradation pathway regulating plant survival and stress response. *Frontiers in Plant Science, 11*, 164. DOI: 10.3389/fpls.2020.00164

Suman, S., Bagal, D., Jain, D., Singh, R., Singh, I. K., & Singh, A. (2021). Biotic stresses on plants: reactive oxygen species generation and antioxidant mechanism. In *Frontiers in plant-soil interaction* (pp. 381–411). Elsevier. DOI: 10.1016/B978-0-323-90943-3.00014-6

Sun, Z., Li, S., Chen, W., Zhang, J., Zhang, L., Sun, W., & Wang, Z. (2021). Plant dehydrins: Expression, regulatory networks, and protective roles in plants challenged by abiotic stress. *International Journal of Molecular Sciences, 22*(23), 12619. DOI: 10.3390/ijms222312619

Teper, D., White, F. F., & Wang, N. (2023). The dynamic transcription activator-like effector family of Xanthomonas. *Phytopathology, 113*(4), 651–666. DOI: 10.1094/PHYTO-10-22-0365-KD

Tessema, N., Yadeta, D., Kebede, A., & Ayele, G. T. (2022). Soil and irrigation water salinity, and its consequences for agriculture in Ethiopia: A systematic review. *Agriculture, 13*(1), 109. DOI: 10.3390/agriculture13010109

Thilakarathne, A. S., Liu, F., & Zou, Z. (2025). Plant Signaling Hormones and Transcription Factors: Key Regulators of Plant Responses to Growth, Development, and Stress. *Plants, 14*(7), 1070. Advance online publication. DOI: 10.3390/plants14071070

Tiwari, S., Patel, A., Singh, M., & Prasad, S. M. (2020). Regulation of temperature stress in plants. In *Plant life under changing environment* (pp. 25–45). Elsevier. DOI: 10.1016/B978-0-12-818204-8.00002-3

Tong, H., Wang, H., Wang, X., Liu, N., Li, G., Wu, D., & Wei, Y. (2024). Development of deaminase-free T-to-S base editor and C-to-G base editor by engineered human uracil DNA glycosylase. *Nature Communications, 15*(1), 4897. DOI: 10.1038/s41467-024-49343-5

Trump, B., Cummings, C., Klasa, K., Galaitsi, S., & Linkov, I. (2023). Governing biotechnology to provide safety and security and address ethical, legal, and social implications. *Frontiers in Genetics*, *13*, 1052371. DOI: 10.3389/fgene.2022.1052371

Turnbull, C., Lillemo, M., & Hvoslef-Eide, T. A. (2021). Global regulation of genetically modified crops amid the gene edited crop boom–a review. *Frontiers in Plant Science*, *12*, 630396. DOI: 10.3389/fpls.2021.630396

Umar, O. B., Ranti, L. A., Abdulbaki, A. S., Bola, A. u. L., Abdulhamid, A. K., Biola, M. R., & Victor, K. O. (2021). Stresses in plants: Biotic and abiotic. In *Current trends in wheat research*: IntechOpen.

Vågsholm, I., Arzoomand, N. S., & Boqvist, S. (2020). Food security, safety, and sustainability—Getting the trade-offs right. *Frontiers in Sustainable Food Systems*, *4*, 487217. DOI: 10.3389/fsufs.2020.00016

Venegas-Rioseco, J., Ginocchio, R., & Ortiz-Calderón, C. (2021). Increase in phytoextraction potential by genome editing and transformation: A review. *Plants*, *11*(1), 86. DOI: 10.3390/plants11010086

Wang, J., Song, L., Gong, X., Xu, J., & Li, M. (2020). Functions of jasmonic acid in plant regulation and response to abiotic stress. *International Journal of Molecular Sciences*, *21*(4), 1446. DOI: 10.3390/ijms21041446

Wang, Y., Zafar, N., Ali, Q., Manghwar, H., Wang, G., Yu, L., & Wang, G. (2022). CRISPR/Cas genome editing technologies for plant improvement against biotic and abiotic stresses: Advances, limitations, and future perspectives. *Cells*, *11*(23), 3928. DOI: 10.3390/cells11233928

Weih, M., Karley, A. J., Newton, A. C., Kiær, L. P., Scherber, C., Rubiales, D., & Pappagallo, S. (2021). Grain yield stability of cereal-legume intercrops is greater than sole crops in more productive conditions. *Agriculture*, *11*(3), 255. DOI: 10.3390/agriculture11030255

Yan, H., Jiao, X., Chen, Y., Liang, H., Liang, W., & Liu, C. (2024). Knockout of OsHMA3 in an indica rice increases cadmium sensitivity and inhibits plant growth. *Plant Growth Regulation*, *103*(3), 635–646. DOI: 10.1007/s10725-024-01137-x

Zafar, S. A., Zaidi, S. S.-A., Gaba, Y., Singla-Pareek, S. L., Dhankher, O. P., Li, X., Mansoor, S., & Pareek, A. (2020). Engineering abiotic stress tolerance via CRISPR/Cas-mediated genome editing. *Journal of Experimental Botany*, *71*(2), 470–479. DOI: 10.1093/jxb/erz476

Zhang, H., Zhu, J., Gong, Z., & Zhu, J.-K. (2022). Abiotic stress responses in plants. *Nature Reviews. Genetics*, *23*(2), 104–119. DOI: 10.1038/s41576-021-00413-0

Zhang, Y., Berman, A., & Shani, E. (2023). Plant hormone transport and localization: Signaling molecules on the move. *Annual Review of Plant Biology*, *74*(1), 453–479. DOI: 10.1146/annurev-arplant-070722-015329

Zhao, B., Liu, Q., Wang, B., & Yuan, F. (2021). Roles of phytohormones and their signaling pathways in leaf development and stress responses. *Journal of Agricultural and Food Chemistry*, *69*(12), 3566–3584. DOI: 10.1021/acs.jafc.0c07908

Chapter 7
Disease Resistance in Gene-Edited Crops

S. Kanshiya
Department of Microbiology, Sacred Heart College, Tirupattur, India

A. Mubarak Ali
http://orcid.org/0009-0009-6367-7864
Department of Microbiology, Sacred Heart College, Tirupattur, India

G. Anusuya
http://orcid.org/0009-0007-2211-4980
Department of Microbiology, Sacred Heart College, Tirupattur, India

S. Harini
http://orcid.org/0009-0008-0300-5081
Department of Microbiology, Sacred Heart College, Tirupattur, India

U. Subhalakshmi
http://orcid.org/0009-0007-8989-5486
Department of Microbiology, Sacred Heart College, Tirupattur, India

P. Saranraj
http://orcid.org/0000-0002-5052-6532
Department of Microbiology, Sacred Heart College, Tirupattur, India

ABSTRACT

The problem of disease outbreaks and the dearth of crop types resistant to disease have made modern agriculture less reliable. A disease outbreak that damages commercially significant crops have a substantial effect on the economy. However, it is predicted that agriculture will be seriously threatened by the advent of new pathovars or the alteration in pathogen host specificity brought on by climate change. In agricultural crops, genome editing techniques include mutagenesis, knockout, knockdown, modification, repression, and activation of the target gene. As a result, genome editing techniques are being used more frequently. Because of its minimal off-target risk and simplicity of use, the CRISPR/Cas9 system is more efficient for gene editing than other genome editing techniques like TALENs and ZFNs. This review aims to illustrate how gene editing technologies can produce climate-resilient crops.

DOI: 10.4018/979-8-3373-9534-0.ch007

1. INTRODUCTION

Plant diseases, population growth, and climate change are some of the problems that affect global food security. Food crop yields, including those of rice, soybeans, maize, and wheat, are expected to increase continuously by 38–67% by 2050 given the current production capacity. Climate change will surely increase plant diseases, floods, and droughts (Anderson et al., 2004). It is believed that just 40% of the world's food supply, or 15%, is lost due to biological stress. The problem has ultimately been resolved by the extensive use of fertilizers, pesticides, and other chemicals to boost food output, promote plant life, and stop plant illnesses. Nevertheless, the usage of such pesticides damages the environment. These contaminants contaminate plants, water, and soil. Indirect effects include things like fish, birds, insects, and non-target plant life. Chemical Ethyl Methanesulfonate (EMS), biological (T-DNA), and physical (gamma rays) methods have been used for mutagenesis. Plant genomes have experienced point mutations, deletions, and gene duplications as a result of DNA and transposon insertions. Although these methods have increased yields, they are more costly, time-consuming, and sometimes challenging than traditional methods. These methods could change the genomes of crops in unexpected ways. In order for crops to be able to deal with future challenges, they need to be able to develop more plants that are disease-free, high-yielding, and resistant to abiotic stress (Bebber & Gurr, 2015).

The Scientific Community has always been passionate about developing the best crops because it is challenging to improve crops. A trans-kingdom strategy of genetic manipulation is used to introduce desired genes into crops. Scientists have focused on developing genome editing techniques to alter crops' genomes without using transgenes. Since TALENs, ZFNs, and CRISPR/Cas were discovered, the area of genome editing has advanced to previously unimaginable levels. As was previously indicated, it is impossible to overestimate the significance of genome engineering techniques in the creation of modern plants (Ray et al., 2013). The genome engineering revolution is largely the result of "CRISPR" technology, one of the most potent methods now in use. For humans and animals to meet their daily needs, plants should produce food, fuel, and feed. Numerous biological and abiotic elements, including bacteria, viruses, fungi, and insect infestation, will affect the plants as they develop. Plant growth is gradually being impacted by abiotic factors due to human activities and climate change. Abiotic influences include things like heat, cold, salt, drought, waterlogging, heavy metals, and more. These abiotic factors have a major impact on global productivity. To reduce production loss, new crops that are resistant to a variety of stressors must be developed, given the challenges crops face and the expanding demands of the human population. Numerous biotic stressors have complex interactions between the plant host and the causative agent, depending on the plant's susceptibility or resistance to particular diseases (Scheben et al., 2017).

Developments in Molecular Technology have shown how certain crop plants interact with different patterns of pathogen infection. The pathogen invasion strategy is governed by these pathways, which can be altered by indel mutations caused by either natural or artificial events. The interaction may therefore alter. The 20-nucleotide gRNA has the same sequence as the target. Cas9 is attracted to this gRNA. This CRISPR/Cas9 genome editing method is distinct because it selects the target DNA sequence by RNA-DNA interaction rather than DNA-protein interaction. ZFNs and TALENs need two different DNA-binding domains in order to detect two different DNA sequences through DNA-protein interaction. The process could be a little challenging. CRISPR/Cas9 can interact with DNA and RNA molecules with just 18–20 base pairs. Currently, increasing crop productivity without the use of fertilizers or pesticides is the main objective of the research (Chen et al., 2019).

It has been established that conventional breeding is difficult, labor-intensive, and time-consuming. Genetic engineering has made it extremely easy to develop new and improved varieties of crops with better agronomic traits, such as enhanced agricultural production, longer shelf life, and resistance to disease and abiotic stress. It has become possible to precisely modify the genetic composition of crops without adding foreign DNA by recent advancements in plant breeding technologies, such as genome editing. The most promising application of recent plant breeding technologies has been identified as the development of elite germplasm, which has the ability to produce high yields and has resistance to biotic and abiotic stressors. It has been established that conventional breeding problems can be addressed by using recent plant breeding technologies to develop new and improved varieties of crops (Tyagi et al., 2021).

2. GENOME EDITING STRATEGY

A further method for enhancing crop plants and learning more about their genomes is genome editing. The fundamental elements of the genetic modification mediated by programmable nucleases include the identification of the genomic locus to be targeted, the binding of the DNA binding domain, the restriction endonucleases FokI and Cas, which make double-strand breaks in the target DNA, and the double-strand break repair via homologous recombination or non-homologous end joining (Kaiser et al., 2020). Several genome editing techniques can be used to add new features and remove undesired ones from an organism's DNA. Site-Specific Recombination (SSR) and Site-Specific Nucleases (SSN) are used in many genome editing methods, including TALEN, ZFN, and CRISPR/Cas. Both parties should be aware of the order. The SSN method breaks single or double strands of DNA to start endogenous DNA repair. The orientation of the sites determines whether genes are knocked down or knocked into the eukaryotic genome surrounding the target. Flippase Recombination Enzyme / Flippase Recognition Target and Causes Recombination / Locus of X-over P1 are the methods used in SSR. Base editing, prime editing, mobile CRISPR, meganucleases, ZFNs, TALENs, and DSB-free genome editing are four distinct plant genome editing techniques that employ site-specific endonucleases. Each of these methods is explained in depth in the next section (Foley et al., 2011).

Figure 1. Schematic Diagram of Major Genome Editing Technologies

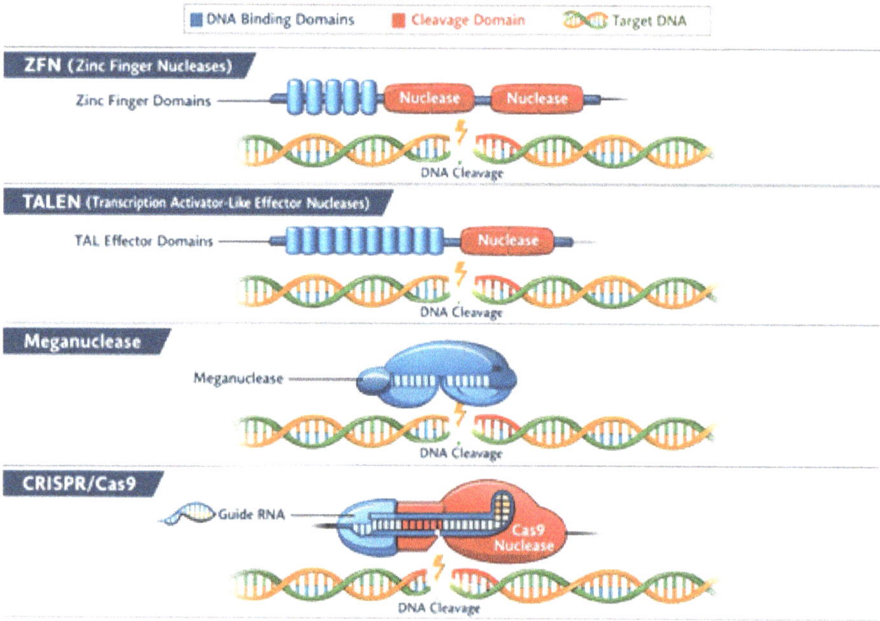

2.1. Zinc-Finger Nucleases (ZFNs)

Zinc fingers and the restriction enzyme FokI's DNA cleavage domain comprise ZFNs' DNA recognition modules. This process can result in a variety of genetic alterations to the organism's DNA, including point mutations, deletions, insertions, inversions, and translocations. There are two pieces to these artificial restriction enzymes: a domain that binds to DNA and a domain that cuts DNA. The DNA-binding domain includes three to six zinc finger repeats and can find a specific DNA sequence that is nine to eighteen nucleotides long. The ZFN-1: At this point, when the ZFN is introduced to the plant genome, there is no template for repair (Prado et al., 2014). It creates DSBs in the plant genome, which causes Non-Homologous End Joining (NHEJ) in the DNA of the host cell. Random modifications, together with little additions and deletions, are made at a particular location. Unlike ZFN-I, ZFN-2 interacts with the crop genome through the ZFN enzyme to offer a quick repair template and Homologous Recombination Deficiency (HDR). When the target DNA connects to specific locations because of its similarity to the template DNA, a double-strand break takes place. When the template uses an endogenous repair mechanism during Homologous Recombination (HR) to target site-specific point mutations, the repair process begins. ZFN 3: The transcription gene encoding ZFN3 is located next to the large repair template in the plant genome. More powerful and specific ZFNs have been developed to induce deletion and gain-of-function mutations that prevent the gene from functioning (Pallarès Masmitjà et al., 2019).

2.2. Meganucleases

Those endonucleases that have the ability to identify and cleave nucleotide sequences of more than twelve nucleotides are called Mega nucleases. This has been observed in bacteria, fungi, algae, yeast, archaebacteria, and even plants, as stated by Borrelli et al. (2018). It has been observed that meganucleases are able to endure minor variations at their target site. Meganucleases are categorized into five classes on the basis of their structure and sequence, namely: His-Cys box, GIY-YIG, LAGLIDADG, HNH, and PD-(D/E) XK. The protein family of LAGLIDADG, referred to as LMNs, has been observed to be accountable for genome editing activities. The nomenclature of the protein is based on its amino acid sequence, which is found in its primary motif. These are usually found in the mitochondria and chloroplasts of unicellular organisms. The gene for the enzyme I-SceI is found in the 21S rRNA gene, whereas the I-CreI enzyme, which is a meganuclease, is found in the 23S rRNA gene, specifically in the chloroplast of the unicellular alga *Chlamydomonas reinhardtii*. Meganucleases' utility for genome editing has been limited since it has been challenging to "tune" them to a particular location (Dong & Ronald, 2019).

2.3. Transcription Activator-Like Effector Nucleases (TALENs)

Restriction enzymes are known as transcription activator-like effector nucleases, or TALENs. Transcription activator-like effector nucleases, consisting of a nuclease and a free, flexible DNA-binding domain, have been the subject of numerous investigations. The TALE domain of transcription activator-like effector nucleases is designed to mimic naturally occurring transcription activator-like effector proteins. Many animal species have had their genomes altered using the state-of-the-art

TALEN technique (Dracatos et al., 2018). Many plant species, including Arabidopsis, Nicotiana, Brachypodium, barley, potatoes, tomatoes, sugarcane, flax, rapeseed, soybeans, rice, maize, wheat, and many more, have had their genomes altered using TALEN technology. Rice is the first crop to benefit from TALEN technology. The main blight disease, *Xanthomonas oryzae*, seriously impairs rice production each year. After the fatty acid desaturase genes were eliminated, soybeans with high levels of oleic acid and low levels of linoleic acid were generated. *Xanthomonas species* have been referred to as TALENs since they encode the native type III effectors. As effectors, the *Xanthomonas species* regulate how genes are expressed in plants. The nuclear localization, activation, and DNA-binding components comprise the TALENs. The activation component initiates the transcription machinery to generate the gene, while the nuclear localization component assists the TALENs in reaching the nucleus (Manghwar et al., 2019).

2.4. CRISPR-Cas9 Gene Editing Technology

Most bacterial and archaeal organisms include CRISPR/Cas9 repeating units. The plant kingdom is protected from bacteria, viruses, and other invasive particles by the remarkably potent CRISPR/Cas9 system and associated proteins. The adaptive immune system was initially reported in 2007. The CRISPR/Cas9 pathway is the foundation of the adaptive immune system. Gene editing technology has dramatically altered the way, scientists conduct research into the biology of animals and plants since its discovery as an effective tool for altering genes in mammalian cells in 2012 (Nerkar et al., 2018). The original CRISPR/Cas9 gene editing tool allows for the alteration of genes in various RNA guides through simple cloning mechanisms. This allows for the alteration of various loci within a given species. Crop species can be accurately altered to create agricultural germplasm that is beneficial for sustainable agriculture through this mechanism. Recent developments in the application of the CRISPR/Cas9 system have accelerated agricultural development through the genetic modification of crops to previously unimaginable levels of precision. Thus far, the method has only been applied to a limited range of crops. Technology has improved yield, quality, disease resistance, and temperature resilience for both dicots and monocots (Altpeter et al., 2016). This method has been used on vegetables like potatoes and tomatoes, as well as cereal crops like cotton, wheat, corn, and rice.

Table 1. Comparison of Major Genome Editing Technologies

Feature	ZFNs	TALENs	Meganucleases	CRISPR/Cas9
DNA Recognition Mechanism	Protein-DNA interaction	Protein-DNA interaction	Protein-DNA interaction	RNA-DNA interaction
Target Recognition	9-18 bp per ZFN pair	24-40 bp per TALEN pair	12-40 bp	20-22 bp (sgRNA) + PAM
Nuclease Domain	FokI	FokI	Endogenous nuclease	Cas9 or Cas12
Design Complexity	High; requires protein engineering	Moderate; requires repeat assembly	Very high; difficult to reprogram	Low; simple sgRNA design
Multiplexing Capability	Limited	Limited	Limited	High; multiple sgRNAs possible
Off-Target Effects	Moderate	Low to moderate	Very low	Low to moderate (can be minimized)
Cost	High	Moderate to high	Very high	Low to moderate
Ease of Use	Difficult	Moderate	Very difficult	Easy
Applications in Plants	Rice, maize, Arabidopsis	Rice, wheat, soybean, barley	Limited applications	Wide range of crops

3. CRISPR/CAS-MEDIATED GENOME EDITING FOR DISEASE RESISTANCE

Plant productivity is seriously threatened by diseases caused by biotic agents such as bacteria, fungi, viruses, and oomycetes. Plant diseases have a detrimental effect on fruit, food plants, and agricultural products. 20–40% of agricultural losses are caused by biotic factors. Plant disease resistance has traditionally been achieved by introducing the wild cousins of the domesticated plant. Pathogen Avr genes, the most common type of plant disease resistance, are identified using nucleotide-binding leucine-rich repeat proteins (Fiaz et al., 2020). When this kind of plant disease resistance is found, the hypersensitive response causes host cells to undergo programmed cell death. The inhibition of the susceptibility (S) genes required for pathogen infection is the second form of plant disease resistance. In order to infect and spread, plant diseases frequently need particular host plant genes. The long-term association with host cells is required for the development of various biotrophic fungal diseases of plants, such as the powdery mildews. The role of the S-genes, also called susceptibility genes, is to fulfill the structural and metabolic needs of the pathogen and to avoid the host's defensive mechanisms, as well as to identify the pathogen and infect the host cells (Zhang & Showalter, 2020).

Resistances that are long-lasting, recessive, and heritable can be brought about by mutations in S-genes. When considering the relationship between plants and infectious agents, the S-genes are of critical importance. The susceptibility of the host to disease is brought about by the S-genes. There are three major molecular mechanisms for the activation of S-genes: basic compatibility, long-lasting compatibility, and the negative control of immune signals in plants. The penetration of viruses and the recognition of the host plants is a demonstration of the basic compatibility of S-genes. The development and advancement of the infections in the host plants is an indication of the long-lasting compatibility of S-genes. The immune signals in the host plants are negatively controlled and function to reduce the reaction of the host to the disease. The molecular mechanisms of S-gene activation are of critical importance for the development of effective means of increasing resistance in plants to infectious agents (Gao, 2021).

The Mole (Mlo) gene is conserved in the kingdom of plants, as there is the presence of numerous copies of the gene in higher plants, as researched in the cloning of the gene in barley. Plants are immune to powdery mildew due to a mutation in the Mlo gene, according to research. The Mlo gene product facilitates the entry of powdery mildew-causing fungal spores into the epidermis of host cells. The CRISPR-Cas9 technology has altered the Mlo gene, which codes for a membrane-attached protein with seven transmembrane domains (Carroll, 2011). The outcome is transgene-free resistant plants. It has been demonstrated that the Mlo-like gene increases the susceptibility of several plant species to powdery mildew infection. These plant species include bread wheat, grapevine, peach, tomato, pea, pepper, tobacco, and Arabidopsis. Mlo-like genes have been grouped using phylogenetic connections. Monocot plant species are associated with Class IV Mlo-like genes, while dicot plant species are associated with Class V Mlo-like genes. S-genes also do not include class I, II, III, or VI Mlo-like genes (Urnov et al., 2010). CRISPR technology and next-generation sequencing have simplified the identification and use of S-genes, which are important for enhancing resistance to infections in plants.

CRISPR/Cas9 has been used to edit S-genes such as the Mlo gene for enhancing resistance in agriculture against lethal diseases. CRISPR/Cas9 technology has the potential for developing persistent resistance to diseases by making the pathogen and the host incompatible. CRISPR/Cas9 has also shown potential in reducing powdery mildew in wheat crops by editing the Enhanced Disease Resistance 1 (EDR1) gene using the CRISPR/Cas9 technology. The ability to edit many genes at once using a single construct is another feature of CRISPR/Cas9 technology (Lusser et al., 2011). According to research, one structure within an organism such as Arabidopsis can change 14 different genes at once. Multiplex genome editing has also been applied to crops, introducing multiple gRNAs into a single construct using a U3/U6 promoter. Excision and segregation were formerly accomplished through selfing or backcrossing to the original paternal line.

The Cas9 method can be used to choose transgene-free plants with the desired phenotype from the segregating populations if the chromosomal site's effective mutation has been confirmed. Viruses are responsible for the spread of diseases that damage a range of commercially important crops. For plants, viruses continue to be a significant biotic stressor. The production of single-stranded positive sense RNA viruses belonging to the Potyviridae family requires eukaryotic translation initiation factor 4E (Khalil, 2020). It has been shown that innate resistance to the virus exists at the species level if the eIF4E gene is connected to the virus protein genome-linked region of the potyviruses' 5-terminal. By modifying their eIF4E gene utilizing CRISPR Cas9 gene editing technology, Arabidopsis and Cucumis sativus have successfully generated resistance to potyviruses and ipomoviruses (Arnould et al., 2011). The genome-edited crops' lack of transgenes shows how well CRISPR technology works to create crops that are resistant to viruses. Research is being done on the CRISPR-Cas system's potential to fight viral and bacterial diseases. Citrus canker is a dangerous disease that severely affects citrus crops all over the world. The bacteria *Xanthomonas citri subsp.* citri is responsible for the sickness. The γ-proteobacterium *Xanthomonas oryzae* pv. oryzae causes rice blight by modifying the OsSWEET13 gene utilizing the CRISPR/Cas9 system (Kaur et al., 2022).

Table 2. CRISPR/Cas9-Mediated Disease Resistance in Major Crops

Crop	Disease	Pathogen	Target Gene	Strategy	Outcome
Wheat	Powdery mildew	*Blumeria graminis* f. sp. *tritici*	*Mlo* (three homeologs)	S-gene knockout	Broad-spectrum resistance; transgene-free lines
Barley	Powdery mildew	*Blumeria graminis* f. sp. *hordei*	*Mlo*	S-gene knockout	Durable resistance
Rice	Bacterial blight	*Xanthomonas oryzae* pv. *oryzae*	*OsSWEET13*, *OsSWEET14*	Promoter editing	Broad-spectrum resistance to multiple Xoo strains
Rice	Rice blast	*Magnaporthe oryzae*	*OsERF922*	S-gene knockout	Enhanced blast resistance
Tomato	Powdery mildew	*Oidium neolycopersici*	*Mlo*	S-gene knockout	Resistance to powdery mildew
Cucumber	Potyviruses	Potyviruses (e.g., CVYV, ZYMV)	*eIF4E*	S-gene knockout	Broad-spectrum virus resistance

continued on following page

Table 2. Continued

Crop	Disease	Pathogen	Target Gene	Strategy	Outcome
Arabidopsis	Potyviruses	Turnip Mosaic Virus (TuMV)	*eIF4E*	S-gene knockout	Recessive resistance to virus
Citrus	Citrus canker	*Xanthomonas citri* subsp. *citri*	*CsLOB1*	S-gene knockout	Enhanced resistance to canker
Soybean	Phytophthora root rot	*Phytophthora sojae*	*GmRps* genes	R-gene modification	Enhanced resistance
Maize	Southern leaf blight	*Bipolaris maydis*	*ZmMlo*	S-gene knockout	Reduced susceptibility

4. CROP DISEASES: ECONOMIC IMPACT

Ten years ago, Streptococcus pyogenes produced class 2 type II CRISPR/Cas9 system. This is a new technique to change genomes that uses RNA-guided designer nucleases. In contrast to other genome editing techniques like ZFNs, TALENs, and meganucleases, which identify their target regions by interactions between proteins and DNA, CRISPR/Cas9 requires a guide RNA to operate. Short nucleotide sequences known as sgRNAs contain a particular sequence that targets the target genomic sequence (Martínez-Fortún et al., 2017). The resulting RNA/DNA complex is then broken down by the Cas9 nuclease. In order to build a DSB, a preserved protospacer adjacent motif is required. NHEJ is a repair technique that introduces indels in the protein coding region of certain genes, resulting in a frameshift and knockdown. The CRISPR/Cas9 reagents are frequently and quickly used in genome editing because they are simple to base pair with DNA. The Protospacer Adjacent Motif (PAM) or "NGG" in the target DNA is also necessary for Cas9, which is derived from the bacteria Streptococcus pyogenes. SpCas9 and other Cas9 variants with distinct PAM requirements have also been employed in plant genome editing. VQR-"NGA," SpCas9-EQR-"NGAG," Cas9-NG-"NG," and xCas9 3.7-"NG/GAA/GAT." Later, the CRISPR toolkit was expanded to include type-V CRISPR systems and class 2 Cas12 nucleases (Li et al., 2012).

Cas12 effectors contain the RuvC-like domain, which cleaves the target DNA site's two strands. A 4-5 nt overhang is also produced by the RuvC-like domain. Cas9 and Cas12 nucleases differ primarily in that the latter only have the RuvC-like domain. The HNH domain is present in Cas9 nucleases. In contrast to Cas9, which is mediated by a 100 nt guide RNA, Cas12 nucleases are primarily mediated by a single 42 nt crRNA. The Cas12 nuclease that is most commonly used for plant genome editing is LbCas12a. The 'TTTV' T-rich PAM can be identified by it. Other

Cas12a variants have higher target ranges and activity (Barrangou et al., 2007). Cas12a's engineering allows it to detect crRNA without Trans-activating CRISPR RNA (tracrRNA), which is useful for transcription, multiplex gene editing, base editing, and epigenetics. The hypercompact type V CRISPR system CRISPR-Cas8 has one 70 kDa Cas8 protein. It is half the weight of Cas9 and Cas12a. It is the latest addition to the CRISPR family. Like Cas12a, it requires a 5'-TBN-3' PAM for a staggered cut with 5' overhangs (Zhou et al., 2020). The efficiency of the CRISPR-Cas8 system for editing the Arabidopsis genome is 0.85%. Base editing has also been used to edit DNA by replacing one DNA nucleotide with another. It has also been used to edit DNA and RNA with single nucleotide precision.

Nuclear and organellar genomes (DNA BEs), proliferating and non-dividing cells, and transcriptomes (RNA BEs) can be edited by base editors. In BEs, a flawed Cas nuclease is fused to DNA repair proteins and a nucleotide deaminase. In particular, it has a flawed dCas9 with H840A and D10A mutations. The flawed NHEJ repairs DSBs caused by SpCas9. In contrast, the accurate base excision repair mechanism repairs single nicks caused by BEs. In this way, the adverse effects of gene editing are minimized (Ricroch et al., 2017). Cereal crops and their wild relatives include graminicolous hyphomycetes that have been linked to crop loss and food insecurity. These phytopathogens have been shown to be highly harmful to crops. Exserohilum turcicum, which causes maize leaf blight in the north, and Bipolaris maydis, which causes it in the south, have both been shown to cause crop losses (Kim et al., 2018). Leaf spots in wheat and barley crops are also caused by the phytopathogen Bipolaris sorokiniana.

The most harmful phytopathogen causing diseases in rice crops is the fungus Pyricularia oryzae, also known as Magnaporthe oryzae. This phytopathogen causes the destruction of up to 30% of the world's rice crops, a condition that has made it difficult to control the diseases caused by it and has resulted in humanitarian crises in Asia. Various researchers have used this phytopathogen to investigate the interaction of the host and phytopathogen; this has resulted in the investigation of the molecular causes of the disease caused by it (Savary et al., 2012). It has since been reported in North America, Europe, Asia, and New Zealand. Defoliation or delayed dieback is the indicator of box wood blight. The fungus Calonectria pseudonaviculata, responsible for the box wood blight disease, was recently reported, causing threats to the growth and cultivation of box wood plants, both in nurseries and the landscape. This has implications for the industry involved in the cultivation of ornamental plants (Jones & Dangl, 2006).

5. GENOME EDITING (GE) TECHNOLOGY FOR IMPROVING PLANT TRAITS

Similar to site-directed insertion, deletion, substitution or targeted mutagenesis, genome editing uses genome targeting techniques to precisely alter the genomic DNA of cells or organisms. This innovative method has great promise for improving agricultural output and gene function (Schie & Takken, 2014). This method uses a set of nucleases that can recognize specific nucleotide sequences, allowing them to cleave the targeted DNA at specific sites and result in a Double-Strand Break (DSB). There are several ways to correct the DSB, including micro-homology-mediated end-joining, non-homologous end-joining, and homology-directed repair. During the initial DSB repair phase, the targeted site of the DSB experiences insertions, deletions, or non-homologous end-joining (Pavan et al., 2010).

Homologous Recombination Deficiency is an error-free method for DSB repair, despite its rarity in plant life. Basic plant research and agricultural development are greatly impacted by CRISPR/Cas9 and other gene editing methods. One of the main benefits of employing gene editing techniques is the effectiveness of genetic segregation in eliminating the initial transgenes employed for genetic engineering (Devoto et al., 2003). When disease genes are targeted using genome editing techniques, disease resistance has diminished. The alterations created by gene editing techniques are similar to those caused by traditional plant breeding techniques. Crop resilience to various diseases has also increased thanks to the application of genome editing techniques. This is accomplished by either cutting the phytopathogens' genetic material to stop their growth or using specific host vulnerability genes (the S gene technique) (Shan et al., 2013).

6. TECHNIQUES TO DEVELOP DISEASE RESISTANCE IN PLANTS

Plant pathologists' primary objective is to find potential targets to take advantage of disease resistance mechanisms to develop disease-resistant cultivars for sustainable agriculture in the future. This implies that knowledge of the molecular processes underpinning plant disease resistance is essential (Nekrasov et al., 2017). Over time, plants have developed an advanced defense system to fend against microbial invasion. The first line of defense against microbial invasion is the physiological barriers that already exist. However, the plant possesses sufficient resistance to microbial invasion thanks to immunologic receptors on the cell wall or membrane, which either directly or indirectly modify the microbe's target. Plant immunity and the signal transduction mechanisms activated in plant cells following infection by various microorganisms have been effectively identified through high-throughput methods (Zaidi et al., 2018). These approaches have been largely centered on immunity-associated genes such as the pathogen's Avr proteins/cffcctors, receptors, Pathogen-Associated Molecular Pattern (PAMPs), channels, and signal molecules.

The two types of immunity in plants, which have been well studied in the past, include PAMP-triggered immunity and effector-triggered immunity. The most cost-effective approach for enhancing the level of disease resistance in the crops is the development or enhancement of the genetic features for resistance (Borrelli et al., 2018). This, however, can be done successfully if the molecular dynamics of the immunity response in the plants, as well as the host-pathogen relationship, are well understood. To develop resistant varieties capable of successfully combating the disease, it is essential to carry out an in-depth analysis of the different ways the virus interacts with the host. For example, it is essential to carry out an in-depth analysis of natural epidemics or "hot spots" in order to identify the genotypes that are susceptible to the illness or can successfully combat it (Bastet et al., 2017). If the genotypes are grown in the "hot spots," they can be efficiently tested for their susceptibility or resistance to the disease. The major concerns that come up during the process of natural screening can be well managed by the genotypes of the crops that have been ascertained to be resistant to the disease. Development of an appropriate and efficient screening method is another fundamental aspect of the breeding for disease resistance project (Pyott et al., 2016). In order to attain the expected results, there is a need to have a comprehensive knowledge of pathogenicity, virulence pattern, type of resistance, and efficient technique. Over the last ten years, sufficient research has been conducted to determine the type and duration of resistance, and an efficient technique for the detection of resistance to various diseases has been established (Stover et al., 2014).

Contemporary methodologies of genomic studies like Quantitative Trait Locus (QTL) sequencing, association mapping, and bi-parental QTL mapping can help elucidate the exact correlation between genes and molecular markers (Zhou et al., 2015). The process of making molecular markers includes making a mapping population, screening for polymorphic markers, phenotyping, genotyping with polymorphic markers, making a linkage map, mapping QTLs, and validating molecular markers. A molecular marker that is closely connected to the QTL or gene of interest is used to screen a candidate. When the pathogen and host share homologous Avr and R genes, respectively, the host imparts total disease resistance on the pathogen in incompatible host-pathogen systems (Ran et al., 2013). British scientist Rowland Biffen showed the potential benefits of gene-mediated resistance in wheat (*Triticum spp.*) breeding in the beginning of the 20th century. Cloned R genes are currently present in many species, genera, and species complexes. Bacterial streak disease resistance was demonstrated in a lab environment by rice (*Oryza sativa*) that acquired the R gene "Rxo1" from maize (Zea mays). Tomatoes containing the pepper gene exhibit a considerable level of resistance to the bacterial illness *Xanthomonas sp.*, according to a multiyear field investigation of tomatoes cultivated in commercial settings (Molla et al., 2021).

Figure 2. Plant Immune System and Gene Editing Targets

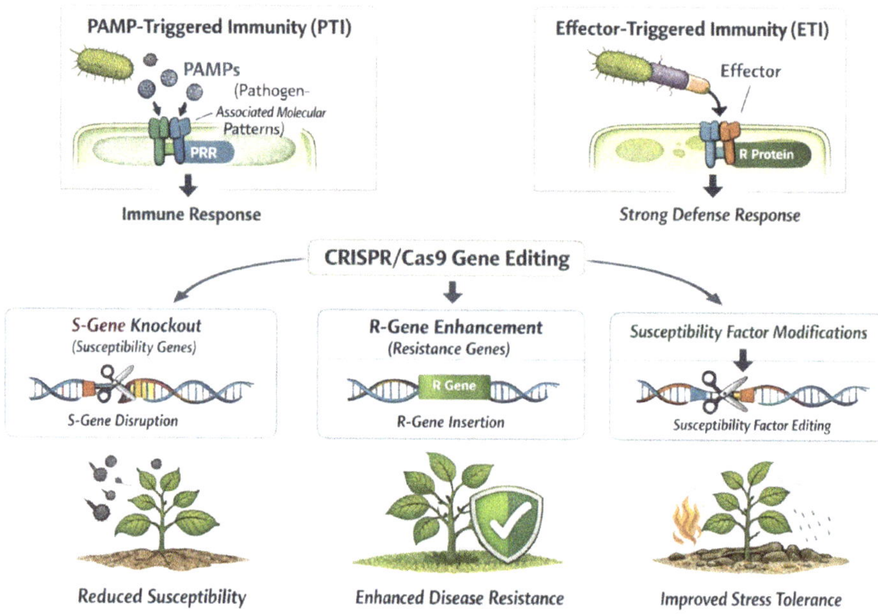

7. PLANT DISEASE MANAGEMENT: BENEFITS AND RISKS

Reducing the loss of production of various crops requires effective treatment of plant diseases. Chemical, biological, and physical methods comprise the majority of illness management strategies (Pausch et al., 2020). The careless use of chemical pesticides that affect plant diseases or their vectors, such as insecticides, fungicides, and bactericides, is one of the most important parts of managing plant diseases. The use of pesticides and the chemical composition of substances pose health risk to human beings and the environment. As a result, it has become more and more important to search for alternative means of controlling diseases in plants. Overuse of the substances mentioned above may lead to the development of diseases that are resistant to treatment. The habitats of wildlife may also be interfered with by the use of chemical substances. Searching for alternative pesticides and fungicides has become more and more important. It has also become more and more important to research the development of safe and efficient insecticides and fungicides.

On the other hand, plant diseases can be managed using biological management techniques, which involve the use of beneficial organisms. Plant diseases can, to some extent, be managed using natural enemies such as bacteria, parasitoids, and predators. The management of plant diseases using this approach is environmentally friendly because it reduces the use of pesticides and the occurrence of resistant populations. Therefore, this approach is sustainable (Zetsche et al., 2015). This approach, which involves the gradual growth of natural enemies, leads to unforeseen ecological effects. Another important approach for controlling plant diseases, especially in the agricultural environment, involves the use of cultural management techniques. Some of the techniques include crop rotation, hygiene, and irrigation. However, cultural customs change farming methods and/or demand a large labor force, which could raise production costs (Manzar et al., 2022). This approach uses less hazardous insecticides while maintaining human safety. It could take some time for genetic engineering or conventional breeding to produce resistant plants. If viruses are able to overcome the plants' resistance, it could not be particularly successful in the long run. It is now possible to quickly change the traits of agricultural plants thanks to genome editing techniques, and this has become a practical means of producing disease-resistant crops (Kashyap et al., 2023).

8. CRISPR/CAS SYSTEMS: ADVANCES AND LIMITATIONS

There are two methods for delivering CRISPR/Cas: direct and indirect. Examples of direct procedures include biolistics and Polyethylene Glycol (PEG) distribution. On the contrary, floral dip and the distribution of Agrobacterium tumefaciens are considered indirect procedures. Transient expression of a gene is often obtained by direct gene delivery techniques (Savary et al., 2019). Plants use Agrobacterium tumefaciens-mediated genetic transformation as their main indirect gene delivery method. The foundation of almost all GE techniques applied to plants is tissue culture and plant regeneration. The difficult, expensive, time-consuming, and genotype-dependent approach for regenerating plants is tissue culture (LeBlanc et al., 2018). Forest woody plants have proven to be challenging for CRISPR/Cas GE in plants due to their low efficiency in genetic transformation, long period for propagation, and lack of mutations. The objective for delivery can be considered when selecting the vector, which will allow CRISPR/Cas to spread rapidly to plants. Plasmids, viruses, or nonviral agents could be used as delivery vectors. Among the viral vectors that plants employ is tobacco mosaic virus, cowpea mosaic virus, bean yellow dwarf virus, and potato virus X (Ahmad et al., 2020).

Due to their limitations, these vectors cannot be used with large sequence fragments or the Cas proteins themselves, but doing so will likely increase the plant immune system's resistance. Inorganic nanoparticles, polymeric nanoparticles, liposomes, carbon nanotubes, and protein and peptide nanoparticles are examples of non-viral vectors (Molla & Yang, 2020). The newly developed non-viral vectors have a lot of potential for future GE applications due to their compact size, non-toxicity, and capacity to endure a variety of physical challenges in the field. Ribonucleoprotein (RNP), transgene-free editing techniques, and transient CRISPR gene expression allow for efficient gene editing in citrus and grape plant species, it is difficult to create new technologies and enhance those that already exist, and efficiency is quite poor (Zhang et al., 2017). Algae, bryophyta, and pteridophyta are among the various members of the plant kingdom that can be targeted by CRISPR/Cas9 technology.

The leaves of liverwort, or Marchantia polymorpha L., have been molecularly cultivated using CRISPR/Cas9-mediated targeted mutagenesis (Zaidi et al., 2018). These days, the model organism is utilized to show the evolution of plants. It is possible for new species to emerge or for existing bacterial, viral, or fungal species to rationally evolve. The potential of CRISPR/Cas9 to modify plants in new ways has been demonstrated. Because mismatched regions of the genome can be altered by off-target chromosomal rearrangements, the application of GE for therapeutic reasons is restricted (Jacob et al., 2013). Reduced gene expression brought on by off-target effects may also lead to several physiological or signaling issues. There has also been a notable reduction in CRISPR/Cas9's off-target consequences. Sniper Cas9, eSpCas9, HypaCas9, and HF-Cas9, for instance, exhibit robust on-target activity and dramatically lower off-target levels (Ye et al., 2002).

Figure 3. Delivery Methods for CRISPR/Cas Systems in Plants

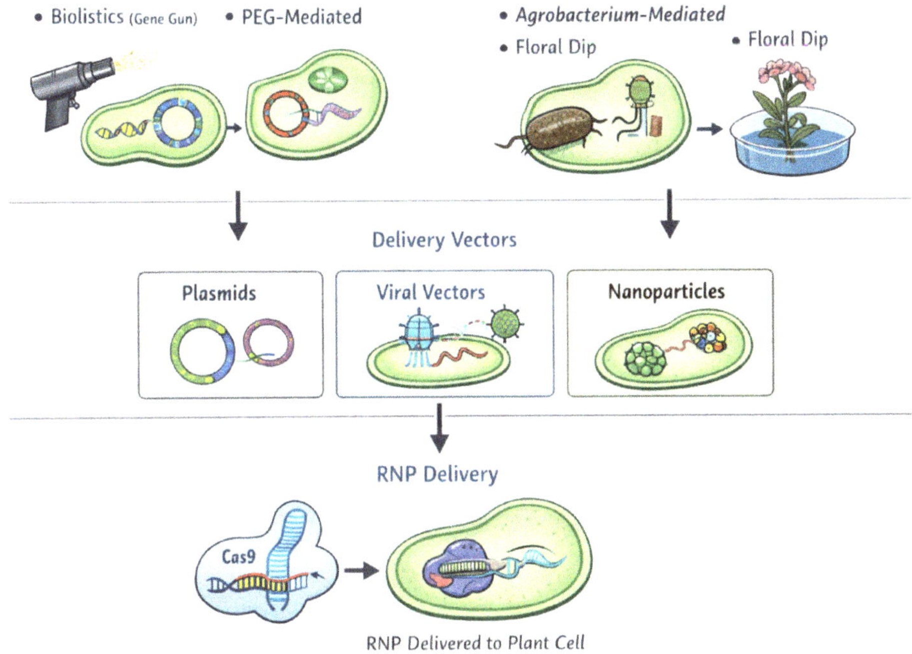

9. APPLICATIONS OF GE TECHNOLOGIES

These days, single gene knockdown and multiplex gene knockout research in plants make extensive use of genome editing. The most popular and widespread application of genome editing in plants is gene knockout (Collard et al., 2005). This program is very useful when one wants to study a gene in depth and how it interacts with other biological processes. This is achieved by modifying genes. Genome editing, through a technique called multiplex gene knockout, has been used to knock out multiple genes in various plant species simultaneously. To appreciate the intricacies of the plant kingdom, one must appreciate the intricacies of genes through the application of multiplex gene knockdown techniques (Takagi et al., 2013). Research has been able to achieve a significant reduction in genes, as identified by the analysis of genes and non-coding RNA. The achievement of genome editing is further identified by the production of crops that are able to use fewer resources, as observed in polyploid plants. The successful editing of genes from triploid fruits, such as apples and oranges, is one example of genome editing (Kunwar et al., 2018).

Therefore, complex genomes such as those of polyploid plants can now be accurately altered. "Gene targeting" is the intentional use of genetic engineering methods to add a new DNA fragment to a specific area of the genome or to precisely swap out one DNA fragment for another. The targeted strategy of gene targeting offers the promise of precision genetic fine-tuning of crops or other genetic modifications that could transform the process of crop improvement (Miedaner, 2016). Many crops, including rice, wheat, sorghum, soybean, maize, potatoes, and model crops like Arabidopsis thaliana and Nicotiana benthamiana, have benefited from genome editing in terms of stress tolerance, growth, and yield attributes. Genome editing has also shown potential in bolstering crop defense mechanisms against diseases by modifying genes associated with diseases (Köhl et al., 2019).

10. MECHANISM OF GENOME EDITING IN CROP MANAGEMENT

A contemporary technique for improving crop quality and, consequently, vegetable crop output is Genetic Engineering (GE). Compared to other horticultural crops, tomatoes have received over 42% more scientific attention through GE, whilst potatoes have only received about 13%. Vegetable crops make up over 72% of horticultural genetic engineering, yet GE has been effectively utilized for the development of various leafy and underground vegetables (Jun et al., 2013). The application of CRISPR-Cas9 gene editing technology has dramatically altered the field of plant breeding through the achievement of unprecedented precision in altering certain genes of plants. Genetic engineering (GE) can improve vegetable crops through enhanced nutritional value, shelf life, resistance to certain diseases, and tolerance to abiotic stress, among other advantages (Xiao et al., 2014).

Using the concepts of genetic engineering, which involves the manipulation of certain genes of the plant used to produce certain qualities in vegetables, it is possible to develop breeds of vegetables with certain qualities (Emebiri et al., 2016). It is essential to conduct research on the ethical, legal, and environmental impacts of GE use in agriculture to ensure the safe and effective use of GE in the future. GE is capable of increasing the yields of vegetable harvest and transforming vegetable breeding. In the future, vegetable breeding using GE is likely to be precise, effective, and dynamic to target certain complex traits and vegetable genomes. Using the concepts of GE, superior breeds of vegetables can be produced to serve the needs of humans and the environment through the regulation of certain genes or certain areas of the vegetable genomes (Li et al., 2013). Vegetable breeding with GE technology or GE organisms could help solve global problems while still being widely accepted as ethical and socially acceptable. This could lead to strong, nutrient-rich vegetable crops in the future. One of the main variables influencing how much food crops can grow in both natural and controlled situations is the arms race between insect and plant pests. Crop plants have a number of ways to protect themselves from insect herbivory so that they don't lose their ability to produce food. Plants have a number of direct and indirect ways to protect themselves from insect attacks (Chen et al., 2017).

Plants specialized morphological structures are examples of direct defense mechanisms. Plants activated secondary metabolism is an example of an indirect defense mechanism. Genome-editing technology, which is seen as a significant tool, has changed the way crop plants are made to be able to handle abiotic stress in the last few decades. Genome-editing techniques could illustrate how tolerant plants are to abiotic pressures by focusing on genes that respond to stress, are susceptible to it, or are negatively controlled, and then turning on genes that positively regulate abiotic stress responses (Slaymaker et al., 2016). The ideal things to aim for when trying to improve crops are known positive and negative regulators of a specific trait. Knockouts in structural genes, regulatory genes, transcription factors, and promoter regions change biochemical pathways, metabolite profiles, and physiology. This leads to proteins that are broken and changed and stress-tolerant phenotypes. For example, changing the way proteins are expressed and how ROS are removed in rice might lead to curled leaves and drought tolerance in maize. SlMAPK3 stops tomatoes from responding to heat stress by lowering their defensive response. Deleting it makes the plants better able to handle heat stress by keeping ROS levels stable (Lee et al., 2018).

11. CONCLUSION

The domains of disease-resistant breeding and agricultural enhancement have undergone substantial change due to the rapid development of genome editing techniques like CRISPR/Cas9 systems. In the face of hitherto unheard-of challenges to global food security brought on by population growth, climate change, and emerging crop diseases, such cutting-edge technologies have been providing promising solutions for crop improvement and disease management. By modifying susceptibility (S) genes and controlling immune responses, genome editing techniques including ZFNs, TALENs, meganucleases, and CRISPR/Cas systems have shown significant promise for producing disease-resistant crops. The CRISPR/Cas9 system is regarded as the most promising of these methods due to its low off-target effects, high efficiency, and simplicity of use. Moreover, the technology can mediate multiplex gene editing. Numerous significant agricultural plants, including rice, wheat, maize, soybean, potato, and tomato, have demonstrated successful CRISPR/Cas9 applications. It has been discovered that modified crops are more resistant to a variety of bacterial, viral, and fungal diseases. For instance, the editing of susceptibility genes such as Mlo for powdery mildew and virus resistance genes such as eIF4E has been recognized as having tremendous potential for crop improvement.

REFERENCES

Ahmad, N., Rahman, M. U., Mukhtar, Z., Zafar, Y., & Zhang, B. (2020). A critical look on CRISPR-based genome editing in plants. *Journal of Cellular Physiology, 235*(2), 666–682. DOI: 10.1002/jcp.29052 PMID: 31317541

Altpeter, F., Springer, N. M., Bartley, L. E., Blechl, A., Brutnell, T. P., Citovsky, V., Conrad, L., Gelvin, S. B., Jackson, D., Kausch, A. P., Lemaux, P. G., Medford, J. I., Orozo-Cardenas, M., Tricoli, D., VanEck, J., Voytas, D. F., Walbot, V., Wang, K., Zhang, Z. J., & Stewart, C. N.Jr. (2016). Advancing crop transformation in the era of genome editing. *The Plant Cell, 28*(7), 1510–1520. DOI: 10.1105/tpc.16.00196 PMID: 27335450

Anderson, P. K., Cunningham, A. A., Patel, N. G., Morales, F. J., Epstein, P. R., & Daszak, P. (2004). Emerging infectious diseases of plants: Pathogen pollution, climate change and agrotechnology drivers. *Trends in Ecology & Evolution, 19*(10), 535–544. DOI: 10.1016/j.tree.2004.07.021 PMID: 16701319

Arnould, S., Delenda, C., Grizot, S., Desseaux, C., Pâques, F., Silva, G. H., & Smith, J. (2011). The I-CreI meganuclease and its engineered derivatives: Applications from cell modification to gene therapy. *Protein Engineering, Design & Selection : PEDS, 24*(1-2), 27–31. DOI: 10.1093/protein/gzq083 PMID: 21047873

Barrangou, R., Fremaux, C., Deveau, H., Richards, M., Boyaval, P., Moineau, S., Romero, D. A., & Horvath, P. (2007). CRISPR provides acquired resistance against viruses in prokaryotes. *Science, 315*(5819), 1709–1712. DOI: 10.1126/science.1138140 PMID: 17379808

Bastet, A., Robaglia, C., & Gallois, J. L. (2017). eIF4E resistance: Natural variation should guide gene editing. *Trends in Plant Science, 22*(5), 411–419. DOI: 10.1016/j.tplants.2017.01.008 PMID: 28258958

Bebber, D. P., & Gurr, S. J. (2015). Crop-destroying fungal and oomycete pathogens challenge food security. *Fungal Genetics and Biology : FG & B, 74*, 62–64. DOI: 10.1016/j.fgb.2014.10.012 PMID: 25459533

Borrelli, V. M., Brambilla, V., Rogowsky, P., Marocco, A., & Lanubile, A. (2018). The enhancement of plant disease resistance using CRISPR/Cas9 technology. *Frontiers in Plant Science, 9*, 1245. DOI: 10.3389/fpls.2018.01245 PMID: 30197654

Borrelli, V. M. G., Brambilla, V., Rogowsky, P., Marocco, A., & Lanubile, A. (2018). The enhancement of plant disease resistance using CRISPR/Cas9 technology. *Frontiers in Plant Science, 9*, 1245. DOI: 10.3389/fpls.2018.01245 PMID: 30197654

Carroll, D. (2011). Genome engineering with zinc-finger nucleases. *Genetics*, *188*(4), 773–782. DOI: 10.1534/genetics.111.131433 PMID: 21828278

Chen, J. S., Dagdas, Y. S., Kleinstiver, B. P., Welch, M. M., Sousa, A. A., Harrington, L. B., Sternberg, S. H., Joung, J. K., Yildiz, A., & Doudna, J. A. (2017). Enhanced proofreading governs CRISPR-Cas9 targeting accuracy. *Nature*, *550*(7676), 407–410. DOI: 10.1038/nature24268 PMID: 28931002

Chen, K., Wang, Y., Zhang, R., Zhang, H., & Gao, C. (2019). CRISPR/Cas genome editing and precision plant breeding in agriculture. *Annual Review of Plant Biology*, *70*(1), 667–697. DOI: 10.1146/annurev-arplant-050718-100049 PMID: 30835493

Collard, B. C., Jahufer, M. Z. Z., Brouwer, J. B., & Pang, E. C. K. (2005). An introduction to markers, quantitative trait loci (QTL) mapping and marker-assisted selection for crop improvement: The basic concepts. *Euphytica*, *142*(1), 169–196. DOI: 10.1007/s10681-005-1681-5

Devoto, A., Hartmann, H. A., Piffanelli, P., Elliott, C., Simmons, C., Taramino, G., & Panstruga, R. (2003). Molecular phylogeny and evolution of the plant-specific seven-transmembrane MLO family. *Journal of Molecular Evolution*, *56*(1), 77–88. DOI: 10.1007/s00239-002-2382-5 PMID: 12569425

Dong, O. X., & Ronald, P. C. (2019). Genetic engineering for disease resistance in plants: Recent progress and future perspectives. *Plant Physiology*, *180*(1), 26–38. DOI: 10.1104/pp.18.01224 PMID: 30867331

Dracatos, P. M., Haghdoust, R., Singh, D., & Park, R. F. (2018). Exploring and exploiting the boundaries of host specificity using the cereal rust and mildew models. *The New Phytologist*, *218*(2), 453–462. DOI: 10.1111/nph.15044 PMID: 29464724

Emebiri, L. C., Tan, M. K., El-Bouhssini, M., Wildman, O., Jighly, A., Tadesse, W., & Ogbonnaya, F. C. (2016). QTL mapping identifies a major locus for resistance in wheat to Sunn pest (*Eurygaster integriceps*) feeding at the vegetative growth stage. *Theoretical and Applied Genetics*, *130*(2), 1–10. DOI: 10.1007/s00122-016-2805-1 PMID: 27744491

Fiaz, S., Xiukang, W., Afifa, Y., Alharthi, B., & Ali, H. (2020). Apomixis and strategies for induce apomixis to preserve hybrid seed vigor for multiple generations. *GM Crops & Food*, *12*(1), 57–70. DOI: 10.1080/21645698.2020.1808423 PMID: 32877304

Foley, J. A., Ramankutty, N., Brauman, K. A., Cassidy, E. S., Gerber, J. S., Johnston, M., & Zaks, D. P. (2011). Solutions for a cultivated planet. *Nature*, *478*(7369), 337–342. DOI: 10.1038/nature10452 PMID: 21993620

Gao, C. (2021). Genome engineering for crop improvement and future agriculture. *Cell, 184*(6), 1621–1635. DOI: 10.1016/j.cell.2021.01.005 PMID: 33581057

Jacob, F., Vernaldi, S., & Markawa, T. (2013). Evolution and conservation of plant NLR functions. *Frontiers in Immunology, 4*, 297. DOI: 10.3389/fimmu.2013.00297 PMID: 24093022

Jones, J. D., & Dangl, J. L. (2006). The plant immune system. *Nature, 444*(7117), 323–329. DOI: 10.1038/nature05286 PMID: 17108957

Jun, T. H., Mian, M. R., & Michel, A. P. (2013). Genetic mapping of three quantitative trait loci for soybean aphid resistance in PI 567324. *Heredity, 111*(1), 16–22. DOI: 10.1038/hdy.2013.10 PMID: 23486080

Kaiser, N., Douches, D., Dhingra, A., Glenn, K. C., Herzig, P. R., Stowe, E. C., & Swarup, S. (2020). The role of conventional plant breeding in ensuring safe levels of naturally occurring toxins in food crops. *Trends in Food Science & Technology, 100*, 51–66. DOI: 10.1016/j.tifs.2020.03.042

Kashyap, A. S., Manzar, N., Maurya, A., Mishra, D. D., Singh, R. P., & Sharma, P. K. (2023). Development of diagnostic markers and applied for genetic diversity study and population structure of *Bipolaris sorokiniana* associated with leaf blight complex of wheat. *Journal of Fungi (Basel, Switzerland), 9*(2), 153. DOI: 10.3390/jof9020153 PMID: 36836268

Kaur, N., Sharma, S., Hasanuzzaman, M., & Pati, P. K. (2022). Genome editing: A promising approach for achieving abiotic stress tolerance in plants. *International Journal of Genomics, 5547231*, 1–12. Advance online publication. DOI: 10.1155/2022/5547231 PMID: 35465040

Khalil, A. M. (2020). The genome editing revolution [review]. *Journal of Genetic Engineering and Biotechnology, 18*(1), 68. DOI: 10.1186/s43141-020-00078-y PMID: 33123803

Kim, D., Alptekin, B., & Budak, H. (2018). CRISPR/Cas9 genome editing in wheat. *Functional & Integrative Genomics, 18*(1), 31–41. DOI: 10.1007/s10142-017-0572-x PMID: 28918562

Köhl, J., Kolnaar, R., & Ravensberg, W. J. (2019). Mode of action of microbial biological control agents against plant diseases: Relevance beyond efficacy. *Frontiers in Plant Science, 10*, 845. DOI: 10.3389/fpls.2019.00845 PMID: 31379891

Kunwar, S., Iriarte, F., Fan, Q., Evaristo da Silva, E., Ritchie, L., Nguyen, N. S., Freeman, J. H., Stall, R. E., Jones, J. B., Minsavage, G. V., Colee, J., Scott, J. W., Vallad, G. E., Zipfel, C., Horvath, D., Westwood, J., Hutton, S. F., & Paret, M. L. (2018). Transgenic expression of EFR and Bs2 genes for field management of bacterial wilt and bacterial spot of tomato. *Phytopathology*, *108*(12), 1402–1411. DOI: 10.1094/PHYTO-12-17-0424-R PMID: 29923802

LeBlanc, N., Salgado-Salazar, C., & Crouch, J. A. (2018). Boxwood blight: An ongoing threat to ornamental and native boxwood. *Applied Microbiology and Biotechnology*, *102*(10), 4371–4380. DOI: 10.1007/s00253-018-8936-2 PMID: 29610965

Lee, J. K., Jeong, E., Lee, J., Jung, M., Shin, E., Kim, Y. H., Lee, K., Jung, I., Kim, D., Kim, S., & Kim, J. S. (2018). Directed evolution of CRISPR-Cas9 to increase its specificity. *Nature Communications*, *9*(1), 3048. DOI: 10.1038/s41467-018-05477-x PMID: 30082838

Li, C., Chen, M., Chao, S., Yu, J., & Bai, G. (2013). Identification of a novel gene, H34, in wheat using recombinant inbred lines and single nucleotide polymorphism markers. *Theoretical and Applied Genetics*, *126*(8), 2065–2071. DOI: 10.1007/s00122-013-2118-5 PMID: 23689741

Li, T., Liu, B., Spalding, M. H., Weeks, D. P., & Yang, B. (2012). High-efficiency TALEN-based gene editing produces disease-resistant rice. *Nature Biotechnology*, *30*(5), 390–392. DOI: 10.1038/nbt.2199 PMID: 22565958

Lusser, M., Parisi, C., Plan, D., & Rodriguez-Cerezo, E. (2011). New plant breeding techniques. In *State-of-the-art and prospects for commercial development* (pp. 1–220). Joint Research Centre-Institute for Prospective Technological Studies, Publications Office of the European Union., DOI: 10.2791/54761

Manghwar, H., Lindsey, K., Zhang, X. L., & Jin, S. X. (2019). CRISPR/Cas system: Recent advances and future prospects for genome editing. *Trends in Plant Science*, *24*(12), 1102–1125. DOI: 10.1016/j.tplants.2019.09.006 PMID: 31727474

Manzar, N., Kashyap, A. S., Maurya, A., Rajawat, M. V. S., Sharma, P. K., Srivastava, A. K., Roy, M., Saxena, A. K., & Singh, H. V. (2022). Multi-gene phylogenetic approach for identification and diversity analysis of *Bipolaris maydis* and *Curvularia lunata* isolates causing foliar blight of *Zea mays*. *Journal of Fungi (Basel, Switzerland)*, *8*(8), 802. DOI: 10.3390/jof8080802 PMID: 36012790

Martínez-Fortún, J., Phillips, D. W., & Jones, H. D. (2017). Potential impact of genome editing in world agriculture. *Emerging Topics in Life Sciences*, *1*(2), 117–133. DOI: 10.1042/ETLS20170010 PMID: 33525764

Miedaner, T. (2016). Breeding strategies for improving plant resistance to diseases. In *Advances in plant breeding strategies: agronomic, abiotic and biotic stress traits* (pp. 561–599). Springer., DOI: 10.1007/978-3-319-22518-0_15

Molla, K. A., Sretenovic, S., Bansal, K. C., & Qi, Y. (2021). Precise plant genome editing using base editors and prime editors. *Nature Plants*, *7*(9), 1166–1187. DOI: 10.1038/s41477-021-00991-1 PMID: 34518669

Molla, K. A., & Yang, Y. (2020). Predicting CRISPR/Cas9-induced mutations for precise genome editing. *Trends in Biotechnology*, *38*(2), 136–141. DOI: 10.1016/j.tibtech.2019.08.002 PMID: 31526571

Nekrasov, V., Wang, C., Win, J., Lanz, C., Weigel, D., & Kamoun, S. (2017). Rapid generation of a transgene-free powdery mildew resistant tomato by genome deletion. *Scientific Reports*, *7*(1), 482. DOI: 10.1038/s41598-017-00578-x PMID: 28352080

Nerkar, G. A., Purankar, M. V., Sheelavantmath, S., & Devarumath, R. M. (2018). Biotechnological approach: a new dimension for sugarcane improvement. In K. Sengar (Ed.), *Biotechnology to enhance sugarcane productivity and stress tolerance* (pp. 1–24). CRC Press, Taylor & Francis Group. https://doi.org/DOI: 10.1201/9781315152776-1

Pallarès Masmitjà, M., Knödlseder, N., & Güell, M. (2019). CRISPR-gRNA design. In *CRISPR gene editing: methods and protocols* (pp. 3–11). Humana., DOI: 10.1007/978-1-4939-9170-9_1

Pausch, P., Al-Shayeb, B., Bisom-Rapp, E., Tsuchida, C. A., Li, Z., Cress, B. F., Knott, G. J., Jacobsen, S. E., Banfield, J. F., & Doudna, J. A. (2020). CRISPR-casΦ from huge phages is a hypercompact genome editor. *Science*, *369*(6501), 333–337. DOI: 10.1126/science.abb1400 PMID: 32675376

Pavan, S., Jacobsen, E., Visser, R. G., & Bai, Y. (2010). Loss of susceptibility as a novel breeding strategy for durable and broad-spectrum resistance. *Molecular Breeding*, *25*(1), 1–12. DOI: 10.1007/s11032-009-9323-6 PMID: 20234841

Prado, J. R., Segers, G., Voelker, T., Carson, D., Dobert, R., Phillips, J., & Martino-Catt, S. (2014). Genetically engineered crops: From idea to product. *Annual Review of Plant Biology*, *65*(1), 769–790. DOI: 10.1146/annurev-arplant-050213-040039 PMID: 24579994

Pyott, D. E., Sheehan, E., & Molnar, A. (2016). Engineering of CRISPR/Cas9-mediated potyvirus resistance in transgene-free Arabidopsis plants. *Molecular Plant Pathology*, *17*(8), 1276–1288. DOI: 10.1111/mpp.12417 PMID: 27103354

Ran, F. A., Hsu, P. D., Lin, C. Y., Gootenberg, J. S., Konermann, S., Trevino, A. E., Scott, D. A., Inoue, A., Matoba, S., Zhang, Y., & Zhang, F. (2013). Double nicking by RNA-guided CRISPR Cas9 for enhanced genome editing specificity. *Cell, 154*(6), 1380–1389. DOI: 10.1016/j.cell.2013.08.021 PMID: 23992846

Ray, D. K., Mueller, N. D., West, P. C., & Foley, J. A. (2013). Yield trends are insufficient to double global crop production by 2050. *PLoS One*, *8*(6), e66428. DOI: 10.1371/journal.pone.0066428 PMID: 23840465

Ricroch, A., Clairand, P., & Harwood, W. (2017). Use of CRISPR systems in plant genome editing: Toward new opportunities in agriculture. *Emerging Topics in Life Sciences*, *1*(2), 169–182. DOI: 10.1042/ETLS20170085 PMID: 33525765

Savary, S., Ficke, A., Aubertot, J. N., & Hollier, C. (2012). Crop losses due to diseases and their implications for global food production losses and food security. *Food Security*, *4*(4), 519–537. DOI: 10.1007/s12571-012-0200-5

Savary, S., Willocquet, L., Pethybridge, S. J., Esker, P., McRoberts, N., & Nelson, A. (2019). The global burden of pathogens and pests on major food crops. *Nature Ecology & Evolution*, *3*(3), 430–439. DOI: 10.1038/s41559-018-0793-y PMID: 30718852

Scheben, A., Wolter, F., Batley, J., Puchta, H., & Edwards, D. (2017). Towards CRISPR/Cas crops–bringing together genomics and genome editing. *The New Phytologist*, *216*(3), 682–698. DOI: 10.1111/nph.14702 PMID: 28762506

Schie, C. C., & Takken, F. L. (2014). Susceptibility genes 101: How to be a good host. *Annual Review of Phytopathology*, *52*(1), 551–581. DOI: 10.1146/annurev-phyto-102313-045854 PMID: 25001453

Shan, Q., Wang, Y., Li, J., Zhang, Y., Chen, K., Liang, Z., & Gao, C. (2013). Targeted genome modification of crop plants using a CRISPR-Cas system. *Nature Biotechnology*, *31*(8), 686–688. DOI: 10.1038/nbt.2650 PMID: 23929338

Slaymaker, I. M., Gao, L. Y., Zetsche, B., Scott, D. A., Yan, W. X., & Zhang, F. (2016). Rationally engineered Cas9 nucleases with improved specificity. *Science*, *351*(6268), 84–88. DOI: 10.1126/science.aad5227 PMID: 26628643

Stover, E. D., Driggers, R., Richardson, M. L., Hall, D. G., Duan, Y., & Lee, R. F. (2014). Incidence and severity of Asiatic citrus canker on diverse citrus and citrus-related germplasm in a Florida field planting. *HortScience*, *49*(1), 4–9. DOI: 10.21273/HORTSCI.49.1.4

Takagi, H., Abe, A., Yoshida, K., Kosugi, S., Natsume, S., Mitsuoka, C., Uemura, A., Utsushi, H., Tamiru, M., Takuno, S., Innan, H., Cano, L. M., Kamoun, S., & Terauchi, R. (2013). QTL-seq: Rapid mapping of quantitative trait loci in rice by whole genome resequencing of DNA from two bulked populations. *The Plant Journal : For Cell and Molecular Biology*, *74*(1), 174–183. DOI: 10.1111/tpj.12105 PMID: 23289725

Tyagi, S., Kumar, R., Kumar, V., Won, S. Y., & Shukla, P. (2021). Engineering disease resistant plants through CRISPR-Cas9 technology. *GM Crops & Food*, *12*(1), 125–144. DOI: 10.1080/21645698.2020.1831729 PMID: 33079628

Urnov, F. D., Rebar, E. J., Holmes, M. C., Zhang, H. S., & Gregory, P. D. (2010). Genome editing with engineered zinc finger nucleases. *Nature Reviews. Genetics*, *11*(9), 636–646. DOI: 10.1038/nrg2842 PMID: 20717154

Xiao, L., Zhong, Y. P., Wang, B., & Wu, T. L. (2014). Mapping an aphid resistance gene in soybean [*Glycine max* (L.) Merr.] P746. *Genetics and Molecular Research*, *13*(4), 9152–9160. DOI: 10.4238/2014.November.7.2 PMID: 25501137

Ye, G., McNell, D. L., & Hill, G. D. (2002). Breeding for resistance to lentil ascochyta blight. *Plant Breeding*, *121*(3), 185–191. DOI: 10.1046/j.1439-0523.2002.00705.x

Zaidi, S. S. E. A., Mukhtar, M. S., & Mansoor, S. (2018). Genome editing: Targeting susceptibility genes for plant disease resistance. *Trends in Biotechnology*, *36*(9), 898–906. DOI: 10.1016/j.tibtech.2018.04.005 PMID: 29752192

Zaidi, S. S. E. A., Mukhtar, M. S., & Mansoor, S. (2018). Genome editing: Targeting susceptibility genes for plant disease resistance. *Trends in Biotechnology*, *36*(9), 898–906. DOI: 10.1016/j.tibtech.2018.04.005 PMID: 29752192

Zetsche, B., Gootenberg, J. S., Abudayyeh, O. O., Slaymaker, I. M., Makarova, K. S., Essletzbichler, P., Volz, S. E., Joung, J., van der Oost, J., Regev, A., Koonin, E. V., & Zhang, F. (2015). Cpf1 is a single RNA-guided endonuclease of a class 2 CRISPR-Cas system. *Cell*, *163*(3), 759–771. DOI: 10.1016/j.cell.2015.09.038 PMID: 26422227

Zhang, H., Zhang, J., Lang, Z., Botella, J. R., & Zhu, J. K. (2017). Genome editing—Principles and applications for functional genomics research and crop improvement. *Critical Reviews in Plant Sciences*, *36*(4), 291–309. DOI: 10.1080/07352689.2017.1402989

Zhang, Y., & Showalter, A. M. (2020). CRISPR/Cas9 genome editing technology: A valuable tool for understanding plant cell wall biosynthesis and function. *Frontiers in Plant Science*, *11*, 589517. DOI: 10.3389/fpls.2020.589517 PMID: 33329650

Zhou, J., Li, D., Wang, G., Wang, F., Kunjal, M., Joldersma, D., & Liu, Z. (2020). Application and future perspective of CRISPR/Cas9 genome editing in fruit crops. *Journal of Integrative Plant Biology*, *62*(3), 269–286. DOI: 10.1111/jipb.12793 PMID: 30791200

Zhou, J., Peng, Z., Long, J., Sosso, D., Liu, B. O., Eom, J. S., & Yang, B. (2015). Gene targeting by the TAL effector PthXo2 reveals cryptic resistance gene for bacterial blight of rice. *The Plant Journal : For Cell and Molecular Biology*, *82*(4), 632–643. DOI: 10.1111/tpj.12838 PMID: 25824104

Chapter 8
Case Studies:
Successful Gene-Edited Crops

Danial Hassani
http://orcid.org/0000-0001-8024-0837
Institute of Biomaterials and Biomedicine, School of Food and Pharmacy, Shanghai Zhongqiao Vocational and Technical University, Shanghai, China

Sumreen Dawood
University of Mianwali, Mianwali, Pakistan

Mamoona Munir
Rawalpindi Women University, Pakistan

Jazbia Shirin
Peking University Shenzhen Graduate School, China

Ayush Madan
http://orcid.org/0000-0001-5089-1794
Department of Biotechnology, School of Research and Technology, People's University, Bhanpur, India

ABSTRACT

CRISPR/Cas9 gene editing has rapidly advanced agriculture by enabling precise, efficient, and stable crop genome changes without introducing foreign DNA. This chapter reviews key developments and applications for improving yield, nutritional quality, disease resistance, abiotic stress tolerance, and post-harvest traits. Reported examples include drought-tolerant maize, iron/vitamin-enriched rice, low-gluten wheat, and disease-resistant potato and grape. Emerging platforms such as base and prime editing expand options for safer, more accurate edits. Regulatory and public acceptance is growing, with approvals and commercialization increasing in

DOI: 10.4018/979-8-3373-9534-0.ch008

countries such as the United States, Canada, Brazil, and Japan. Beyond biosafety, gene editing can reduce chemical inputs and support environmental and economic gains. The chapter concludes by emphasizing the need for responsible governance, ethics, and equitable access—especially in developing regions—to realize sustainable agriculture and food security.

INTRODUCTION

Gene editing is a controversial technology that allows for precise changes to an organism's genomic sequence. The most recognized approach to create targeted changes in genes is CRISPR-Cas9, which creates site-specific double-stranded breaks with RNA-guided nucleases (Jinek et al., 2012). More recent approaches have involved double-stranded breaks, to promote direct single nucleotide conversions (Komor et al., 2016). Prime editing also represents an expanded capacity for editing by allowing insertion, deletion, and/or base change of targets while avoiding double-stranded breaks (Anzalone et al., 2019). These technologies have disruptively changed the agricultural landscape because they allow for rapid and targeted improvements in crops for yield, nutrition, and stress resilience. Agriculture will have to adapt by considering food safety from gene editing in the face of global challenges, such as climate change and population growth, while using these advancements in gene editing to develop higher productivity in agriculture without compromising farm environmental resilience.

Gene editing is viewed as a trend in agriculture because of its rapid pace and accuracy compared to traditional breeding and transgenic changes. CRISPR-mediated edits generally leave no residual foreign DNA traces and can target specific genomic locations, producing specific transgene expression at those target sites. Such precise editing lowers regulatory hurdles and off-target effects, which will advance commercialization and public acceptance of these technologies.

CRISPR technology has advanced rapidly beyond basic genome editing techniques, such as multiplex editing. Multiplex editing allows for full phenotyping enhancements that were previously difficult to achieve through editing single genes, as several edits can be performed at once. For example, elite cultivars have rapidly emerged for multiple agricultural system needs through multiplex CRISPR editing because wheat productivity, stress tolerance, and disease resistance were all improved simultaneously.

A recent and powerful gene-editing technique is base editing, which allows the insertion of single nucleotides without double-strand breaks. This technique has improved editing accuracy and safety compared to standard editing methods, with significant reductions in cellular damage and unwanted mutations. For example, rice base editing produced point mutations in the *ALS* gene and provided considerable herbicide resistance without the unwanted side effects associated with standard cultivation procedures.

Prime editing is still in the early stages of development but is more flexible than base editing because it provides precise changes (i.e., base swaps, insertions, and deletions) in target DNA regions without a double-strand break. Many of the technical difficulties of CRISPR-Cas9 and base editing methods have been addressed in prime editing. Initial crops, such as rice and maize, have been successfully edited with a prime editor, and precise insertions producing traits, such as improved yield and drought resistance, have been especially promising.

The main limiting factors for the commercial viability of gene-edited crops are regulations and public sentiment. Countries in North America, such as the United States, Canada, and Brazil, as well as Argentina and Japan, have been the first to authorize the commercial production of gene-edited crops.

The reduced regulatory frameworks associated with developing gene-edited crops, particularly those with no transgenic insertions, are generally less complex and more easily attainable than those facing GMOs. If the European Union adopts a more liberal approach to the licensing of gene-edited crops while re-evaluating its regulatory systems, gene-edited crop licensing may also have a home. Collectively, these regulatory frameworks indicate the burgeoning use and potential market growth of gene-editing technologies in agricultural applications worldwide.

On another note, there is no doubt that gene editing will provide environmental sustainability opportunities. As gene-edited crops specifically target traits such as pest resistance and drought tolerance, they have a positive environmental impact by reducing reliance on chemical pesticides and excessive irrigation. Gene-edited pest-resistant crops that rely less on agricultural chemicals, while also resulting in lower farming costs, minimize environmental costs, such as pollution and biodiversity loss.

Finally, from an economic perspective, gene-editing technology represents a possible key innovation pathway for achieving sustainable productivity in agriculture (Figure 1). Economic research comparing gene-edited crops to conventional or GMOs crops has illustrated considerable potential for cost savings and yield improvement. The practical commercial viability and economic appeal of gene editing are demonstrated by the quick entry of gene-edited soybeans with enhanced oil profiles into the U.S. market, which have shown increased profitability and customer acceptability.

Considering all the above, gene-editing technologies, most notably CRISPR-Cas9 and its offspring, base editing and prime editing, are a revolutionary development in agriculture that provides significant benefits in terms of increased yield, improved nutritional value, resistance to disease and pests, and tolerance to abiotic stress. Their continued advancement and acceptance are expected to reshape global agriculture and play a critical role in ensuring food security and environmental sustainability for future generations.

Figure 1. Application of CRISPR/ Cas9 gene editing in crop improvement.

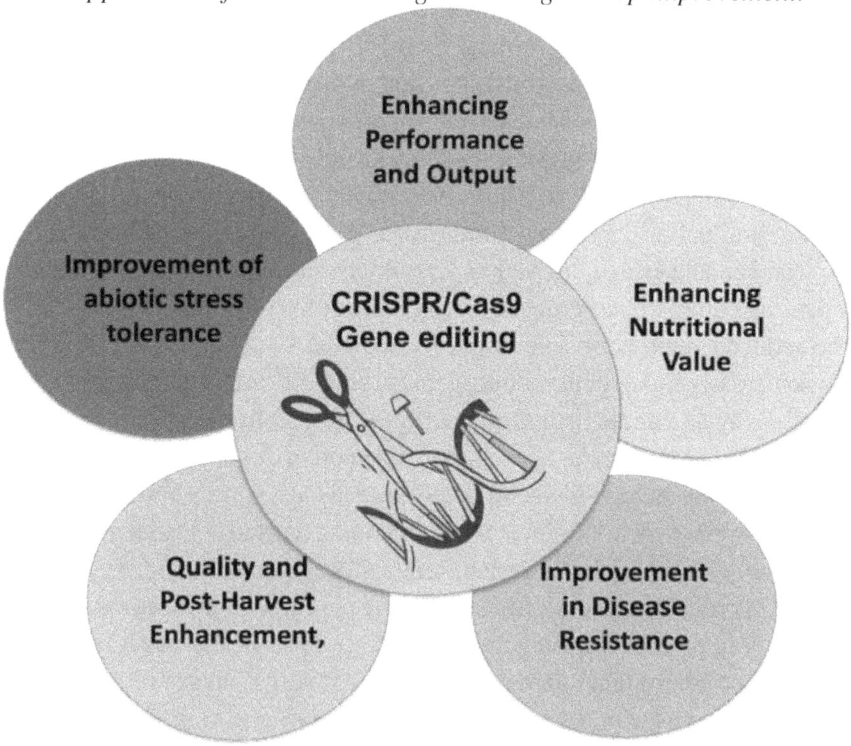

ENHANCING PERFORMANCE AND OUTPUT

Recent advancements in gene-editing technologies, particularly CRISPR/Cas systems, have enabled modern agriculture to maximize crop yield, productivity, and pest resistance to environmental stress. The most notable example is CRISPR-edited rice (*Oryza sativa*), which exhibits significant yield gains and improved resistance to drought stress. For example, it has been shown that specifically altering genes such as *OsGRF4*, *Gn1a*, and *DEP1* significantly increases grain size and number, leading to production increases of 15–20% under a variety of agro-environmental circumstances. Moreover, alterations to transcription factors, including *OsERF3* and *OsDST,* which respond to drought stress, have been shown to significantly improve drought tolerance and reduce yield through water limitation. The successful adoption of these gene-modified rice varieties, particularly in China, highlights farmer and consumer acceptance as a testament to their global market potential (Perochon et al., 2019).

Similarly, maize (*Zea mays*) has undergone precision breeding using the CRISPR system, demonstrating how advances in precision breeding can improve productivity while simultaneously providing resistance to insect pests. In other instances, precision modification of a number of regulatory genes, such as *ZmIPK* and *ARGOS8,* provided maize with higher grain weight and kernel numbers, which corresponded to improvements in productivity of 12% in the field (Shi et al., 2017). Furthermore, CRISPR-based precision modifications to promote insect resistance were also successful in moderating crop loss to the corn earworm (*Helicoverpa zea),* while providing economic and environmental benefits via reduced insecticides (Tian et al., 2021).

Additional case studies of production and yield

Gene editing has been shown to improve yield traits in wheat (Triticum aestivum). Here, the gene *TaGW2* was targeted using CRISPR/Cas9 to develop varieties with improved grain width and thousand-grain weight traits. Field studies have demonstrated wheat yield increases of 10% greater than commercially available or traditional varieties, reinforcing how gene editing could provide a rapidly viable total approach to improve cereal grain yield for evidence-based agricultural consumables with continued demand and sustainability (Chachar et al., 2024).

Another case study involved enhancements to soybean yield productivity using CRISPR technology. Yield traits of soybean (*Glycine max*) have also been improved with targeted CRISPR editing. Targeted edits in growth-regulating genes of soybean (specifically, *GmFT2a* and *GmFT5a*) have led to enhanced flowering time, pod-setting, and increased seed yield. Field trials have reported an average increase in yield of 8-12% all improving its economic and productivity value in the North American and Brazilian markets (Cai et al., 2020).

BARLEY WITH GENE EDITING FOR INCREASED GRAIN YIELD

Barley (Hordeum vulgare) is a globally significant cereal crop that has benefited from precision gene editing to increase production. *HvCKX1*, a cytokinin oxidase gene linked to grain filling and seed yield, is one of the yield-regulating genes in barley that researchers have chosen to target using CRISPR-Cas9 technology. Under controlled field conditions, the resultant barley lines showed enhanced grain filling and an approximate 15% increase in overall production. With very minor crop management changes, CRISPR-edited barley cultivars present farmers with an intriguing opportunity to boost yield sustainably (Gasparis et al., 2019).

Resilient to drought; CRISPR-improved sorghum for increased yield stability

Owing to its natural resistance to drought, sorghum (*Sorghum bicolor)* is a crucial crop for regions that are prone to drought. However, the yields are typically less than ideal. The sorghum *SbBADH2* gene, which controls stress responses and grain quality, was edited using CRISPR/Cas9 to provide significant increases in grain size, weight, and yield stability during drought conditions. An estimated 11–13% increase in yields was observed in field studies conducted in several locations, indicating that CRISPR editing can be a useful strategy for stabilizing and enhancing sorghum production in semi-arid environments, such as sub-Saharan Africa (Cheng et al., 2024).

Yield enhancement in gene-edited oilseed rape (*Brassica napus*)

A critical source of edible oil has been attained by targeted gene editing. Precise alteration of the *BnCLV1* gene, which is linked to yield, resulted in plants with additional pods and higher seed yield per plant. Field trials have demonstrated mean yield increments of approximately 10–15% over traditional cultivars. Gene-edited crops have been welcomed in the market, particularly in Canada and Australia, because they can save input costs while having the potential for increased return on investment (Yang et al., 2018).

CRISPR-BASED IMPROVEMENT OF COTTON YIELD

CRISPR/Cas9 genome editing in cotton (*Gossypium hirsutum*), one of the most economically significant fiber crops globally, has been achieved by targeting genes that affect fiber development and boll growth. Precise editing of the *GhALARP* gene increased the fiber yield and length. In some field trials, it exhibited an average yield increase of approximately 8–10%, and the fiber quality was improved by gene editing. A few such gene-edited cotton varieties have been adopted as part of the textile industry worldwide because of their improved fiber quality and sustainability (Zhu et al., 2018).

Enhancing Nutritional Value

Gene-editing tools have opened new avenues for enhancing the nutritional value of crops, providing practical solutions to targeted dietary and health problems. Recent breakthroughs using CRISPR/Cas9 technology have ushered in dramatic progress, showing the potential of precise gene editing for enhancing crops aimed at counteracting essential nutritional deficiencies and food intolerance, thereby supporting general consumer health.

Biofortified Tomatoes for Vitamin D

Introduction Vitamin D insufficiency affects millions of individuals worldwide, prompting agricultural researchers to explore gene editing as an effective approach for biofortification. The application of CRISPR-Cas9 technology to achieve precise genetic editing of the *Sl7-DR2* gene in tomatoes (*Solanum lycopersicum*) is a high-profile example of the successful application of gene editing, which significantly increases the levels of provitamin D3. The resulting tomatoes have a significantly higher vitamin D content to address nutritional shortfalls, but without any loss of flavor or market acceptability to consumers. Nutritional enhancements were validated in controlled field-and greenhouse-based tests with no adverse impact on fruit yield, shelf-life, or taste. Market acceptance was instant, particularly in Europe, where consumers want their fruit and vegetables to be health-enhancing and enhanced in a naturally occurring setting.

Wheat with Reduced Levels of Gluten by Utilizing CRISPR

Millions of individuals globally endure dietary restrictions due to gluten intolerance and celiac disease; thus, there is a dire necessity for wheat varieties with low gluten levels. CRISPR/Cas9 technology has been successfully utilized to edit target *α-gliadin* genes with precision to reduce the gluten content in wheat (Triticum aestivum). The altered wheat types had significantly decreased gluten proteins associated with immune-mediated responses, thus providing safer options of wheat products for gluten-sensitive consumers. Low-gluten wheat types also have good agronomical characteristics, such as stable yield and general baking quality, and consumer acceptance will be facilitated, particularly in Europe and North America, given the sustained market growth for gluten-free products (Jouanin et al., 2020).

Iron-Biofortified Rice

Iron deficiency anemia is a significant nutritional deficiency on a global scale, with particularly high rates in developing countries such as India. The science of rice biofortification utilizes CRISPR/Cas9 genome editing to alter the genes *OsVIT2* and *OsNAS2*, which store and transport iron in rice. The new rice variety effectively treated iron deficiency anemia with no negative implications on grain quality or yield and increased iron deposition to four times that of the control. Furthermore, field trials in Asia have shown outstanding agronomic performance of this new variety and high consumer acceptance, suggesting an applicable solution for iron deficiency anemia (Rajadurai et al., 2024).

CRISPR-Edited Camelina with Omega-3

Rich Camelina (*Camelina sativa*) is an oilseed crop that is gaining recognition for its favorable oil profile. Editing of fatty acid desaturase genes (*FAD2 and FAD3*) using the CRISPR/Cas9 system has enabled the elevation of omega-3 polyunsaturated fatty acid content, particularly α-linolenic acid (ALA), in camelina seeds. Transgenic camelina lines exhibited elevations in omega-3 fatty acid content by as much as 20–25%, rendering them a viable plant-based source of omega-3 fatty acids with excellent market potential as nutritional supplements and functional foods (Lee et al., 2021).

Lycopene-Enriched Tomatoes via Gene Editing

Tomatoes have been genetically edited to enhance their lycopene content. Lycopene is a valuable antioxidant with well-documented extensive health benefits, including decreased risk of cardiovascular disease and certain cancers. Scientists have utilized CRISPR-Cas9 genome editing of genes participating in carotenoid biosynthesis, such as *SlSGR1* and *SlCRTISO* in tomato plant fruit, to derive lycopene-rich tomatoes. The research increased lycopene levels in tomatoes by 40% compared to normal tomatoes without any loss of fruit quality and yield. Field trials and consumer testing have established that lycopene-enriched tomatoes have potential market value, especially among health-conscious consumers globally.

CRISPR-Edited Potato with Minimized Acrylamide Content

Acrylamide, a potential carcinogen, is produced and accumulates in potatoes subjected to high-temperature cooking processes, such as frying and baking. The use of CRISPR/Cas9 technology for editing genes controlling the production of acrylamide precursors, namely the vacuolar invertase (*VInv*) gene, achieved significant reductions in acrylamide content (over 50%) after cooking in edited potatoes. The genetically modified potato cultivar effectively provided consumers seeking safer and healthier food options with better-processed potato products, with successful acceptance in the North American and European markets (Ly et al., 2023).

Folate-Enriched Rice through CRISPR Editing

CRISPR editing has also been employed to develop Folate-Enriched Rice. Folate deficiency is a worldwide nutritional disorder that leads to severe health risks, particularly in the context of birth defects. Through the application of CRISPR-Cas9 gene editing to edit genes involved in folate biosynthesis, such as *OsGCHI* and *OsADCS*, scientists have created rice with higher folate levels. The gene-edited rice contained up to three times the folate content of the non-edited rice. The rice tested retained desirable agronomic characteristics and seed consumer acceptability, showing great promise for eliminating folate deficiency worldwide.

Improvement in Disease Resistance

Gene editing technologies have been enhancing crop resistance to ravaging diseases, offering sustainable solutions instead of conventional chemical control. By making precise changes in the genome, plant immunity has improved, chemical security has been lowered, and agricultural sustainability has been enhanced. Notable examples are gene-edited potatoes with late blight resistance and powdery mildew-resistant wheat, where the practicability of modifications is exhibited in field applications, all with broad acceptability in agricultural practice.

Disease-resistant CRISPR-edited potatoes

Potato crops (*Solanum tuberosum*) are highly vulnerable to late blight, which is caused by the pathogen Phytophthora infestans. Historically, this disease has caused dramatic reductions in the demand for potato production, and the development of transgenic potato cultivars was originally designed to reduce the economic impact of late blight. With the precise editing features of CRISPR/Cas9 technology used to edit susceptibility (S) genes, namely by the targeted knockout of the susceptibility gene *StDND1*, recent developments have demonstrated that potato cultivars have dramatically enhanced tolerance to late blight. Multigenerational field test results have consistently demonstrated that both the incidence of disease frequency and the requirement for fungicide application can be significantly minimized, resulting in tremendous economic and environmental advantages. These genetically modified potatoes have been gathering momentum for commercial adoption, particularly in Europe, where late blight has been a major factor affecting potato crop productivity for decades (Poudel, 2024).

Wheat varieties engineered for powdery mildew resistance

Powdery mildew, caused by *Blumeria graminis*, is a major limitation to wheat (*Triticum aestivum*) production. Traditional methods have largely relied on fungicide use, resulting in environmental issues and changes in pathogen resistance. The discovery of CRISPR/Cas9 gene editing has enabled the precise modification of susceptibility genes, such as the MLO locus in wheat, to generate varieties with innate resistance to powdery mildew. To date, extensive field trials have demonstrated efficient resistance, considerable yield loss, and substantial reductions in fungicide application, leading to considerable environmental impacts and reduced production costs. Consumer acceptance of CRISPR-edited powdery mildew-resistant wheat has been good, particularly in Europe and China, owing to the consumer demand for sustainably produced goods (Acevedo-Garcia et al., 2017).

CRISPR-edited rice resistant to bacterial blight

CRISPR-edited rice resistant to bacterial blight caused by *Xanthomonas oryzae* pv. *X. Oryzae* poses a great risk to rice cultivation worldwide. Using the CRISPR/Cas9 system, the gene family of SWEET sucrose transporters—namely *OsSWEET11*, *OsSWEET13*, and *OsSWEET14*—were edited to achieve durable and widespread resistance against bacterial blight with simultaneous minimization of crop yield loss and chemical intervention (Figure 2). Field tests have demonstrated strong resistance to multiple strains of Xoo in a wide range of geographic locations, which has gained popularity among both consumers and farmers, particularly in Asia and Africa (Oliva et al., 2019).

Figure 2. Improvement in the disease resistance of crops using CRISPR/Cas9.

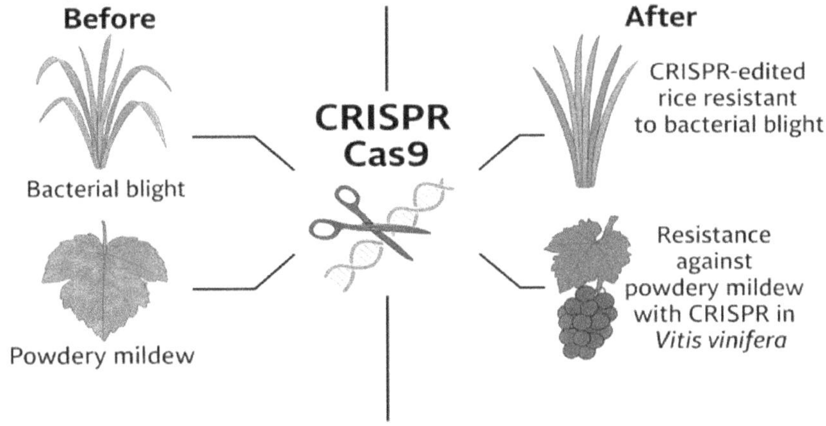

Resistance against Disease with CRISPR in Grapevines

Grapevines (*Vitis vinifera*) are extremely susceptible to fungal pathogens, such as powdery mildew (*Erysiphe necator*), which can severely disrupt production in the global wine and grape industries. Research has shown that CRISPR/Cas9 editing of the susceptibility gene *VvMLO3* greatly increases resistance to powdery mildew. CRISPR-edited grapevines have been shown to confer strong resistance to powdery mildew under both greenhouse and field conditions, resulting in a reduction of fungicide (chemical) applications by more than 80% while maintaining grape quality. With many varietals readily acceptable in wine-producing regions of Europe and North America, development has accelerated based on the economic and environmental advantages of reducing chemical applications (Wan et al., 2020).

CRISPR-edits of Citrus against *Citrus canker*

Citrus canker disease caused by *Xanthomonas citri* subsp. citri has resulted in large losses in global citrus production. Researchers have developed CRISPR/Cas9 edited susceptibility genes, such as *CsLOB1,* to produce citrus cultivars with built-in resistance against canker. Field trials have shown a significant reduction in canker severity and a reduced reliance on chemical controls. The gene-edited citrus plants described above are poised to have major impacts in areas of the world that produce the largest levels of citrus, including the United States, Brazil, and China, which will lower agricultural costs and improve the sustainability of commodities (Peng et al., 2017).

CRISPR-improved Bananas' Resistance to Fusarium Wilt

Millions of people worldwide depend on bananas (*Musa acuminata*) as a staple food; however, the destructive Fusarium wilt poses a serious threat to them. High levels of resistance to Fusarium wilt have been successfully produced through targeted CRISPR editing of susceptibility genes, such as MaDND1, which has significantly decreased the severity of the disease and growers' financial losses. These banana types maintained their productivity, quality, and consumer acceptance, according to trials conducted in tropical field locations. The potential of gene editing to reduce the risks to the world's banana output is demonstrated by this significant example.

CRISPR-modified Cucumber with Virus Resistance

Viral infections, particularly cucumber mosaic viruses (CMVs), restrict the productivity of cucumbers (*Cucumis sativus*) and other crops. Cucumber lines resistant to CMV were identified using CRISPR/Cas9-edited *eIF4E* (eukaryotic initiation factor 4E), which is necessary for viral replication. Ecological trials under controlled conditions showed enhanced fruit yield and health, as well as a lower incidence of disease, and they were also unaffected by the use of pesticides to manage disease. Growers who choose pesticide-free, sustainable vegetable crops, particularly in East Asia and Europe, have shown a rapid market uptake (Fidan et al., 2023).

Improvement of abiotic stress tolerance

Abiotic stresses, including salinity and drought, greatly diminish crop yields and pose a risk to global food security. Gene-editing technologies, especially CRISPR-mediated approaches, have emerged as effective tools for engineering crops that are resilient to environmental stresses. Examples of such approaches have already been successfully implemented to develop drought-resistant maize cultivars and salt-tolerant soybean cultivars. Both examples illustrate that precision breeding offers a sustainable means of enhancing crop performance in the context of agricultural challenges associated with environmental conditions.

Drought-resistant CRISPR-edited maize

Zea mays often suffers from drought stress, leading to considerable economic losses globally. Subsequently, researchers have deployed CRISPR/Cas9 editing technologies on maize to target genes that play a role in the drought response process, including ZmWRKY transcription factors and genes that regulate stomatal closure, such as *ARGOS8*. Drought-resistant maize cultivars have demonstrated improved water-use efficiency, reduced leaf wilting, and improved grain yield during prolonged drought periods. Furthermore, a field trial generated utility data and appearance of yield stability of approximately 10-15% in comparison to non-edited controls, indicating substantial potential for development and expansion of crops into drought-prone regions. The market acceptance of cultivars was also enhanced in drought-prone regions, notably Sub-Saharan Africa and North America, because of decreased irrigation costs and consistent performance during drought periods (Varotto 2023).

CRISPR Edited Soybeans with Increased Salt Tolerance

Salinity is an important abiotic stress that limits soybean (Glycine max) production worldwide, particularly in arid and semi-arid regions. Salt tolerance has been enhanced using CRISPR/Cas9 gene-editing technology by knocking out genes known to negatively regulate sodium transport and accumulation, such as *GmSOS1* and *GmNAC*. Edited plant lines had significantly lower sodium uptake and improved plant vigor, thereby sustaining yield under ebb high-salinity conditions. Controlled and open-field trials provided evidence of yield increases of up to 18% over conventional lines in saline soils. These varieties are perceived positively by growers in areas affected by saline soils across Asia, North America, and South America because of their ability to maintain productivity in formerly underfit soils (Rasheed et al., 2022).

CRISPR-Edited Rice Exhibiting Improved Cold Tolerance

Cold stress has a major influence on global rice (*Oryza sativa*) production and is especially deleterious in the temperate regions of the world. The authors investigated the potential of CRISPR/Cas9 technology to edit cold-sensitive genes in rice. The two genes studied were *OsMYB30* and *OsCOLD1*, both of which have been associated with cold tolerance in rice. The CRISPR-edited rice lines showed significant reductions in cold-induced damage, improved survival, and consistently high grain yields under low-temperature field conditions. These novel rice varieties hold excellent promise for increasing rice production worldwide, especially in colder climates, which will benefit the sustainability of rice production (Zeng et al., 2020).

CRISPR-Edited Tomato Showing Improved Heat Tolerance

Heat stress can significantly reduce tomato production by reducing fruit-setting, growth, and yield. The authors used CRISPR/Cas9 gene editing of two heat stress-related transcription factors, *SlAGL6* and *SlMAPK3*, to create improved heat tolerance in tomatoes. Field trials have demonstrated that CRISPR-edited tomato plants suffer less yield loss and fruit quality reduction under heat stress conditions, which reduces economic loss. This technology is becoming commercially viable and significant, particularly in regions suffering from high temperatures (Wang et al., 2017).

CRISPR-Cas9 Methodology for Drought Tolerance in Wheat

Drought is one of the most prevalent abiotic stresses affecting wheat (*Triticum aestivum*) globally. For example, scientists have used CRISPR/Cas9 methodology to edit genes that respond to drought stress, including wheat TaDREB2 transcription factor genes, and enhance drought tolerance. Under field conditions, researchers have found improvements in plant survival, yield stability, and water-use efficiencies under very high drought stress, demonstrating substantial promise for using CRISPR technology to improve wheat production in drought-prone and semi-arid regions (Abdallah et al., 2025).

CRISPR-Cas9 Methodology for Flooding Tolerance in Soybeans

Flooding stress dramatically reduces soybean yields globally, and CRISPR/Cas9 editing can be employed to edit flooding-sensitive regulatory genes, demonstrating the ability to improve flooding tolerance in soybeans by reducing hypoxic stress damage to flood-edited genes, such as *GmERF3*. After editing, the soybean cultivars showed improved plant survivorship, vigor, and yield stability during prolonged flooding, which has important implications in smaller flood-prone areas, such as the Mississippi Delta region of the United States and other parts of South America (Yijun et al., 2022).

CRISPR-Edited Potato with Improved Heat and Drought Resilience

Potato crops (*Solanum tuberosum*) are extremely vulnerable to combined heat and drought stress, which is becoming increasingly common phenomenon associated with global climate change. The authors edited potato genes for stress responses, specifically *StDREB* and *StCBF* genes, to improve potato tolerance to combined abiotic stressors. Field trials have shown significantly less yield loss, more efficient use of water, and maintained tuber quality during both heat and drought events, demonstrating the influence of gene editing in support of global potato production (Ndudzo et al., 2024).

CRISPR-Enhanced Salt-Resistance in Rice

Salinity poses an increasingly challenging threat to global rice production, particularly in coastal environments. The authors utilized CRISPR/Cas9 technology to edit salt-responsive regulatory genes, including *OsRR22* and *OsDST*. The edited rice varieties exhibited significant enhancements in salt tolerance, limited sodium uptake, growth under saline conditions, and maintained grain yield in saline-damaged soils. These gene-edited varieties have tremendous potential for expansion of rice production in coastal portions of Asia and Africa and reinforce viable agriculture in saline-prone areas (Farhat et al., 2019)

Nutritional Gene-Edited Rice (An Alternative to Golden Rice)

Vitamin deficiencies, including vitamin A deficiency (VAD), continue to threaten the health of millions of people, especially children in early childhood, in developing parts of the world. Gene editing is now an option that allows the targeted and customary enhancement of rice (*Oryza sativa*) nutritional content without the regulatory complications associated with traditional transgenic Golden Rice. By editing endogenous genes involved in carotenoid biosynthesis, namely, *OsPsy1* and *OsOr*, researchers have increased the levels of provitamin A (β-carotene) in rice grains. Previous field trials have shown that these edited rice varieties have significantly higher β-carotene content, while correcting for nutrition without negative effects on agronomic performance or human consumption. This gene-edited rice has now been successfully introduced to the market at a rapid speed, especially in Asian countries, owing to the consumer notion of being a safer and less natural alternative to GMO varieties. Another point worth mentioning is that gene-edited rice can reproduce naturally for economic benefits (Dong et al., 2020).

Gene-Edited Tomatoes, Enhanced for Their Vitamin Content

Tomatoes (Solanum lycopersicum) are among the most widely consumed vegetables in the world and represent an excellent source for enhancing nutrition via gene editing. Most recently, gene edits for the CRISPR/Cas9 gene editing technique that activated vitamin pathways, especially for genes associated with the synthesis of provitamin D3(Sl7-DR2) and biosynthetic pathways for vitamin C (SlGGP1), have been carried out in tomatoes. Gene-edited tomatoes were deemed enriched with vitamin levels and provided an experimental food product that addressed widespread dietary deficiencies without altering tomato flavor, storage quality, or shelf life. This type of tomato was evaluated for agronomical and consumer acceptance attributes and was well-accepted in sustainable food markets, and among consumers who expressed an increased preference for appealing foods that are optimized for nutrition but are not considered transgenic(Jie et al. 2022).

Iron Bio-fortified Rice Cultivars via CRISPR Gene Editing

Iron deficiency anemia affects millions of people worldwide, particularly in developing countries where rice is the primary commodity. CRISPR/Cas9 gene editing technology has been used by researchers to develop community-supported iron-enriched rice grain products. Using available rice cultivars as targets for gene edits, the researchers obtained iron-enhanced rice products by targeting genetic loci *OsNAS2* and *OsYSL2* that have been shown to regulate iron transport and homeostasis. The researchers showed that the CRISPR gene-edited rice contained up to four times higher iron accumulation levels in the grain than conventional cultivars, and the trials reported that the bioactive iron was significantly enhanced compared to conventional rice. The rice grains were field-tested for traits, and the results confirmed a reasonable analysis of trait inheritance. These iron-optimized rice products have been widely accepted, and positive results have been reported in the target community in Asia and neighboring regions, where iron deficiency is prevalent (Jie et al., 2022).

Zinc Biofortification in Rice Using CRISPR-Based Technology

Almost two billion people worldwide suffer from zinc deficiency, which can result in impaired immune responses or growth disorders. Utilizing CRISPR-Cas9 technology, researchers have targeted *OsZIP* genes involved in zinc uptake and transport in rice. After editing the *OsZIP* genes, rice cultivars showed a significantly elevated zinc concentration in the kernels (2–3 times higher), thus enhancing the nutrients of edible kernels without changing the growth or yield of the plants. In field trials, rice cultivars demonstrated high agronomic performance and consumer acceptance, supporting the use of CRISPR technology as a global strategy for nutritional biofortification, particularly in regions of South and Southeast Asia, where zinc deficiency is a concern (Ankanagari et al., 2024).

CRISPR-Edited Rice with Enhanced Protein Concentration

Protein malnutrition is a significant health issue in many developing areas of the world where rice is the primary source of dietary calories but typically has very low protein levels. Researchers targeted the *OsAAP6* gene (an amino acid transporter) through CRISPR/Cas9 editing to downregulate the expression of this gene and increase grain protein levels by 10–15%. Field experiments indicated that gene-edited rice lines did not yield less and maintained their agronomic characteristics while simultaneously enhancing dietary protein levels. Consumer acceptance trials have shown that consumers are highly accepting of rice cultivars with enhanced protein content, especially in areas where dietary protein availability is highly limited (Li et al., 2022).

CRISPR-Edited Rice with Increased Antioxidant Activity via Increased Anthocyanin Content

Anthocyanins are potent antioxidants that offer potential health benefits; however, they are often limited in conventional rice varieties. Researchers have used CRISPR/Cas9 on rice by activating the anthocyanin regulatory gene (*OsC1*), thereby substantially increasing anthocyanin accumulation in the rice endosperm. The lines of rice that were subsequently developed showed enhanced antioxidant capacity, which can also lead to a decreased risk of chronic diseases. Controlled research studies have confirmed sustained anthocyanin enhancement, positive agronomic performance, and consumer acceptability, especially in markets that focus on functional foods and health-beneficial foods (Wu et al., 2025).

CRISPR-modified Rice Grains with Elevated Lysine Levels

Lysine is an amino acid that is important for human nutrition, but is often limiting in cereals, including rice. Scientists have utilized CRISPR/Cas9 technology to edit two genes associated with lysine biosynthesis (*OsAK* and *OsDHPS*) to produce rice with significantly elevated levels of lysine in the grains (up to 25% more than controls). Data from the researchers showed that this rice is viable in terms of yield and plant health. The lysine-enhanced rice variety was subjected to rigorous field testing, leading to stable inheritance and consumer acceptance characteristics, to help enhance protein quality for locations heavily reliant on rice as a primary staple (Chen et al., 2024).

Quality Enhancement, and Post-Harvest Attributes and Organizer of Shelf Life

Gene-editing technologies have been highly successful in improving the quality attributes of certain crops (including shelf-life and nutritional characteristics) that correlate with less food waste and a more satisfactory consumer experience. Quality deterioration normally observed after harvest is generally determined by whether or not crops were allowed to achieve full maturation and whether intrinsic traits facilitate or impede the storage process (Table 1). Publicized gene-editing news on quality improvement is highly visible in the commercial development of tomatoes with enhanced shelf-life properties and low-browning apples and potatoes. Specifically, the quality traits attributed to high-profile tomatoes and apples are noteworthy.

Table 1. Gene Editing in Agriculture.

Section	Main Focus	Genes Involved	Traits Improved	Regions Applied	Effectiveness
1. Performance Enhancement	Crop yield, productivity, pest resistance	CRISPR-Cas9; OsGRF4, Gn1a, DEP1 (rice); ARGOS8 (maize)	Yield ↑, Drought resistance ↑, Pest resistance ↑	China, North America, Brazil	10–20% yield increases; stress tolerance proven
2. Nutritional Enhancement	Biofortification and allergen reduction	Sl7-DR2 (tomato), α-gliadin (wheat), OsVIT2/OsNAS2 (rice)	Vitamin D ↑, Gluten ↓, Iron ↑, Lycopene ↑	Europe, Asia, North America	2–4× nutritional increase; no yield loss
3. Disease Resistance	Resistance to pathogens and reduced fungicide use	StDND1 (potato), MLO (wheat), OsSWEET11/13/14 (rice)	Late blight ↓, Mildew ↓, Bacterial blight ↓	Europe, Africa, Asia	Disease incidence ↓80%; pesticide ↓
4. Abiotic Stress Tolerance	Coping with drought, salinity, heat, cold	ZmWRKY, OsMYB30, GmSOS1, SlMAPK3	Drought resistance ↑, Salt tolerance ↑, Heat resilience ↑	Sub-Saharan Africa, Asia, North America	10–18% yield under stress; better water-use
5. Post-harvest & Quality	Shelf-life, browning, and processing quality	SlPG2a (tomato), PPO (apple/ mushroom), MaACO1 (banana)	Shelf-life ↑, Browning ↓, Texture maintained	USA, EU	Shelf-life ↑ 40–100%; food waste ↓
6. Regulatory & Economic Aspects	Policy, market adoption, public perception	No transgene, CRISPR edits only	↓regulatory hurdles, higher profit	USA, Canada, Argentina, Japan	Faster commercialization; public trust ↑
7.Technological Innovation	Advances in precision and safety	CRISPR-Cas9, Base editing, Prime editing	Multiplex traits, High fidelity edits	Global	Editing efficiency 30–75%; low off-targets

Gene-Edited Tomatoes with Extended Shelf Life

Tomatoes (*Solanum lycopersicum*) suffer significant losses due to post-harvest management practices. Tomatoes require advanced post-harvest management methods to protect against bruising and rapid softening. Researchers have utilized CRISPR/Cas9 editing at specific genetic loci associated with traits such as *SlPG2a* and *SlPL*, which are codified areas in the genomes for fruit softening and cell wall degradation, respectively. These tomatoes exhibited significant advancements in shelf life, reductions in loss, and maintenance of firmness and nutritional quality for multiple air and refrigeration temperatures and durations. Field and commercial experiences have revealed high consumer acceptance based on quality criteria related to these sensory properties (texture and eyesight), which reduced the spoilage rate (40%) pre-sale (Ortega-Salazar et al., 2024).

Nutritionally Enhanced Gene-Edited Golden Rice Alternative

Traditional Golden Rice has faced regulatory challenges associated with its transgenic origin. CRISPR/Cas9 technology, a powerful precision gene-editing method, has been utilized as an alternative to biofortified rice with increased levels of vitamin A (β-carotene), an important nutritional enhancement for areas impacted by vitamin A deficiency (VAD). Editing genes from the endogenous carotenoid biosynthetic pathway genes (*OsLCYe* and *OsOr*) significantly increased β-carotene levels in rice grains without inserting any foreign DNA. From a field trial standpoint, CRISPR-derived rice cultivars alleviating dietary vitamin A deficiency contribute to enhanced nutritional levels. In addition to this nutritional enhancement, the CRISPR-derived rice plants did not reduce crop yield or taste and garnered robust U.S. market acceptance in Asia and parts of Africa, which had traditionally been much more cautious towards GMOs (Kim et al., 2022).

CRISPR-Modified Non-Browning Apples

The browning of apples (*Malus domestica*) results in a loss of consumer appeal, leading to food wastage. Arctic apples have been developed using gene-editing technology to silence the gene(s) that produce polyphenol oxidase (PPO), which is responsible for enzymatic browning of apples. Edited apple varieties demonstrated markedly less browning upon cutting or mechanical damage and remained fresher and visually more appealing for longer than leading conventional apple varieties. Market studies show strong consumer acceptance and rapid commercialization, particularly in North America, where consumers regard less food waste and improved longevity in the fresh market as positive attributes (Maioli et al., 2020).

CRISPR-Modified Browned Mushrooms

Mushrooms (*Agaricus bisporus*) undergo rapid browning post-harvest, which creates a barrier to their shelf-life and marketability. Researchers have utilized CRISPR-Cas9 to target and knock out polyphenol oxidase (PPO) genes responsible for enzymatic browning. Edited varieties of the mushroom have demonstrated significant delays in browning, extended shelf life, and a decrease of approximately 30% in overall post-harvest losses. As for Arctic apples, market adoption was good because of consumer preference for fresh produce that is visually appealing in markets in the United States and Europe, ultimately developing a consumer base for modified mushrooms (Zhang et al., 2023).

Enhancing the Postharvest Sustainability of Bananas using CRISPR Technology

Fresh market bananas (*Musa acuminata*) have a rapid postharvest life owing to their accelerated senescence process, resulting in large-scale losses. CRISPR/Cas9 editing was performed on the banana ripening-associated (*MaACO1*) gene to retard the ripening process. Gene-edited bananas had a twofold longer shelf life, while the flavor, texture, and overall nutrient value remained unchanged. Real-world field trials and market acceptance validations found high acceptance of gene-edited bananas for farms and consumers (consumers/enhanced production/industry sectors affected) in tropical countries that produce and export bananas (Hu et al., 2021).

Gene-Edited Potatoes to Increase Storage Life

When grown for a long time, potatoes convert starch to sugars, resulting in poor-quality tubers and waste, which is undesirable. Researchers chose the potato *StVInv* gene for CRISPR editing; to delay sugar accumulation in stored potatoes ('the word accumulation was found in the previous sentence). The end product was genetically modified potatoes with improved storage quality and reduced sugar content, which in turn produced potentially higher-quality processing potatoes and reduced acrylamide cooking levels. A high level of consumer acceptance has occurred, mainly in Europe and North America, owing to the safety and flavor quality attributes (Danieli et al., 2021).

Lettuce (*Lactuca sativa*) subjected to CRISPR gene editing led to decreased post-harvest browning

Post-harvest browning is a major impediment to the market longevity of lettuce (Lactuca sativa) and its eventual acceptance by consumers. Successful experiments using CRISPR/Cas9 have utilized targeted mutagenesis to silence important browning genes, such as *LsPPO* and *LsPAL*. The gene-edited lettuce showed significant delays in browning and increased shelf life in the marketplace, with 7 to 10 days of additional marketable shelf life, which has implications for food waste in the supply chain. The commercial side of the experiment in both Europe and the United States illustrated the market's acceptance of CRISPR-created lettuce in all its use states owing to improved freshness, reduced food waste, and overall better quality (Guo et al., 2025).

CONCLUSION

Gene-editing technologies, especially CRISPR tools, provide clear real-world benefits for agriculture in terms of crop yield, nutritional quality, pest/disease resistance, abiotic stress resistance, and shelf life. The demonstrated benefits of gene-edited crops, including enhanced yield during times of environmental stressors, enriched nutritional value that helps address deficiencies worldwide, and lowering of pesticide use, identify the potential to be truly disruptive to food security and sustainability worldwide. Landmark examples, including drought-resistant maize, enhanced vitamin content in rice, low-gluten wheat, and disease resistance in potatoes, show both the potential for performance and consumer perception of the acceptance of these novel crops (Figure 3).

The ability to alter a plant's endogenous trait without the addition of foreign DNA has made it feasible for consumers to understand, considering that CRISPR-edited canola and soybean have been commercialized in North America, and maize varieties have been commercialized in Brazil and Argentina. The regulatory framework differentiating gene-edited crops from traditional genetically modified organisms (GMO's) has further supported a positive reception and has been adopted commercially worldwide.

From an environmental perspective, the extra quantitative benefit is reduced chemical input (e.g., pesticides and synthetic fertilizers). Gene editing has been shown to be efficacious in lowering agrochemical inputs through specific improvements in pest/disease resistance traits. Decreased environmental contamination and the benefits of biodiversity advancement, in the name of sustainability, support the need for agriculture to be sustainable and "greener."

From an economic perspective, gene-edited crops provide increased profitability to farmers through increases in crop yield, quality, and marketability. Cost-benefit analyses indicate that by growing either traditional or CRISPR-derived varieties, an economically viable system is created because of the increased performance of crops, lower production costs, and reduction in post-harvest losses. For example, the newly developed shelf life and disease resistance of gene-edited potatoes propelled rapid market development and consumer acceptance.

Despite the apparent benefits of previous developments in gene editing research and technology, some caveats warrant attention to gain the complete utility of gene editing in agriculture. Public perception remains a barrier to acceptance; more Westernized regions that have always embraced biotechnology remain hesitant to embrace gene-edited crops (GECs). Public perception will rely on clear, honest communication about the positives of gene-editing benefits for farmers and the environmental advantages it offers, all while enforcing regulatory policies to separate GECs from GMO.

Separately, summarizing the ethical and social issues connected to gene editing (i.e., social equity concerns about technology use, smallholder farmers in lower income countries, access to technology, and if benefits of developing countries come to those groups), and who will get the advantages is an area of future research and policy development. The potential of international work or committees could address and provide solutions to concerns related to building capacity and technology handing down to achieve inclusive benefits from the knowledge transfer.

Technologies adapted around CRISPR-based tools are starting to proliferate and range from multiplex gene editing, base editing, and prime editing, all of which will increase precision, flexibility, and efficiency. Continued research and development will assess the ability to synergistically improve multiple traits simultaneously, not only optimizing the breeding process but also improving a cultivar's overall resilience to biotic and abiotic stresses at any plant stage. Continued work on gene-editing innovations will not only contribute positively to ensuring food security but also alleviate some of the unpredictability related to climate change.

Figure 3. Key Outcomes of Next-Generation Crop Innovation

REFERENCES

Abdallah, N. A., Hany, E., Abulela, H. A., Thilmony, R., Abdelhadi, A. A., & Elarabi, N. I. (2025). Multiplex CRISPR/Cas9-mediated genome editing to address drought tolerance in wheat. *GM Crops & Food, 16*(1), 1–17. DOI: 10.1080/21645698.2022.2120313 PMID: 36200515

Acevedo-Garcia, J., Spencer, D., Thieron, H., Reinstädler, A., Hammond-Kosack, K., Phillips, A. L., & Panstruga, R. (2017). mlo-based powdery mildew resistance in hexaploid bread wheat generated by a non-transgenic TILLING approach. *Plant Biotechnology Journal, 15*(3), 367–378. DOI: 10.1111/pbi.12631 PMID: 27565953

Ankanagari, S., Lingfa, L., Kavi Kishor, P. B., & Singam, P. (2024). Biofortification of iron and zinc in cereal crops: A genetic engineering and genome editing perspective. In *Harnessing crop biofortification for sustainable agriculture* (pp. 73–99). Springer. DOI: 10.1007/978-981-97-3438-2_5

Anzalone, A. V., Randolph, P. B., Davis, J. R., Sousa, A. A., Koblan, L. W., Levy, J. M., Chen, P. J., Wilson, C., Newby, G. A., Raguram, A., & Liu, D. R. (2019). Search-and-replace genome editing without double-strand breaks or donor DNA. *Nature, 576*(7785), 149–157. DOI: 10.1038/s41586-019-1711-4 PMID: 31634902

Cai, Y., Wang, L., Chen, L., Wu, T., Liu, L., Sun, S., Wu, C., Yao, W., Jiang, B., Yuan, S., Han, T., & Hou, W. (2020). Mutagenesis of GmFT2a and GmFT5a mediated by CRISPR/Cas9 contributes for expanding the regional adaptability of soybean. *Plant Biotechnology Journal, 18*(1), 298–309. DOI: 10.1111/pbi.13199 PMID: 31240772

Chachar, Z., Fan, L., Chachar, S., Ahmed, N., Narejo, M.-N., Ahmed, N., Lai, R., & Qi, Y. (2024). Genetic and genomic pathways to improved wheat (Triticum aestivum L.) yields: A review. *Agronomy (Basel), 14*(6), 1201. DOI: 10.3390/agronomy14061201

Chen, F., Chen, L., Yan, Z., Xu, J., Feng, L., He, N., Guo, M., Zhao, J., Chen, Z., Chen, H., Yao, G., & Liu, C. (2024). Recent advances of CRISPR-based genome editing for enhancing staple crops. *Frontiers in Plant Science, 15*, 1478398. DOI: 10.3389/fpls.2024.1478398 PMID: 39376239

Cheng, Z., Li, K., Liu, H., Wei, X., Yin, T., Xing, X., Han, L., & Sui, Y. (2024). Establishment of a genome-editing system to create fragrant germplasm in sweet sorghum. *aBIOTECH, 5*(4), 502–506. DOI: 10.1007/s42994-024-00180-6 PMID: 39650140

Danieli, R., Assouline, S., Salam, B. B., Teper-Bamnolker, P., Belausov, E., Friedmann, Y., Granot, D., & Eshel, D. (2021). Sugar units, not chilling units, control endodormancy duration in plants: Potato tuber as a case study. *bioRxiv*. DOI: 10.1101/2021.05.10.443356

Dong, O. X., Yu, S., Jain, R., Zhang, N., Duong, P. Q., Butler, C., Li, Y., Lipzen, A., Martin, J. A., Barry, K. W., Schmutz, J., Tian, L., & Ronald, P. C. (2020). Marker-free carotenoid-enriched rice generated through targeted gene insertion using CRISPR-Cas9. *Nature Communications*, *11*(1), 1178. DOI: 10.1038/s41467-020-14981-y PMID: 32132530

Farhat, S., Jain, N., Singh, N., Sreevathsa, R., Dash, P. K., Rai, R., Yadav, S., Kumar, P., Sarkar, A. K., Jain, A., Singh, N. K., & Rai, V. (2019). CRISPR-Cas9 directed genome engineering for enhancing salt stress tolerance in rice. *Seminars in Cell & Developmental Biology*, *96*, 91–99. DOI: 10.1016/j.semcdb.2019.05.003 PMID: 31075379

Fidan, H., Calis, O., Ari, E., Atasayar, A., Sarikaya, P., Tek, M. I., Izmirli, A., Oz, Y., & Firat, G. (2023). Knockout of eIF4E using CRISPR/Cas9 for large-scale production of resistant cucumber cultivar against WMV, ZYMV, and PRSV. *Frontiers in Plant Science*, *14*, 1143813. DOI: 10.3389/fpls.2023.1143813 PMID: 37008503

Gasparis, S., Przyborowski, M., Kała, M., & Nadolska-Orczyk, A. (2019). Knockout of the HvCKX1 or HvCKX3 gene in barley (Hordeum vulgare L.) by RNA-guided Cas9 nuclease affects the regulation of cytokinin metabolism and root morphology. *Cells*, *8*(8), 782. DOI: 10.3390/cells8080782 PMID: 31357516

Guo, M., Tang, Y., Yang, Y., Luo, J., & Gao, J. (2025). Identification and expression analysis of polyphenol oxidase gene family members in response to wound stress in lettuce (Lactuca sativa L.). *Plants*, *14*(6), 972. DOI: 10.3390/plants14060972 PMID: 40265893

Hu, C., Sheng, O., Deng, G., He, W., Dong, T., Yang, Q., Dou, T., Li, C., Gao, H., Liu, S., Yi, G., & Bi, F. (2021). CRISPR/Cas9-mediated genome editing of MaACO1 (aminocyclopropane-1-carboxylate oxidase 1) promotes the shelf life of banana fruit. *Plant Biotechnology Journal*, *19*(4), 654–666. DOI: 10.1111/pbi.13534 PMID: 33369835

Jie, A. S., Mora Gonzalez, N., D'Orso, F., Yue, Y., Nemeth, K., Saalbach, G., Hill, L., de Oliveira Martins, C., Moran, R., Santino, A., & Martin, C. (2022). Biofortified tomatoes provide a new route to vitamin D sufficiency. *Nature Plants*, *8*(6), 611–616. DOI: 10.1038/s41477-022-01154-6 PMID: 35606499

Jinek, M., Chylinski, K., Fonfara, I., Hauer, M., Doudna, J. A., & Charpentier, E. (2012). A programmable dual-RNA-guided DNA endonuclease in adaptive bacterial immunity. *Science*, *337*(6096), 816–821. DOI: 10.1126/science.1225829 PMID: 22745249

Jouanin, A., Gilissen, L. J. W. J., Schaart, J. G., Leigh, F. J., Cockram, J., Wallington, E. J., Boyd, L. A., van den Broeck, H. C., van der Meer, I. M., America, A. H. P., Visser, R. G. F., & Smulders, M. J. M. (2020). CRISPR/Cas9 gene editing of gluten in wheat to reduce gluten content and exposure—Reviewing methods to screen for coeliac safety. *Frontiers in Nutrition*, *7*, 51. DOI: 10.3389/fnut.2020.00051 PMID: 32391373

Kim, J. H., Yu, J., Kim, H. K., Kim, J. Y., Kim, M.-S., Cho, Y.-G., Bae, S., Kang, K. K., & Jung, Y. J. (2022). Genome editing of golden SNP-carrying lycopene epsilon-cyclase (LcyE) gene using the CRISPR-Cas9/HDR and geminiviral replicon system in rice. *International Journal of Molecular Sciences*, *23*(18), 10383. DOI: 10.3390/ijms231810383 PMID: 36142294

Komor, A. C., Kim, Y. B., Packer, M. S., Zuris, J. A., & Liu, D. R. (2016). Programmable editing of a target base in genomic DNA without double-stranded DNA cleavage. *Nature*, *533*(7603), 420–424. DOI: 10.1038/nature17946 PMID: 27096365

Lee, K.-R., Jeon, I., Yu, H., Kim, S.-G., Kim, H.-S., Ahn, S.-J., Lee, J., Lee, S.-K., & Kim, H. U. (2021). Increasing monounsaturated fatty acid contents in hexaploid Camelina sativa seed oil by FAD2 gene knockout using CRISPR-Cas9. *Frontiers in Plant Science*, *12*, 702930. DOI: 10.3389/fpls.2021.702930 PMID: 34267775

Li, Z., Huang, Q., Lin, B., Guo, B., Wang, J., Huang, C., Liao, J., & Zhuo, K. (2022). CRISPR/Cas9-targeted mutagenesis of a representative member of a novel PR10/Bet v1-like protein subfamily significantly reduces rice plant height and defense against Meloidogyne graminicola. *Phytopathology Research*, *4*(1), 38. DOI: 10.1186/s42483-022-00143-z

Ly, D. N. P., Iqbal, S., Fosu-Nyarko, J., Milroy, S., & Jones, M. G. K. (2023). Multiplex CRISPR-Cas9 gene-editing can deliver potato cultivars with reduced browning and acrylamide. *Plants*, *12*(2), 379. DOI: 10.3390/plants12020379 PMID: 36679094

Maioli, A., Gianoglio, S., Moglia, A., Acquadro, A., Valentino, D., Milani, A. M., Prohens, J., Orzaez, D., Granell, A., Lanteri, S., & Comino, C. (2020). Simultaneous CRISPR/Cas9 editing of three PPO genes reduces fruit flesh browning in Solanum melongena L. *Frontiers in Plant Science*, *11*, 607161. DOI: 10.3389/fpls.2020.607161 PMID: 33343607

Ndudzo, A., Makuvise, A. S., Moyo, S., & Bobo, E. D. (2024). CRISPR-Cas9 genome editing in crop breeding for climate change resilience: Implications for smallholder farmers in Africa. *Journal of Agriculture and Food Research, 18*, 101132. DOI: 10.1016/j.jafr.2024.101132

Oliva, R., Ji, C., Atienza-Grande, G., Huguet-Tapia, J. C., Perez-Quintero, A., Li, T., Eom, J.-S., Li, C., Nguyen, H., Liu, B., Auguy, F., Sciallano, C., Luu, V. T., Dossa, G. S., Cunnac, S., Schmidt, S. M., Slamet-Loedin, I. H., Vera Cruz, C., Szurek, B., & Yang, B. (2019). Broad-spectrum resistance to bacterial blight in rice using genome editing. *Nature Biotechnology, 37*(11), 1344–1350. DOI: 10.1038/s41587-019-0267-z PMID: 31659337

Ortega-Salazar, D. C., Sbodio, A. O., Sugiyama, Y., Adaskaveg, A., Wang, D., Seymour, G. B., Li, X., Wang, S. C., & Blanco-Ulate, B. (2024). Double CRISPR knockout of pectin degrading enzymes improves tomato shelf-life while ensuring fruit quality. *Plants, People, Planet, 6*(2), 330–340. DOI: 10.1002/ppp3.10445

Peng, A., Chen, S., Lei, T., Xu, L., He, Y., Wu, L., Yao, L., & Zou, X. (2017). Engineering canker-resistant plants through CRISPR/Cas9-targeted editing of the susceptibility gene CsLOB1 promoter in citrus. *Plant Biotechnology Journal, 15*(12), 1509–1519. DOI: 10.1111/pbi.12733 PMID: 28371200

Perochon, A., Kahla, A., Vranić, M., Jia, J., Malla, K. B., Craze, M., Wallington, E., & Doohan, F. M. (2019). A wheat NAC interacts with an orphan protein and enhances resistance to Fusarium head blight disease. *Plant Biotechnology Journal, 17*(10), 1892–1904. DOI: 10.1111/pbi.13105 PMID: 30821405

Poudel, B. (2024). CRISPR/Cas9 based gene editing in wheat and potato to enhance multiple disease resistance. [Unpublished manuscript].

Rajadurai, G., Varanavasiappan, S., Arul, L., Kokiladevi, E., & Kumar, K. K. (2024). Biofortification of rice: Enhancing nutritional value by genetic manipulation. In *Harnessing crop biofortification for sustainable agriculture* (pp. 101–124). Springer. DOI: 10.1007/978-981-97-3438-2_6

Rasheed, A., Raza, A., Jie, H., Mahmood, A., Ma, Y., Zhao, L., Xing, H., Li, L., Hassan, M. U., Qari, S. H., & Jie, Y. (2022). Molecular tools and their applications in developing salt-tolerant soybean (Glycine max L.) cultivars. *Bioengineering (Basel, Switzerland), 9*(10), 495. DOI: 10.3390/bioengineering9100495 PMID: 36290463

Shi, J., Gao, H., Wang, H., Lafitte, H. R., Archibald, R. L., Yang, M., Hakimi, S. M., Mo, H., & Habben, J. E. (2017). ARGOS8 variants generated by CRISPR-Cas9 improve maize grain yield under field drought stress conditions. *Plant Biotechnology Journal, 15*(2), 207–216. DOI: 10.1111/pbi.12603 PMID: 27442592

Tian, L., Chen, X., Jia, X., Wang, S., Wang, X., Zhang, J., Zhang, Y., Wu, S., Chen, Y., & Wu, L. (2021). First report of antifungal activity conferred by non-conventional peptides. *Plant Biotechnology Journal*, *19*(11), 2147–2149. DOI: 10.1111/pbi.13691 PMID: 34455677

Varotto, S. (2023). Current status and future prospective of genome editing application in maize. In *A roadmap for plant genome editing* (pp. 165–182). Springer Nature.

Wan, D.-Y., Guo, Y., Cheng, Y., Hu, Y., Xiao, S., Wang, Y., & Wen, Y.-Q. (2020). CRISPR/Cas9-mediated mutagenesis of VvMLO3 results in enhanced resistance to powdery mildew in grapevine (Vitis vinifera). *Horticulture Research*, *7*(1), 116. DOI: 10.1038/s41438-020-0339-8 PMID: 32821399

Wang, L., Chen, L., Li, R., Zhao, R., Yang, M., Sheng, J., & Shen, L. (2017). Reduced drought tolerance by CRISPR/Cas9-mediated SlMAPK3 mutagenesis in tomato plants. *Journal of Agricultural and Food Chemistry*, *65*(39), 8674–8682. DOI: 10.1021/acs.jafc.7b02745 PMID: 28873302

Wu, W., Miao, R., Li, Z., & Fang, Z. (2025). CRISPR/Cas9-mediated editing of BADH2 and Wx genes for the development of novel aromatic and soft-textured black and red rice. *Physiologia Plantarum*, *177*(2), e70194. DOI: 10.1111/ppl.70194 PMID: 40171918

Yang, Y., Zhu, K., Li, H., Han, S., Meng, Q., Khan, S. U., Fan, C., Xie, K., & Zhou, Y. (2018). Precise editing of genes in Brassica napus L. regulates multilocular silique development. *Plant Biotechnology Journal*, *16*(7), 1322–1335. DOI: 10.1111/pbi.12872 PMID: 29250878

Yijun, G., Zhiming, X., Jianing, G., Qian, Z., Rasheed, A., Hussain, M. I., Ali, I., Shuheng, Z., Hassan, M. U., Hashem, M., Mostafa, Y. S., Wang, Y., Chen, L., Xiaoxue, W., & Jian, W. (2022). The intervention of classical and molecular breeding approaches to enhance flooding stress tolerance in soybean: A review. *Frontiers in Plant Science*, *13*, 1085368. DOI: 10.3389/fpls.2022.1085368 PMID: 36643298

Zeng, Y., Wen, J., Zhao, W., Wang, Q., & Huang, W. (2020). Rational improvement of rice yield and cold tolerance by editing the three genes OsPIN5b, GS3, and OsMYB30 with the CRISPR-Cas9 system. *Frontiers in Plant Science*, *10*, 1663. DOI: 10.3389/fpls.2019.01663 PMID: 31993066

Zhang, Y., Chen, S., Yang, L., & Zhang, Q. (2023). Application progress of CRISPR/Cas9 genome-editing technology in edible fungi. *Frontiers in Microbiology*, *14*, 1169884. DOI: 10.3389/fmicb.2023.1169884 PMID: 37303782

Zhu, X. Y., Li, Y., Sun, Y., Zhu, Q., & Sun, J. (2018). Highly efficient targeted gene editing in upland cotton using the CRISPR/Cas9 system. *International Journal of Molecular Sciences*, *19*(10), 3000. DOI: 10.3390/ijms19103000 PMID: 30275376

Compilation of References

Abbasi, A., Anas, M., Elahi, M., Khan, A., Khattak, W. A., Saleem, M. H., & Quraishi, U. M. (2025). Restoring Wheat Productivity and Nutrient Balance Under Cadmium Stress Through Reducing Toxicity, Metal Uptake, and Oxidative Damage Using Selenium Nanoparticles. *Journal of Trace Elements in Medicine and Biology : Organ of the Society for Minerals and Trace Elements (GMS), 127644*, 127644. Advance online publication. DOI: 10.1016/j.jtemb.2025.127644

Abdallah, N. A., Hany, E., Abulela, H. A., Thilmony, R., Abdelhadi, A. A., & Elarabi, N. I. (2025). Multiplex CRISPR/Cas9-mediated genome editing to address drought tolerance in wheat. *GM Crops & Food*, *16*(1), 1–17. DOI: 10.1080/21645698.2022.2120313 PMID: 36200515

Abdelaal, K., Alsubeie, M. S., Hafez, Y., Emeran, A., Moghanm, F., Okasha, S., & Ibrahim, M. F. (2022). Physiological and biochemical changes in vegetable and field crops under drought, salinity and weeds stresses: Control strategies and management. *Agriculture*, *12*(12), 2084. DOI: 10.3390/agriculture12122084

Abdelrahman, M., Wei, Z., Rohila, J. S., & Zhao, K. (2021). Multiplex genome-editing technologies for revolutionizing plant biology and crop improvement. *Frontiers in Plant Science*, *12*, 721203. DOI: 10.3389/fpls.2021.721203

Abdul Aziz, M., Brini, F., Rouached, H., & Masmoudi, K. (2022). Genetically engineered crops for sustainably enhanced food production systems. *Frontiers in Plant Science*, *13*, 1027828. DOI: 10.3389/fpls.2022.1027828

Acevedo-Garcia, J., Spencer, D., Thieron, H., Reinstädler, A., Hammond-Kosack, K., Phillips, A. L., & Panstruga, R. (2017). mlo-based powdery mildew resistance in hexaploid bread wheat generated by a non-transgenic TILLING approach. *Plant Biotechnology Journal*, *15*(3), 367–378. DOI: 10.1111/pbi.12631 PMID: 27565953

Adhikari, P., & Poudel, M. (2020). CRISPR-Cas9 in agriculture: Approaches, applications, future perspectives, and associated challenges. *Malaysian Journal of Halal Research*, *3*(1), 6–12. DOI: 10.2478/mjhr-2020-0002

Afzal, S., Sirohi, P., & Singh, N. K. (2020). A review of CRISPR associated genome engineering: Application, advances and future prospects of genome targeting tool for crop improvement. *Biotechnology Letters*, *42*(9), 1611–1632. DOI: 10.1007/s10529-020-02950-w

Aglawe, S. B., Barbadikar, K. M., Mangrauthia, S. K., & Madhav, M. S. (2018). New breeding technique "genome editing" for crop improvement: Applications, potentials and challenges. *3 Biotech*, *8*(8), 336. https://doi.org/DOI: 10.1007/s13205-018-1355-3

Ahmadikhah, A., Zarabizadeh, H., Nayeri, S., & Abbasi, M. S. (2025). Advancements in genome editing tools for genetic studies and crop improvement. *Frontiers in Plant Science*, *15*, 1370675. DOI: 10.3389/fpls.2024.1370675

Ahmad, M. (2023). Plant breeding advancements with "CRISPR-Cas" genome editing technologies will assist future food security. *Frontiers in Plant Science*, *14*, 1133036. DOI: 10.3389/fpls.2023.1133036 PMID: 36993865

Ahmad, N., Rahman, M. U., Mukhtar, Z., Zafar, Y., & Zhang, B. (2020). A critical look on CRISPR-based genome editing in plants. *Journal of Cellular Physiology*, *235*(2), 666–682. DOI: 10.1002/jcp.29052 PMID: 31317541

Ahmar, S., Gill, R. A., Jung, K.-H., Faheem, A., Qasim, M. U., Mubeen, M., & Zhou, W. (2020). Conventional and molecular techniques from simple breeding to speed breeding in crop plants: Recent advances and future outlook. *International Journal of Molecular Sciences*, *21*(7), 2590. DOI: 10.3390/ijms21072590

Ahmar, S., Saeed, S., Khan, M. H. U., Ullah Khan, S., Mora-Poblete, F., Kamran, M., & Saleem, S. (2020). A revolution toward gene-editing technology and its application to crop improvement. *International Journal of Molecular Sciences*, *21*(16), 5665. DOI: 10.3390/ijms21165665

Akram, F., Sahreen, S., Aamir, F., Haq, I., Malik, K., Imtiaz, M., Naseem, W., Nasir, N., & Waheed, H. M. (2023). An insight into modern targeted genome-editing technologies with a special focus on CRISPR/Cas9 and its applications. *Molecular Biotechnology*, *65*(2), 227–242. DOI: 10.1007/s12033-022-00501-4

Ali, Q., Yu, C., Hussain, A., Ali, M., Ahmar, S., Sohail, M. A., Riaz, M., Ashraf, M. F., Abdalmegeed, D., Wang, X., Imran, M., Manghwar, H., & Zhou, L. (2022). Genome engineering technology for durable disease resistance: Recent progress and future outlooks for sustainable agriculture. *Frontiers in Plant Science*, *13*, 1025145. DOI: 10.3389/fpls.2022.860281 PMID: 35371164

Alotaibi, M. (2023). Climate change, its impact on crop production, challenges, and possible solutions. *Notulae Botanicae Horti Agrobotanici Cluj-Napoca*, *51*(1), 13020. DOI: 10.15835/nbha51113020

Altpeter, F., Springer, N. M., Bartley, L. E., Blechl, A. E., Brutnell, T. P., Citovsky, V., Conrad, L., Gelvin, S. B., Jackson, D., Kausch, A. P., Lemaux, P. G., Medford, J. I., Orozo-Cardenas, M., Tricoli, D., VanEck, J., Voytas, D. F., Walbot, V., Wang, K., Zhang, Z. J., & Stewart, C. N.Jr. (2016). Advancing crop transformation in the era of genome editing. *The Plant Cell*, *28*(7), 1510–1520. DOI: 10.1105/tpc.16.00196

Ambrin, G., & Eram, R. (2024). Perspective chapter: Major insights into CRISPR-Cas9 in edible oilseeds research. In *IntechOpen eBooks*. IntechOpen., DOI: 10.5772/intechopen.114967

Amoah, P., Oumarou Mahamane, A. R., Byiringiro, M. H., Mahula, N. J., Manneh, N., Oluwasegun, Y. R., Assfaw, A. T., Mukiti, H. M., Garba, A. D., Chiemeke, F. K., Bernard Ojuederie, O., & Olasanmi, B. (2024). Genome editing in Sub-Saharan Africa: A game-changing strategy for climate change mitigation and sustainable agriculture. *GM Crops & Food*, *15*(1), 233–248. DOI: 10.1080/21645698.2024.2411767 PMID: 39481911

Anas, M., Bibi, S., Farwa, U., Khan, A., & Quraishi, U. M. (2024). Manganese oxide nanoparticles elicit cadmium tolerance in wheat seedlings by ionomic and phenomic adjustment, regulation of AsA–GSH pathway, cellular thickness recovery, and antioxidant modulation. *Environmental Science. Nano*, *11*(11), 4588–4605. DOI: 10.1039/D4EN00490F

Anas, M., Khattak, W. A., Fahad, S., Alrawiq, N., Alrawiq, H. S., Abdelsalam, N. R., Jaremko, M., & Quraishi, U. M. (2025). Mitigating heavy metal pollution in agriculture: A multi-omics and nanotechnology approach to safeguard global wheat production. *Journal of Hazardous Materials Advances*, *17*, 100584. DOI: 10.1016/j.hazadv.2024.100584

Anderson, P. K., Cunningham, A. A., Patel, N. G., Morales, F. J., Epstein, P. R., & Daszak, P. (2004). Emerging infectious diseases of plants: Pathogen pollution, climate change and agrotechnology drivers. *Trends in Ecology & Evolution*, *19*(10), 535–544. DOI: 10.1016/j.tree.2004.07.021 PMID: 16701319

Angon, P. B., & Habiba, U. (2023). Application of the CRISPR/Cas9 gene-editing system and its participation in plant and medical science. *Current Applied Science and Technology, 23*(3). Advance online publication. DOI: 10.55003/cast.2022.03.23.003

Anjali, K., Kumar, S., Korra, T., Thakur, R., Arutselvan, R., Kashyap, A. S., Nehela, Y., Chaplygin, V., Minkina, T., & Keswani, C. (2023). Role of plant secondary metabolites in defence and transcriptional regulation in response to biotic stress. *Plant Stress, 8*, 100154. DOI: 10.1016/j.stress.2023.100154

Ankanagari, S., Lingfa, L., Kavi Kishor, P. B., & Singam, P. (2024). Biofortification of iron and zinc in cereal crops: A genetic engineering and genome editing perspective. In *Harnessing crop biofortification for sustainable agriculture* (pp. 73–99). Springer. DOI: 10.1007/978-981-97-3438-2_5

Ansari, W. A., Chandanshive, S. U., Bhatt, V., Nadaf, A. B., Vats, S., Katara, J. L., Sonah, H., & Deshmukh, R. (2020). Genome editing in cereals: Approaches, applications and challenges. *International Journal of Molecular Sciences, 21*(11), 4040. DOI: 10.3390/ijms21114040

Ansori, A. N., Antonius, Y., Susilo, R. J., Hayaza, S., Kharisma, V. D., Parikesit, A. A., Zainul, R., Jakhmola, V., Saklani, T., Rebezov, M., Ullah, M. E., Maksimiuk, N., Derkho, M., & Burkov, P. (2023). Application of CRISPR-Cas9 genome editing technology in various fields: A review. *Narra J, 3*(2), e184. DOI: 10.52225/narra.v3i2.184

Anyshchenko, A. (2022). Aligning policy design with science to achieve food security: The contribution of genome editing to sustainable agriculture. *Frontiers in Sustainable Food Systems, 6*, 937742. DOI: 10.3389/fsufs.2022.897643

Anzalone, A. V., Randolph, P. B., Davis, J. R., Sousa, A. A., Koblan, L. W., Levy, J. M., Chen, P. J., Wilson, C., Newby, G. A., Raguram, A., & Liu, D. R. (2019). Search-and-replace genome editing without double-strand breaks or donor DNA. *Nature, 576*(7785), 149–157. DOI: 10.1038/s41586-019-1711-4

Arif, M., Haroon, M., Nawaz, A. F., Abbas, H., Xu, R., & Li, L. (2025). Enhancing wheat resilience: Biotechnological advances in combating heat stress and environmental challenges. *Plant Molecular Biology, 115*(2), 41. DOI: 10.1007/s11103-025-01569-7

Arnould, S., Delenda, C., Grizot, S., Desseaux, C., Pâques, F., Silva, G. H., & Smith, J. (2011). The I-CreI meganuclease and its engineered derivatives: Applications from cell modification to gene therapy. *Protein Engineering, Design & Selection : PEDS, 24*(1-2), 27–31. DOI: 10.1093/protein/gzq083 PMID: 21047873

Arora, L., & Narula, A. (2017). Gene editing and crop improvement using CRISPR-Cas9 system. *Frontiers in Plant Science*, *8*, 1932. DOI: 10.3389/fpls.2017.01932

Atia, M., Jiang, W., Sedeek, K., Butt, H., & Mahfouz, M. (2024). Crop bioengineering via gene editing: Reshaping the future of agriculture. *Plant Cell Reports*, *43*(4), 98. DOI: 10.1007/s00299-024-03183-1

Aubert, M. K., Coventry, S., Shirley, N. J., Betts, N. S., Würschum, T., Burton, R. A., & Tucker, M. R. (2018). Differences in hydrolytic enzyme activity accompany natural variation in mature aleurone morphology in barley (Hordeum vulgare L.). *Scientific Reports*, *8*(1), 11025. DOI: 10.1038/s41598-018-29068-4

Awulachew, M. T., & Weldehawaria, A. G. (2024). Food security and safety in light of global megatrends: A review. *Asian J. Biol. Sci*, *17*(4), 666–677.

Ayanoğlu, F. B., Elçin, A. E., & Elçin, Y. M. (2020). Bioethical issues in genome editing by CRISPR-Cas9 technology. *Turkish Journal of Biology = Türk Biyoloji Dergisi*, *44*(2), 110–120. DOI: 10.3906/biy-1912-52

Badia-i-Mompel, P., Wessels, L., Müller-Dott, S., Trimbour, R., Ramirez Flores, R. O., Argelaguet, R., & Saez-Rodriguez, J. (2023). Gene regulatory network inference in the era of single-cell multi-omics. *Nature Reviews. Genetics*, *24*(11), 739–754. DOI: 10.1038/s41576-023-00618-5

Balasubramaniam, T., Wijewardene, I., Hu, R., Shen, G., Zhang, J., & Zhang, H. (2022). Co-overexpression of AVP1, PP2A-C5, and AtCLCc in Arabidopsis thaliana greatly increases tolerance to salt and drought stresses. *Environmental and Experimental Botany*, *200*, 104934. DOI: 10.1016/j.envexpbot.2022.104934

Baloch, F. S., Altaf, M. T., Liaqat, W., Bedir, M., Nadeem, M. A., Cömertpay, G., Çoban, N., Habyarimana, E., Barutçular, C., Cerit, I., Ludidi, N., Karaköy, T., Aasim, M., Chung, Y. S., Nawaz, M. A., Hatipoğlu, R., Kökten, K., & Sun, H.-J. (2023). Recent advancements in the breeding of sorghum crop: Current status and future strategies for marker-assisted breeding. *Frontiers in Genetics*, *14*, 1150616. DOI: 10.3389/fgene.2023.1150616 PMID: 37252661

Baloglu, M. C., Celik Altunoglu, Y., Baloglu, P., Yildiz, A. B., Türkölmez, N., & Özden Çiftçi, Y. (2022). Gene-editing technologies and applications in legumes: Progress, evolution, and future prospects. *Frontiers in Genetics*, *13*, 859437. DOI: 10.3389/fgene.2022.859437

Bandyopadhyay, A., Kancharla, N., Javalkote, V. S., Dasgupta, S., & Brutnell, T. P. (2020). CRISPR-Cas12a (Cpf1): A Versatile Tool in the Plant Genome Editing Tool Box for Agricultural Advancement [Review]. *Volume 11 - 2020.* https://doi.org/DOI: 10.3389/fpls.2020.584151

Barrangou, R., Fremaux, C., Deveau, H., Richards, M., Boyaval, P., Moineau, S., Romero, D. A., & Horvath, P. (2007). CRISPR provides acquired resistance against viruses in prokaryotes. *Science*, *315*(5819), 1709–1712. DOI: 10.1126/science.1138140 PMID: 17379808

Bašinskienė, L., & Šeinauskienė, B. (2021). Gene Editing versus Gene Modification: Awareness, Attitudes and Behavioural Intentions of Lithuanian Consumers, Producers and Farmers. *CET Journal-Chemical Engineering Transactions, 87.*

Bastet, A., Robaglia, C., & Gallois, J. L. (2017). eIF4E resistance: Natural variation should guide gene editing. *Trends in Plant Science*, *22*(5), 411–419. DOI: 10.1016/j.tplants.2017.01.008 PMID: 28258958

Bebber, D. P., & Gurr, S. J. (2015). Crop-destroying fungal and oomycete pathogens challenge food security. *Fungal Genetics and Biology : FG & B*, *74*, 62–64. DOI: 10.1016/j.fgb.2014.10.012 PMID: 25459533

Begna, T. (2021). Conventional Breeding Methods Widely used to Improve Self-Pollinated Crops. *International Journal of Research Studies in Agricultural Sciences*, *7*(1), 2454–6224. DOI: 10.20431/2454-6224.0701001

Begna, T. (2021). Conventional breeding methods widely used to improve self-pollinated crops. *International Journal of Research*, *7*(1), 1–16.

Benavente, E., & Giménez, E. (2021). Modern approaches for the genetic improvement of rice, wheat and maize for abiotic constraints-related traits: A comparative overview. *Agronomy (Basel)*, *11*(2), 376. DOI: 10.3390/agronomy11020376

Benitez-Alfonso, Y., Soanes, B. K., Zimba, S., Sinanaj, B., German, L., Sharma, V., Bohra, A., Kolesnikova, A., Dunn, J. A., Martin, A. C., Khashi u Rahman, M., Saati-Santamaría, Z., García-Fraile, P., Ferreira, E. A., Frazão, L. A., Cowling, W. A., Siddique, K. H. M., Pandey, M. K., Farooq, M., & Foyer, C. H. (2023). Enhancing climate change resilience in agricultural crops. *Current Biology : CB*, *33*(23), R1246–R1261. DOI: 10.1016/j.cub.2023.10.028

Bhagtaney, L., & Sundarrajan, P. (2023). An overview of tools for genome editing: ZFNs, mega nucleases, and TALENs. *CRISPR/Cas-mediated genome editing in plants*, 37-64.

Bhandari, M. K., & Paudel, M. (2024). Genetic, biological and sterile insect techniques: Insect pest management strategies: A review. *Agricultural Reviews (Karnal)*, *45*(3), 390–399. DOI: 10.18805/ag.RF-311

Bhardwaj, A., & Nain, V. (2021). TALENs—an indispensable tool in the era of CRISPR: A mini review. *Journal of Genetic Engineering and Biotechnology*, *19*(1), 125. DOI: 10.1186/s43141-021-00225-z

Bharti, D. R., Pramanik, B., Paul, A., & Debnath, S. (2025). CRISPR-based precision breeding for crop improvement: current achievements and future directions. In *Plant Stress Tolerance* (pp. 118–138). CRC Press. DOI: 10.1201/9781003457237-6

Bhatta, B. P., & Malla, S. (2020). Improving horticultural crops via CRISPR/Cas9: Current successes and prospects. *Plants*, *9*(10), 1360. DOI: 10.3390/plants9101360 PMID: 33066510

Bhatta, M., Sandro, P., Smith, M. R., Delaney, O., Voss-Fels, K. P., Gutierrez, L., & Hickey, L. T. (2021). Need for speed: Manipulating plant growth to accelerate breeding cycles. *Current Opinion in Plant Biology*, *60*, 101986. DOI: 10.1016/j.pbi.2020.101986

Bhavanee, K. N. D., Krishnamoorthi, A., Rathva, H. M., Mareguddikar, S. C., Singh, A., Singh, B. P., Nageshwar, , & Chittibomma, K. (2024). Advancements in genetic engineering for enhanced traits in horticulture crops: A comprehensive review. *Journal of Advances in Biology & Biotechnology*, *27*(4), 90–105. DOI: 10.9734/jabb/2024/v27i2702

Bhavani, S., Singh, P., Qureshi, N., He, X., Biswal, A. K., Juliana, P., . . . Mourad, A. M. (2021). Globally important wheat diseases: status, challenges, breeding and genomic tools to enhance resistance durability. *Genomic designing for biotic stress resistant cereal crops*, 59-128.

Bhushan, K., Pratap, D., & Sharma, P. K. (2016). Transcription activator-like effector nucleases (TALENs): An efficient tool for plant genome editing. *Engineering in Life Sciences*, *16*(4), 330–337. DOI: 10.1002/elsc.201500126

Bibikova, M., Beumer, K., Trautman, J. K., & Carroll, D. (2003). Enhancing gene targeting with designed zinc finger nucleases. *Science*, *300*(5620), 764. DOI: 10.1126/science.1079512

Boch, J., Scholze, H., Schornack, S., Landgraf, A., Hahn, S., Kay, S., Lahaye, T., Nickstadt, A., & Bonas, U. (2009). Breaking the code of DNA binding specificity of TAL-type III effectors. *Science*, *326*(5959), 1509–1512. DOI: 10.1126/science.1178811

Bohra, A., Chand Jha, U., Godwin, I. D., & Kumar Varshney, R. (2020). Genomic interventions for sustainable agriculture. *Plant Biotechnology Journal, 18*(12), 2388–2405. DOI: 10.1111/pbi.13472

Borg, E., & Policante, A. (2024). The Gene Editing Business: Rent Extraction in the Biotech Industry. *Review of Political Economy*, •••, 1–36.

Borrelli, V. M., Brambilla, V., Rogowsky, P., Marocco, A., & Lanubile, A. (2018). The enhancement of plant disease resistance using CRISPR/Cas9 technology. *Frontiers in Plant Science*, *9*, 1245. DOI: 10.3389/fpls.2018.01245 PMID: 30197654

Brookes, G., & Smyth, S. J. (2024). Risk-appropriate regulations for gene-editing technologies. *GM Crops & Food*, *15*(1), 1–14. DOI: 10.1080/21645698.2023.2293510

Brower-Toland, B., Stevens, J. L., Ralston, L., Kosola, K., & Slewinski, T. L. (2024). A crucial role for technology in sustainable agriculture. *ACS Agricultural Science & Technology*, *4*(2), 173–180. DOI: 10.1021/acsagscitech.3c00426

Bullock, D., Wilson, W., & Neadeau, J. (2021). Gene Editing Versus Genetic Modification in the Research and Development of New Crop Traits: An Economic Comparison. *American Journal of Agricultural Economics*, *103*(5), 1700–1719. Advance online publication. DOI: 10.1111/ajae.12201

Burgess, A. J., Masclaux-Daubresse, C., Strittmatter, G., Weber, A. P., Taylor, S. H., Harbinson, J., & Westhoff, P. (2023). Improving crop yield potential: Underlying biological processes and future prospects. *Food and Energy Security*, *12*(1), e435. DOI: 10.1002/fes3.435

Burmistrz, M., Krakowski, K., & Krawczyk-Balska, A. (2020). RNA-targeting CRISPR–Cas systems and their applications. *International Journal of Molecular Sciences*, *21*(3), 1122. DOI: 10.3390/ijms21031122

Cai, Y., Wang, L., Chen, L., Wu, T., Liu, L., Sun, S., Wu, C., Yao, W., Jiang, B., Yuan, S., Han, T., & Hou, W. (2020). Mutagenesis of GmFT2a and GmFT5a mediated by CRISPR/Cas9 contributes for expanding the regional adaptability of soybean. *Plant Biotechnology Journal*, *18*(1), 298–309. DOI: 10.1111/pbi.13199 PMID: 31240772

Camerlengo, F., Frittelli, A., & Pagliarello, R. (2022). CRISPR towards a sustainable agriculture. *Encyclopedia*, *2*(2), 650–664. DOI: 10.3390/encyclopedia2020044

Caradus, J. R. (2023). Processes for regulating genetically modified and gene edited plants. *GM Crops & Food*, *14*(1), 1–41. DOI: 10.1080/21645698.2023.2252947

Carroll, D. (2011). Genome engineering with zinc-finger nucleases. *Genetics*, *188*(4), 773–782. DOI: 10.1534/genetics.111.131433

Cermak, T., Doyle, E. L., Christian, M., Wang, L., Zhang, Y., Schmidt, C., Baller, J. A., Somia, N. V., Bogdanove, A. J., & Voytas, D. F. (2011). Efficient design and assembly of custom TALEN and other TAL effector-based constructs for DNA targeting. *Nucleic Acids Research, 39*(12), e82. DOI: 10.1093/nar/gkr218

Chachar, Z., Fan, L., Chachar, S., Ahmed, N., Narejo, M.-N., Ahmed, N., Lai, R., & Qi, Y. (2024). Genetic and genomic pathways to improved wheat (Triticum aestivum L.) yields: A review. *Agronomy (Basel), 14*(6), 1201. DOI: 10.3390/agronomy14061201

Chachar, Z., Xue, X., Junteng, F., Chen, M., Chen, W., Hui, L. X., & Liang Chen, Z. (2025). Genetic and Molecular Insights into Tiller Development and Approaches for Crop Yield Improvement. *Frontiers in Plant Science*, *16*, 1532180. DOI: 10.3389/fpls.2025.1532180

Chaffai, R., Ganesan, M., & Cherif, A. (2024). Abiotic Stress in Plants: Challenges and Strategies for Enhancing Plant Growth and Development. In *Plant Adaptation to Abiotic Stress: From Signaling Pathways and Microbiomes to Molecular Mechanisms* (pp. 1-30): Springer.

Chao, G., Wang, M., Li, X., Jiang, Z., Zhang, H., & Gao, D. (2023). TaCHP encoding C1-domain protein stably enhances wheat yield in saline-alkaline field. Journal of Integrative Plant Biology.

Chauhan, P. K., Upadhyay, S. K., Tripathi, M., Singh, R., Krishna, D., Singh, S. K., & Dwivedi, P. (2023). Understanding the salinity stress on plant and developing sustainable management strategies mediated salt-tolerant plant growth-promoting rhizobacteria and CRISPR/Cas9. *Biotechnology & Genetic Engineering Reviews*, *39*(2), 311–347. DOI: 10.1080/02648725.2022.2131958

Chavhan, R. L., Jaybhaye, S. G., Hinge, V. R., Deshmukh, A. S., Shaikh, U. S., Jadhav, P. K., Kadam, U. S., & Hong, J. C. (2025). Emerging applications of gene editing technologies for the development of climate-resilient crops. *Frontiers in Genome Editing*, *7*, 1524767. DOI: 10.3389/fgeed.2025.1524767

Chen, B., Gilbert, L. A., Cimini, B. A., Schnitzbauer, J., Zhang, W., Li, G.-W., Park, J., Blackburn, E. H., Weissman, J. S., Qi, L. S., & Huang, B. (2013). Dynamic imaging of genomic loci in living human cells by an optimized CRISPR/Cas system. *Cell*, *155*(7), 1479–1491. DOI: 10.1016/j.cell.2013.12.001

Chen, F., Chen, L., Yan, Z., Xu, J., Feng, L., He, N., & Chen, H. (2024). Recent advances of CRISPR-based genome editing for enhancing staple crops. *Frontiers in Plant Science*, *15*, 1478398. DOI: 10.3389/fpls.2024.1478398

Cheng, Z., Li, K., Liu, H., Wei, X., Yin, T., Xing, X., Han, L., & Sui, Y. (2024). Establishment of a genome-editing system to create fragrant germplasm in sweet sorghum. *aBIOTECH*, *5*(4), 502–506. DOI: 10.1007/s42994-024-00180-6 PMID: 39650140

Chen, J. S., Dagdas, Y. S., Kleinstiver, B. P., Welch, M. M., Sousa, A. A., Harrington, L. B., Sternberg, S. H., Joung, J. K., Yildiz, A., & Doudna, J. A. (2017). Enhanced proofreading governs CRISPR-Cas9 targeting accuracy. *Nature*, *550*(7676), 407–410. DOI: 10.1038/nature24268 PMID: 28931002

Chen, K., Wang, Y., Zhang, R., Zhang, H., & Gao, C. (2019). CRISPR/Cas genome editing and precision plant breeding in agriculture. *Annual Review of Plant Biology*, *70*(1), 667–697. DOI: 10.1146/annurev-arplant-050718-100049 PMID: 30835493

Chen, L., Park, J. E., Paa, P., Rajakumar, P. D., Prekop, H. T., Chew, Y. T., Mannebach, S. M., & Chew, W. L. (2021). Programmable C:G to G:C genome editing with CRISPR-Cas9-directed base excision repair proteins. *Nature Communications*, *12*(1), 1384. DOI: 10.1038/s41467-021-21559-9

Chen, L., Zhang, S., Xue, N., Hong, M., Zhang, X., Zhang, D., Yang, J., Bai, S., Huang, Y., Meng, H., Wu, H., Gao, C., & Li, D. (2023). Engineering a precise adenine base editor with minimal bystander editing. *Nature Chemical Biology*, *19*(1), 101–110. DOI: 10.1038/s41589-022-01163-8

Chen, M., Zhu, X., Liu, X., Wu, C., Yu, C., Hu, G., Chen, L., Chen, R., Bouzayen, M., Zouine, M., & Hao, Y. (2021). Knockout of Auxin Response Factor SlARF4 Improves Tomato Resistance to Water Deficit. *International Journal of Molecular Sciences*, *22*(7), 3347. DOI: 10.3390/ijms22073347

Chen, X., Cui, X., & Ma, H. (2025). The application of artificial intelligence and gene editing technologies in high-yield crop breeding: Innovations and prospects for future agriculture. *Advances in Resources Research*, *5*(1), 146–166.

Choubey, K., Dey, S. K., Dey, G., & Boruah, T. (2025). Understanding the role of phytohormones in mitigation of biotic and abiotic stress: Current status and future prospects. *Proceedings of the National Academy of Sciences. India. Section B, Biological Sciences*, *95*(1), 9–14. DOI: 10.1007/s40011-024-01566-8

Choudhary, V., Kumar, K. K., Sahu, B., Muthappa, S. K., & Das, A. D. A. A. (2021). Agronomic innovations in biotic stress management and its combined effect with abiotic stresses in crop production. *Indian Journal of Agronomy*, *66*(5), 237–257.

Christian, M., Cermak, T., Doyle, E. L., Schmidt, C., Zhang, F., Hummel, A., Bogdanove, A. J., & Voytas, D. F. (2010). Targeting DNA double-strand breaks with TAL effector nucleases. *Genetics, 186*(2), 757–761. DOI: 10.1534/genetics.110.120717

Clapp, J., & Ruder, S. L. (2020). Precision technologies for agriculture: Digital farming, gene-edited crops, and the politics of sustainability. *Global Environmental Politics*, *20*(3), 49–69. DOI: 10.1162/glep_a_00566

Clarke, R., Heler, R., MacDougall, M. S., Yeo, N. C., Chavez, A., Regan, M., Zhang, F., Doudna, J. A., Cate, J. H. D., & Merrill, B. J. (2018). Enhanced bacterial immunity and mammalian genome editing via RNA-polymerase-mediated dislodging of Cas9 from double-strand DNA breaks. *Molecular Cell, 71*(1), 42–55. DOI: 10.1016/j.molcel.2018.06.005

Collard, B. C., Jahufer, M. Z. Z., Brouwer, J. B., & Pang, E. C. K. (2005). An introduction to markers, quantitative trait loci (QTL) mapping and marker-assisted selection for crop improvement: The basic concepts. *Euphytica, 142*(1), 169–196. DOI: 10.1007/s10681-005-1681-5

Cui, L., Zheng, F., Wang, J., Zhang, C., Zhang, D., Gao, S., & Ouyang, B. (2022). The tomato CONSTANS-LIKE protein SlCOL1 regulates fruit yield by repressing SFT gene expression. *BMC Plant Biology*, *22*(1), 429. DOI: 10.1186/s12870-022-03813-4

Cui, Y., Cao, Q., Li, Y., He, M., & Liu, X. (2023). Advances in cis-element- and natural variation-mediated transcriptional regulation and applications in gene editing of major crops. *Journal of Experimental Botany*, *74*(18), 5441–5457. DOI: 10.1093/jxb/erad248

Curtin, S. J., Zhang, F., Sander, J. D., Haun, W. J., Starker, C., Baltes, N. J., Reyon, D., Dahlborg, E. J., Goodwin, M. J., Coffman, A. P., Dobbs, D., Joung, J. K., Voytas, D. F., & Stupar, R. M. (2011). Targeted mutagenesis of duplicated genes in soybean with zinc-finger nucleases. *Plant Physiology*, *156*(2), 466–473. DOI: 10.1104/pp.111.172981

Danieli, R., Assouline, S., Salam, B. B., Teper-Bamnolker, P., Belausov, E., Friedmann, Y., Granot, D., & Eshel, D. (2021). Sugar units, not chilling units, control endodormancy duration in plants: Potato tuber as a case study. *bioRxiv*. DOI: 10.1101/2021.05.10.443356

Das, P., Dutta, A. K., & Das, A. (2025). CRISPR/Cas9 applications for nutritional enhancement in major oilseed, vegetable, and cereal crops: A comprehensive review. *Plant Molecular Biology Reporter*, *43*(4), 1732–1755. DOI: 10.1007/s11105-025-01592-4

Das, S., Ray, M. K., Panday, D., & Mishra, P. K. (2023). Role of biotechnology in creating sustainable agriculture. *PLOS Sustainability and Transformation*, *2*(5), e0000065. DOI: 10.1371/journal.pstr.0000069

de Dieu Habimana, J., Huang, R., Muhoza, B., Kalisa, Y. N., Han, X., Deng, W., & Li, Z. (2022). Mechanistic insights of CRISPR/Cas nucleases for programmable targeting and early-stage diagnosis: A review. *Biosensors & Bioelectronics*, *203*, 114033. DOI: 10.1016/j.bios.2022.114033

de Melo, B. P., Lourenço-Tessutti, I. T., Paixão, J. F. R., Noriega, D. D., Silva, M. C. M., de Almeida-Engler, J., Fontes, E. P. B., & Grossi-de-Sa, M. F. (2020). Transcriptional modulation of AREB-1 by CRISPRa improves plant physiological performance under severe water deficit. *Scientific Reports*, *10*(1), 16231. DOI: 10.1038/s41598-020-72464-y

Devoto, A., Hartmann, H. A., Piffanelli, P., Elliott, C., Simmons, C., Taramino, G., & Panstruga, R. (2003). Molecular phylogeny and evolution of the plant-specific seven-transmembrane MLO family. *Journal of Molecular Evolution*, *56*(1), 77–88. DOI: 10.1007/s00239-002-2382-5 PMID: 12569425

Dhugga, K. S. (2022). Gene editing to accelerate crop breeding. *Frontiers in Plant Science*, *13*, 889995. DOI: 10.3389/fpls.2022.889995

Dibley, K., Jost, M., McIntosh, R., Lagudah, E., & Zhang, P. (2024). The wheat stripe rust resistance gene YrNAM is Yr10. *Nature Communications*, *15*(1), 3291. DOI: 10.1038/s41467-024-47513-z

Dikilitas, M., Simsek, E., & Roychoudhury, A. (2020). Role of proline and glycine betaine in overcoming abiotic stresses. *Protective chemical agents in the amelioration of plant abiotic stress: biochemical and molecular perspectives*, 1-23.

Dima, O., & Inzé, D. (2021). The role of scientists in policy making for more sustainable agriculture. *Current Biology : CB*, *31*(20), R1235–R1238. DOI: 10.1016/j.cub.2021.01.090 PMID: 33689712

Ding, G., Li, Z., Iqbal, Z., Zhao, M., Cui, Z., Cao, L., & Bai, L. (2025). Identifications of genes involved in ABA and MAPK signaling pathways positively regulating cold tolerance in rice. *Plants*, *14*(4), 498. DOI: 10.3390/plants14040498

Dixit, S., Sivalingam, P. N., Baskaran, R. M., Senthil-Kumar, M., & Ghosh, P. K. (2024). Plant responses to concurrent abiotic and biotic stress: Unravelling physiological and morphological mechanisms. *Plant Physiology Reports*, *29*(1), 6–17. DOI: 10.1007/s40502-023-00766-0

Dong, G., & Fan, Z. (2024). CRISPR/Cas-mediated germplasm improvement and new strategies for crop protection. *Crop Health*, *2*(1), 2. DOI: 10.1007/s44297-023-00020-x PMID: 41649714

Dong, O. X., & Ronald, P. C. (2019). Genetic engineering for disease resistance in plants: Recent progress and future perspectives. *Plant Physiology*, *180*(1), 26–38. DOI: 10.1104/pp.18.01224 PMID: 30867331

Dong, O. X., Yu, S., Jain, R., Zhang, N., Duong, P. Q., Butler, C., Li, Y., Lipzen, A., Martin, J. A., Barry, K. W., Schmutz, J., Tian, L., & Ronald, P. C. (2020). Marker-free carotenoid-enriched rice generated through targeted gene insertion using CRISPR-Cas9. *Nature Communications*, *11*(1), 1178. DOI: 10.1038/s41467-020-14981-y PMID: 32132530

Dong, S. S., Wang, X. R., Han, K. K., Tang, J. Q., Lu, W. T., Kong, X. Y., & Zhai, H. (2025). Creation of dual-purpose soybean germplasm for grain and forage by CRISPR/Cas9-mediated targeting mutation of GmFT2a and GmFT5a. *International Journal of Biological Macromolecules*, •••, 142234.

Dormatey, R., Antwi-Boasiako, A., Chimsah, F. A., Kojo, K. Y. K., Mensah, J., Keteku, A. K., & Dapaah, H. K. (2025). Applications of CRISPR in Crop Improvement and Global Food Security. In *CRISPR Technologies-Advances in Genome Editing*. Applications, and Ethical Implications. IntechOpen. DOI: 10.5772/intechopen.1011717

Dormatey, R., Sun, C., Ali, K., Coulter, J. A., Bi, Z., & Bai, J. (2020). Gene pyramiding for sustainable crop improvement against biotic and abiotic stresses. *Agronomy (Basel)*, *10*(9), 1255. DOI: 10.3390/agronomy10091255

Dos Santos, T. B., Ribas, A. F., de Souza, S. G. H., Budzinski, I. G. F., & Domingues, D. S. (2022). Physiological responses to drought, salinity, and heat stress in plants: A review. *Stresses*, *2*(1), 113–135. DOI: 10.3390/stresses2010009

Dossa, K., Diouf, D., Wang, L., Wei, X., Zhang, Y., Niang, M., & Yehouessi, L. W. (2017). The emerging oilseed crop Sesamum indicum enters the "Omics" era. *Frontiers in Plant Science*, *8*, 1154. DOI: 10.3389/fpls.2017.01154

Doudna, J. A., & Charpentier, E. (2014). The new frontier of genome engineering with CRISPR-Cas9. *Science*, *346*(6213), 1258096. DOI: 10.1126/science.1258096

Dracatos, P. M., Haghdoust, R., Singh, D., & Park, R. F. (2018). Exploring and exploiting the boundaries of host specificity using the cereal rust and mildew models. *The New Phytologist*, *218*(2), 453–462. DOI: 10.1111/nph.15044 PMID: 29464724

Durai, S., Mani, M., Kandavelou, K., Wu, J., Porteus, M. H., & Chandrasegaran, S. (2005). Zinc finger nucleases: Custom-designed molecular scissors for genome engineering of plant and mammalian cells. *Nucleic Acids Research*, *33*(18), 5978–5990. DOI: 10.1093/nar/gki912

Ebeling, S., & Gebhard, U. (2022). Ideas, hopes, and fears: What young adults think about genome editing, nature, and society. *Cultural Studies of Science Education*, *17*(3), 745–764. DOI: 10.1007/s11422-022-10102-x

Eckerstorfer, M. F., Dolezel, M., Engelhard, M., Giovannelli, V., Grabowski, M., Heissenberger, A., & Staiano, G. (2023). Recommendations for the assessment of potential environmental effects of genome-editing applications in plants in the EU. *Plants*, *12*(9), 1764. DOI: 10.3390/plants12091764

El-Mounadi, K., Morales-Floriano, M. L., & García-Ruiz, H. (2020). Principles, applications, and biosafety of plant genome editing using CRISPR-Cas9. *Frontiers in Plant Science*, *11*, 56. DOI: 10.3389/fpls.2020.00056 PMID: 32117392

Emebiri, L. C., Tan, M. K., El-Bouhssini, M., Wildman, O., Jighly, A., Tadesse, W., & Ogbonnaya, F. C. (2016). QTL mapping identifies a major locus for resistance in wheat to Sunn pest (*Eurygaster integriceps*) feeding at the vegetative growth stage. *Theoretical and Applied Genetics*, *130*(2), 1–10. DOI: 10.1007/s00122-016-2805-1 PMID: 27744491

Erkes, A., Reschke, M., Boch, J., & Grau, J. (2017). Evolution of Transcription Activator-Like Effectors in Xanthomonas oryzae. *Genome Biology and Evolution*, *9*(6), 1599–1615. DOI: 10.1093/gbe/evx108

Faizal, A., Nugroho, S., Sembada, A. A., Theda, Y., Komariyah, T., & Esyanti, R. R. (2024). Genome editing in future crop protection: Utilizing CRISPR/Cas9 to improve crop resistance against diseases, pests, and weeds. *Discover Agriculture*, *2*(1), 104. DOI: 10.1007/s44279-024-00124-0

Falak, A., Anas, M., Khan, A., Hayat, A., Shaheen, Z., Saleem, M. H., Fahad, S., & Quraishi, U. M. (2025). Efficacy of ascorbic acid coated quantum dots in alleviating lead-induced oxidative damage and enhancing growth parameters in rice (Oryza sativa L.) for sustainable cultivation. *Journal of Trace Elements in Medicine and Biology : Organ of the Society for Minerals and Trace Elements (GMS)*, *88*, 127603. DOI: 10.1016/j.jtemb.2025.127603

Farhat, S., Jain, N., Singh, N., Sreevathsa, R., Dash, P. K., Rai, R., Yadav, S., Kumar, P., Sarkar, A. K., Jain, A., Singh, N. K., & Rai, V. (2019). CRISPR-Cas9 directed genome engineering for enhancing salt stress tolerance in rice. *Seminars in Cell & Developmental Biology*, *96*, 91–99. DOI: 10.1016/j.semcdb.2019.05.003 PMID: 31075379

Farinati, S., Draga, S., Betto, A., Palumbo, F., Vannozzi, A., Lucchin, M., & Barcaccia, G. (2023). Current insights and advances into plant male sterility: New precision breeding technology based on genome editing applications. *Frontiers in Plant Science*, *14*, 1223861. DOI: 10.3389/fpls.2023.1223861

Farooq, M., Khan, A., Hassan, A., & Shah, M. M. (2026). Advances in CRISPR/Cas systems for engineering abiotic stress tolerance in plants: Mechanisms and future prospects. *Planta*, *263*(3), 72. DOI: 10.1007/s00425-026-04942-y PMID: 41665762

Feng, W., Newbigging, A. M., Tao, J., Cao, Y., Peng, H., Le, C., Wu, J., Pang, B., Li, J., Qiu, D. L., Huang, J., Zhang, X., Zhang, X. C., & Le, X. C. (2021). CRISPR technology incorporating amplification strategies: Molecular assays for nucleic acids, proteins, and small molecules. *Chemical Science (Cambridge)*, *12*(13), 4683–4698. DOI: 10.1039/D0SC06973F

Feng, X., Hu, Y., Zhang, W., Xie, R., Guan, H., Xiong, H., & Zheng, D. (2022). Revisiting the role of delta-1-pyrroline-5-carboxylate synthetase in drought-tolerant crop breeding. *The Crop Journal*, *10*(4), 1213–1218. DOI: 10.1016/j.cj.2022.04.002

Fernandes, P. M. B., Favaratto, L., Fernandes, A. A. R., Vicien, C., Capalbo, D. M. F., & Zerbini, F. M. (2022). To become more sustainable organic agriculture needs genome editing technology. *Frontiers in Bioengineering and Biotechnology*, *10*, 860051. DOI: 10.3389/fbioe.2022.912793 PMID: 36061445

Fiaz, S., Xiukang, W., Afifa, Y., Alharthi, B., & Ali, H. (2020). Apomixis and strategies for induce apomixis to preserve hybrid seed vigor for multiple generations. *GM Crops & Food*, *12*(1), 57–70. DOI: 10.1080/21645698.2020.1808423 PMID: 32877304

Fidan, H., Calis, O., Ari, E., Atasayar, A., Sarikaya, P., Tek, M. I., Izmirli, A., Oz, Y., & Firat, G. (2023). Knockout of eIF4E using CRISPR/Cas9 for large-scale production of resistant cucumber cultivar against WMV, ZYMV, and PRSV. *Frontiers in Plant Science*, *14*, 1143813. DOI: 10.3389/fpls.2023.1143813 PMID: 37008503

Fidler, J., Graska, J., Gietler, M., Nykiel, M., Prabucka, B., Rybarczyk-Płońska, A., Muszyńska, E., Morkunas, I., & Labudda, M. (2022). PYR/PYL/RCAR receptors play a vital role in the abscisic-acid-dependent responses of plants to external or internal stimuli. *Cells*, *11*(8), 1352. DOI: 10.3390/cells11081352

Filyushin, M. A., Kochieva, E. Z., & Shchennikova, A. V. (2022). ZmDREB2. 9 gene in maize (Zea mays L.): Genome-wide identification, characterization, expression, and stress response. *Plants*, *11*(22), 3060. DOI: 10.3390/plants11223060

Foley, J. A., Ramankutty, N., Brauman, K. A., Cassidy, E. S., Gerber, J. S., Johnston, M., & Zaks, D. P. (2011). Solutions for a cultivated planet. *Nature*, *478*(7369), 337–342. DOI: 10.1038/nature10452 PMID: 21993620

Fudjoe, S. K., Shi, S., Li, L., Anwar, S., Xie, J., Chimsah, F., & Wang, L. (2025). Gene expression through comparative transcriptome analysis unravels the molecular mechanisms of fertilizer-induced hormonal regulation and leaf senescence in maize for enhanced yield in semiarid regions. *BMC Plant Biology*, *25*(1), 327. DOI: 10.1186/s12870-025-06303-5

Fujita, M., & Hasanuzzaman, M. (2022). *Approaches to enhancing antioxidant defense in plants* (Vol. 11). MDPI. DOI: 10.3390/books978-3-0365-4185-3

Gao, C. (2021). Genome engineering for crop improvement and future agriculture. *Cell*, *184*(6), 1621–1635. DOI: 10.1016/j.cell.2021.01.005

Gasiunas, G., Barrangou, R., Horvath, P., & Siksnys, V. (2012). Cas9–crRNA ribonucleoprotein complex mediates specific DNA cleavage for adaptive immunity in bacteria. *Proceedings of the National Academy of Sciences of the United States of America*, *109*(39), E2579–E2586. DOI: 10.1073/pnas.1208507109

Gasparis, S., Przyborowski, M., Kała, M., & Nadolska-Orczyk, A. (2019). Knockout of the HvCKX1 or HvCKX3 gene in barley (Hordeum vulgare L.) by RNA-guided Cas9 nuclease affects the regulation of cytokinin metabolism and root morphology. *Cells*, *8*(8), 782. DOI: 10.3390/cells8080782 PMID: 31357516

Gatahi, D. M. (2020). Challenges and opportunities in tomato production chain and sustainable standards. *International Journal of Horticultural Science and Technology*, *7*(3), 235–262.

Gaudelli, N. M., Komor, A. C., Rees, H. A., Packer, M. S., Badran, A. H., Bryson, D. I., & Liu, D. R. (2017). Programmable base editing of A•T to G•C in genomic DNA without DNA cleavage. *Nature*, *551*(7681), 464–471. DOI: 10.1038/nature24644

Gerashchenkov, G., Rozhnova, N., Kuluev, B., Kiryanova, O. Y., Gumerova, G., Knyazev, A., & Matniyazov, R. (2020). Design of guide RNA for CRISPR/Cas plant genome editing. *Molecular Biology*, *54*(1), 24–42. DOI: 10.1134/S0026893320010069

Gerelchuluun, A. (2020). DNA Damage, Repair Mechanisms, and Chromosomal Aberrations. *Proton Beam Radiotherapy: Physics and Biology*, 183-208.

Gonzalez Guzman, M., Cellini, F., Fotopoulos, V., Balestrini, R., & Arbona, V. (2022). New approaches to improve crop tolerance to biotic and abiotic stresses. *Physiologia Plantarum, 174*(1), e13547. DOI: 10.1111/ppl.13547

Gordon, D. R., Jaffe, G., Doane, M., Glaser, A., Gremillion, T. M., & Ho, M. D. (2021). Responsible governance of gene editing in agriculture and the environment. *Nature Biotechnology, 39*(9), 1055–1057. DOI: 10.1038/s41587-021-01023-1 PMID: 34381209

Guo, M., Tang, Y., Yang, Y., Luo, J., & Gao, J. (2025). Identification and expression analysis of polyphenol oxidase gene family members in response to wound stress in lettuce (Lactuca sativa L.). *Plants, 14*(6), 972. DOI: 10.3390/plants14060972 PMID: 40265893

Guo, Y., Zhao, G., Gao, X., Zhang, L., Zhang, Y., Cai, X., Yuan, X., & Guo, X. (2023). CRISPR/Cas9 gene editing technology: A precise and efficient tool for crop quality improvement. *Planta, 258*(2), 36. DOI: 10.1007/s00425-023-04187-z

Gupta, R., Gupta, D., Ahmed, K. T., Dey, D., Singh, R., Swarnakar, S., & Ghosh, D. (2021). Modification of Cas9, gRNA and PAM: Key to further regulate genome editing and its applications. *Progress in Molecular Biology and Translational Science, 178*, 85–98. DOI: 10.1016/bs.pmbts.2020.12.001

Gupta, S., Kumar, A., Patel, R., & Kumar, V. (2021). Genetically modified crop regulations: Scope and opportunity using the CRISPR-Cas9 genome editing approach. *Molecular Biology Reports, 48*(5), 4851–4863. DOI: 10.1007/s11033-021-06477-9

Haider, S., Singh, A. P., Panthi, B., Sindhu, S. R., Safa, N. T., Malik, S., & Rahimi, M. (2026). Advances in CRISPR/Cas9 genome editing for crop improvement and global food security. *Current Plant Biology, 100593*, 100593. Advance online publication. DOI: 10.1016/j.cpb.2026.100593

Haldeman, J. M., Conway, A. E., Arlotto, M. E., Slentz, D. H., Muoio, D. M., Becker, T. C., & Newgard, C. B. (2019). Creation of versatile cloning platforms for transgene expression and dCas9-based epigenome editing. *Nucleic Acids Research, 47*(4), e23. DOI: 10.1093/nar/gky1286

Hamdan, M. F., Hensel, G., Alok, A., & Tan, B. C. (2023). Editorial: Genome editing and biotechnological advances for crop improvement and future agriculture. *Frontiers in Plant Science, 14*, 1132821. DOI: 10.3389/fpls.2023.1132821 PMID: 36798708

Hamdan, M. F., & Tan, B. C. (2024). Genetic modification techniques in plant breeding: A comparative review of CRISPR/Cas and GM technologies. *Horticultural Plant Journal*. Advance online publication. DOI: 10.1016/j.hpj.2024.02.012

Han, X., Li, S., Zeng, Q., Sun, P., Wu, D., Wu, J., & Kang, Z. (2025). Genetic engineering, including genome editing, for enhancing broad-spectrum disease resistance in crops. *Plant Communications*, *6*(2), 101195. DOI: 10.1016/j.xplc.2024.101195

Haque, E., Taniguchi, H., Hassan, M. M., Bhowmik, P., Karim, M. R., Śmiech, M., Zhao, K., Rahman, M., & Islam, T. (2018). Application of CRISPR/Cas9 genome editing technology for the improvement of crops cultivated in tropical climates: Recent progress, prospects, and challenges. *Frontiers in Plant Science*, *9*, 617. DOI: 10.3389/fpls.2018.00617

Harfouche, A. L., Petousi, V., Meilan, R., Sweet, J., Twardowski, T., & Altman, A. (2021). Promoting ethically responsible use of agricultural biotechnology. *Trends in Plant Science*, *26*(6), 546–559. DOI: 10.1016/j.tplants.2020.12.015

Hasan, N., Choudhary, S., Naaz, N., Sharma, N., & Laskar, R. A. (2021). Recent advancements in molecular marker-assisted selection and applications in plant breeding programmes. *Journal of Genetic Engineering and Biotechnology*, *19*(1), 128. DOI: 10.1186/s43141-021-00231-1 PMID: 34448979

Hernandes-Lopes, J., Yassitepe, J., Koltun, A., Pauwels, L., Silva, V., Dante, R. A., Gerhardt, I. R., & Arruda, P. (2023). Genome editing in maize: Toward improving complex traits in a global crop. *Genetics and Molecular Biology*, *46*(1, Suppl 1), e20220217. DOI: 10.1590/1678-4685-gmb-2022-0217

Hossain, A., Skalicky, M., Brestic, M., Maitra, S., Ashraful Alam, M., Syed, M. A., & Bhadra, P. (2021). Consequences and mitigation strategies of abiotic stresses in wheat (Triticum aestivum L.) under the changing climate. *Agronomy (Basel)*, *11*(2), 241. DOI: 10.3390/agronomy11020241

Hosseini, S. Y., Mallick, R., Mäkinen, P., & Ylä-Herttuala, S. (2024). Insights into Prime Editing Technology: A Deep Dive into Fundamentals, Potentials, and Challenges. *Human Gene Therapy*, *35*(17-18), 649–668. DOI: 10.1089/hum.2024.043

Huang, L., Chao, Y., Zhao, W., Rao, Y., Shen, H., Gu, Z., & Liu, Q. (2023). Creating high-resistant starch rice by simultaneous editing of SS3a and SS3b. *Plant Biotechnology Journal*, *22*(4), 787–789. DOI: 10.1111/pbi.14053

Huang, W. (2024). The current situation and future of using GWAS strategies to accelerate the improvement of crop stress resistance traits. *Molecular Plant Breeding*, •••, 15. DOI: 10.5376/mpb.2024.15.0007

Huang, Y., Sun, Z., & Zhou, X. (2024). WRKY Transcription Factors in Response to Metal Stress in Plants: A Review. *International Journal of Molecular Sciences*, *25*(20), 10952. DOI: 10.3390/ijms252010952

Hu, B., Wang, W., Chen, J., Liu, Y., & Chu, C. (2023). Genetic improvement toward nitrogen-use efficiency in rice: Lessons and perspectives. *Molecular Plant, 16*(1), 64–74. DOI: 10.1016/j.molp.2022.11.007

Hu, C., Sheng, O., Deng, G., He, W., Dong, T., Yang, Q., Dou, T., Li, C., Gao, H., Liu, S., Yi, G., & Bi, F. (2021). CRISPR/Cas9-mediated genome editing of MaACO1 (aminocyclopropane-1-carboxylate oxidase 1) promotes the shelf life of banana fruit. *Plant Biotechnology Journal, 19*(4), 654–666. DOI: 10.1111/pbi.13534 PMID: 33369835

Idris, S. H., Mat Jalaluddin, N. S., & Chang, L. W. (2023). Ethical and legal implications of gene editing in plant breeding: A systematic literature review. *Journal of Zhejiang University. Science. B, 24*(12), 1093–1105. DOI: 10.1631/jzus.B2200601

Ilyas, M. (2025). Improving resistance of traditional Italian tomato cultivars against biotic stress.

Ishino, Y., Shinagawa, H., Makino, K., Amemura, M., & Nakata, A. (1987). Nucleotide sequence of the iap gene, responsible for alkaline phosphatase isozyme conversion in Escherichia coli, and identification of the gene product. *Journal of Bacteriology, 169*(12), 5429–5433. DOI: 10.1128/jb.169.12.5429-5433.1987

Jabłoński, B., Ogonowska, H., Szala, K., Bajguz, A., Orczyk, W., & Nadolska-Orczyk, A. (2020). Silencing of TaCKX1 Mediates Expression of Other TaCKX Genes to Increase Yield Parameters in Wheat. *International Journal of Molecular Sciences, 21*(13), 4809. Advance online publication. DOI: 10.3390/ijms21134809

Jacob, F., Vernaldi, S., & Markawa, T. (2013). Evolution and conservation of plant NLR functions. *Frontiers in Immunology, 4*, 297. DOI: 10.3389/fimmu.2013.00297 PMID: 24093022

Jiang, L. (2020). Commercialization of the gene-edited crop and morality: Challenges from the liberal patent law and the strict GMO law in the EU. *New Genetics and Society, 39*(2), 191–218. DOI: 10.1080/14636778.2019.1686968

Jinek, M., Chylinski, K., Fonfara, I., Hauer, M., Doudna, J. A., & Charpentier, E. (2012). A programmable dual-RNA-guided DNA endonuclease in adaptive bacterial immunity. *Science, 337*(6096), 816–821. DOI: 10.1126/science.1225829 PMID: 22745249

Ji, Z., Ji, C., Liu, B., Zou, L., Chen, G., & Yang, B. (2016). Interfering TAL effectors of Xanthomonas oryzae neutralize R-gene-mediated plant disease resistance. *Nature Communications, 7*(1), 13435. DOI: 10.1038/ncomms13435

Jogawat, A., Yadav, B., Chhaya, , Lakra, N., Singh, A. K., & Narayan, O. P. (2021). Crosstalk between phytohormones and secondary metabolites in the drought stress tolerance of crop plants: A review. *Physiologia Plantarum*, *172*(2), 1106–1132. DOI: 10.1111/ppl.13328

Jones, J. D., & Dangl, J. L. (2006). The plant immune system. *Nature*, *444*(7117), 323–329. DOI: 10.1038/nature05286 PMID: 17108957

Joshi, A., Yang, S.-Y., Song, H.-G., Min, J., & Lee, J.-H. (2023). Genetic databases and gene editing tools for enhancing crop resistance against abiotic stress. *Biology (Basel)*, *12*(11), 1400. DOI: 10.3390/biology12111400

Jouanin, A., Gilissen, L. J. W. J., Schaart, J. G., Leigh, F. J., Cockram, J., Wallington, E. J., Boyd, L. A., van den Broeck, H. C., van der Meer, I. M., America, A. H. P., Visser, R. G. F., & Smulders, M. J. M. (2020). CRISPR/Cas9 gene editing of gluten in wheat to reduce gluten content and exposure—Reviewing methods to screen for coeliac safety. *Frontiers in Nutrition*, *7*, 51. DOI: 10.3389/fnut.2020.00051 PMID: 32391373

Juma, B. S., Mukami, A., Mweu, C., Ngugi, M. P., & Mbinda, W. (2022). Targeted mutagenesis of the CYP79D1 gene via CRISPR/Cas9-mediated genome editing results in lower levels of cyanide in cassava. *Frontiers in Plant Science*, *13*, 1009860. DOI: 10.3389/fpls.2022.1009860

Jun, T. H., Mian, M. R., & Michel, A. P. (2013). Genetic mapping of three quantitative trait loci for soybean aphid resistance in PI 567324. *Heredity*, *111*(1), 16–22. DOI: 10.1038/hdy.2013.10 PMID: 23486080

Ju, Y., Jia, Y., Cheng, B., Wang, D., Gu, D., Jing, W., Zhang, H., Chen, X., & Li, G. (2024). NRT1.1B mediates rice plant growth and soil microbial diversity under different nitrogen conditions. *AMB Express*, *14*(1), 39. DOI: 10.1186/s13568-024-01683-7

Kabato, W., Getnet, G. T., Sinore, T., Nemeth, A., & Molnár, Z. (2025). Towards climate-smart agriculture: Strategies for sustainable agricultural production, food security, and greenhouse gas reduction. *Agronomy (Basel)*, *15*(3), 565. DOI: 10.3390/agronomy15030565

Kafle, S. (2023). CRISPR/CAS9: A new paradigm for crop improvement revolutionizing agriculture. *Journal of Agriculture and Food Research*, *11*, 100484. DOI: 10.1016/j.jafr.2022.100484

Kaiser, N., Douches, D., Dhingra, A., Glenn, K. C., Herzig, P. R., Stowe, E. C., & Swarup, S. (2020). The role of conventional plant breeding in ensuring safe levels of naturally occurring toxins in food crops. *Trends in Food Science & Technology*, *100*, 51–66. DOI: 10.1016/j.tifs.2020.03.042

Kaniganti, S., Saini, H., Chaitanya, A. K., Hegde, N., Shah, P., Magar, N. D., Rijal, R., Kaushik, J. J., Nanda, D., Sachan, S., Kumar, A., Bhoite, R., & Jamedar, H. R. (2026). CRISPR/Cas Genome Editing and Its Applications in Cereal Crop Improvement. *Plant-Environment Interactions*, *7*(2), e70133. DOI: 10.1002/pei3.70133

Kantor, A., McClements, M. E., & MacLaren, R. E. (2020). CRISPR-Cas9 DNA base-editing and prime-editing. *International Journal of Molecular Sciences*, *21*(17), 6240. DOI: 10.3390/ijms21176240

Karavolias, N. G., Horner, W., & Abugu, M. N.. (2021). Gene editing applications for climate-resilient agriculture and sustainable food systems. *Frontiers in Sustainable Food Systems*, *5*, 685801. DOI: 10.3389/fsufs.2021.685801

Karunarathne, S. D., Han, Y., Zhang, X. Q., & Li, C. (2022). CRISPR/Cas9 gene editing and natural variation analysis demonstrate the potential for HvARE1 in improvement of nitrogen use efficiency in barley. *Journal of Integrative Plant Biology*, *64*(3), 756–770. DOI: 10.1111/jipb.13214

Kashyap, A. S., Manzar, N., Maurya, A., Mishra, D. D., Singh, R. P., & Sharma, P. K. (2023). Development of diagnostic markers and applied for genetic diversity study and population structure of *Bipolaris sorokiniana* associated with leaf blight complex of wheat. *Journal of Fungi (Basel, Switzerland)*, *9*(2), 153. DOI: 10.3390/jof9020153 PMID: 36836268

Kato-Nitta, N., Tachikawa, M., Inagaki, Y., & Maeda, T. (2022). Public Perceptions of Risks and Benefits of Gene-edited Food Crops: An International Comparative Study between the US, Japan, and Germany. *Science, Technology & Human Values*, *48*(6), 1360–1392. DOI: 10.1177/01622439221123830

Kaur, N., Qadir, M., Francis, D. V., Alok, A., Tiwari, S., & Ahmed, Z. F. R. (2025). CRISPR/Cas9: A sustainable technology to enhance climate resilience in major staple crops. *Frontiers in Genome Editing*, *7*, 1533197. DOI: 10.3389/fgeed.2025.1533197 PMID: 40171546

Kaur, N., Sharma, S., Hasanuzzaman, M., & Pati, P. K. (2022). Genome editing: A promising approach for achieving abiotic stress tolerance in plants. *International Journal of Genomics*, *5547231*, 1–12. Advance online publication. DOI: 10.1155/2022/5547231 PMID: 35465040

Kettenburg, A. T., Lopez, M. A., Yogendra, K., Prior, M. J., Rose, T., Bimson, S., Heuer, S., Roy, S. J., & Bailey-Serres, J. (2023). PHOSPHORUS-STARVATION TOLERANCE 1 (OsPSTOL1) is prevalent in upland rice and enhances root growth and hastens low phosphate signaling in wheat. *Plant, Cell & Environment*, *46*(7), 2187–2205. DOI: 10.1111/pce.14588

Khalid, S., Shahid, M. N., Nawaz, K., Naeem, M., & Jabeen, K. (2025). Structure and Mechanism of Transcription Activator-like Effector Nuclease-mediated Genome Editing in Plants. 47–55. https://doi.org/DOI: 10.1079/9781800622517.0004

Khalil, A. M. (2020). The genome editing revolution [review]. *Journal of Genetic Engineering and Biotechnology*, *18*(1), 68. DOI: 10.1186/s43141-020-00078-y PMID: 33123803

Khan, Z., & Shahwar, D. (2020). Role of heat shock proteins (HSPs) and heat stress tolerance in crop plants. *Sustainable agriculture in the era of climate change*, 211-234.

Khan, Z., Ali, Z., Khan, A. A., Sattar, T., Zeshan, A., Saboor, T., & Binyamin, B. (2022). History and classification of CRISPR/Cas system. *The CRISPR/Cas Tool Kit for Genome Editing*, 29-52.

Khan, N., Ray, R. L., Sargani, G. R., Ihtisham, M., Khayyam, M., & Ismail, S. (2021). Current progress and future prospects of agriculture technology: Gateway to sustainable agriculture. *Sustainability (Basel)*, *13*(9), 4883. DOI: 10.3390/su13094883

KhokharVoytas, A., Shahbaz, M., Maqsood, M. F., Zulfiqar, U., Naz, N., Iqbal, U. Z., Sara, M., Aqeel, M., Khalid, N., Noman, A., Zulfiqar, F., Al Syaad, K. M., & AlShaqhaa, M. A.KhokharVoytas. (2023). Genetic modification strategies for enhancing plant resilience to abiotic stresses in the context of climate change. *Functional & Integrative Genomics*, *23*(3), 283. DOI: 10.1007/s10142-023-01202-0

Kim, D., Alptekin, B., & Budak, H. (2018). CRISPR/Cas9 genome editing in wheat. *Functional & Integrative Genomics*, *18*(1), 31–41. DOI: 10.1007/s10142-017-0572-x PMID: 28918562

Kim, J. H., Yu, J., Kim, H. K., Kim, J. Y., Kim, M.-S., Cho, Y.-G., Bae, S., Kang, K. K., & Jung, Y. J. (2022). Genome editing of golden SNP-carrying lycopene epsilon-cyclase (LcyE) gene using the CRISPR-Cas9/HDR and geminiviral replicon system in rice. *International Journal of Molecular Sciences*, *23*(18), 10383. DOI: 10.3390/ijms231810383 PMID: 36142294

Kim, S.-T., Choi, M., Bae, S.-J., & Kim, J.-S. (2021). The functional association of ACQOS/VICTR with salt stress resistance in Arabidopsis thaliana was confirmed by CRISPR-mediated mutagenesis. *International Journal of Molecular Sciences*, *22*(21), 11389. DOI: 10.3390/ijms222111389

Kiran, F. (2022). Proline and Plant Stress Tolerance: Mechanisms, Applications, and Future Prospects. *Plant and Environment*, *3*(01), 23–36.

Kitenge, S. Y., Nkhabindze, B., Ratemo, B. O., Dada, S. A., Ogbimi, E. R., Karikari, B., Nuaila, V. N. A., Toili, M. E. M., Ndudzo, A., Adetunji, C. O., Fotabong, E., & Akinbo, O. (2025). The role of young scientists in promoting genome editing for sustainable agriculture and food systems in Africa. *Frontiers in Sustainable Food Systems*, *9*, 1523546. DOI: 10.3389/fsufs.2025.1530800

Kleinstiver, B. P., Tsai, S. Q., Prew, M. S., Nguyen, N. T., Welch, M. M., Lopez, J. M., McCawley, T. R., Joung, J. K., & Joung, J. K. (2016). Genome-wide specificities of CRISPR-Cas Cpf1 nucleases in human cells. *Nature Biotechnology*, *34*(8), 869–874. DOI: 10.1038/nbt.3620

Köhl, J., Kolnaar, R., & Ravensberg, W. J. (2019). Mode of action of microbial biological control agents against plant diseases: Relevance beyond efficacy. *Frontiers in Plant Science*, *10*, 845. DOI: 10.3389/fpls.2019.00845 PMID: 31379891

Komor, A. C., Kim, Y. B., Packer, M. S., Zuris, J. A., & Liu, D. R. (2016). Programmable editing of a target base in genomic DNA without double-stranded DNA cleavage. *Nature*, *533*(7603), 420–424. DOI: 10.1038/nature17946

Kopczyńska, M., Saha, U., Romanenko, A., Nojima, T., Gdula, M. R., & Kamieniarz-Gdula, K. (2025). Defining gene ends: RNA polymerase II CTD threonine 4 phosphorylation marks transcription termination regions genome-wide. *Nucleic Acids Research*, *53*(2), gkae1240. DOI: 10.1093/nar/gkae1240

Kopecká, R., Kameniarová, M., Černý, M., Brzobohatý, B., & Novák, J. (2023). Abiotic stress in crop production. *International Journal of Molecular Sciences*, *24*(7), 6603. DOI: 10.3390/ijms24076603

Kovalev, M. A., Davletshin, A. I., & Karpov, D. S. (2024). Engineering Cas9: Next generation of genomic editors. *Applied Microbiology and Biotechnology*, *108*(1), 209. DOI: 10.1007/s00253-024-13056-y

Krishna, R., Ansari, W. A., Soumia, P., Yadav, A., Jaiswal, D. K., Kumar, S., Singh, A. K., Singh, M., & Verma, J. P. (2022). Biotechnological interventions in tomato (Solanum lycopersicum) for drought stress tolerance: Achievements and future prospects. *Biotech*, *11*(4), 48. DOI: 10.3390/biotech11040048

Ku, H. K., & Ha, S. H. (2020). Improving nutritional and functional quality by genome editing of crops: Status and perspectives. *Frontiers in Plant Science*, *11*, 577313. DOI: 10.3389/fpls.2020.577313

Kumari, K., Gusain, S., & Joshi, R. (2025). Engineering cold resilience: Implementing gene editing tools for plant cold stress tolerance. *Planta*, *261*(1), 2. DOI: 10.1007/s00425-024-04578-w

Kumari, P., Jasrotia, P., Kumar, D., Kashyap, P. L., Kumar, S., Mishra, C. N., Kumar, S., & Singh, G. P. (2022). Biotechnological approaches for host plant resistance to insect pests. *Frontiers in Genetics*, *13*, 914029. DOI: 10.3389/fgene.2022.914029 PMID: 35719377

Kumari, V. V., Banerjee, P., Verma, V. C., Sukumaran, S., Chandran, M. A. S., Gopinath, K. A., Venkatesh, G., Yadav, S. K., Singh, V. K., & Awasthi, N. K. (2022). Plant nutrition: An effective way to alleviate abiotic stress in agricultural crops. *International Journal of Molecular Sciences*, *23*(15), 8519. DOI: 10.3390/ijms23158519

Kumari, V., Kumawat, P., Yeri, S., & Rajput, S. S. (2024). CRISPR-Cas9: A genome editing tool in crop plants: A review. *Agricultural Reviews (Karnal)*, *45*(2), 329–334.

Kumar, M., Prusty, M. R., Pandey, M. K., Singh, P. K., Bohra, A., Guo, B., & Varshney, R. K. (2023). Application of CRISPR/Cas9-mediated gene editing for abiotic stress management in crop plants. *Frontiers in Plant Science*, *14*, 1157678. DOI: 10.3389/fpls.2023.1157678

Kumar, R., Khungar, L., Shimphrui, R., Tiwari, L. D., Tripathi, G., Sarkar, N. K., Agarwal, S.-K., Agarwal, M., & Grover, A. (2020). AtHsp101 research sets course of action for the genetic improvement of crops against heat stress. *Journal of Plant Biochemistry and Biotechnology*, *29*(4), 715–732. DOI: 10.1007/s13562-020-00624-2

Kumar, R., Shanker, R., Singh, P., Yadav, M. K., Chaudhary, V., & Kumar, M. (2024). Genome editing towards pests and disease management in agricultural crops: Recent developments, challenges and future prospects. *Physiological and Molecular Plant Pathology*, *134*, 102402. DOI: 10.1016/j.pmpp.2024.102402

Kumlehn, J., Pietralla, J., Hensel, G., Pacher, M., & Puchta, H. (2018). The CRISPR/Cas revolution continues: From efficient gene editing for crop breeding to plant synthetic biology. *Journal of Integrative Plant Biology*, *60*(12), 1127–1153. DOI: 10.1111/jipb.12734

Kunwar, S., Iriarte, F., Fan, Q., Evaristo da Silva, E., Ritchie, L., Nguyen, N. S., Freeman, J. H., Stall, R. E., Jones, J. B., Minsavage, G. V., Colee, J., Scott, J. W., Vallad, G. E., Zipfel, C., Horvath, D., Westwood, J., Hutton, S. F., & Paret, M. L. (2018). Transgenic expression of EFR and Bs2 genes for field management of bacterial wilt and bacterial spot of tomato. *Phytopathology*, *108*(12), 1402–1411. DOI: 10.1094/PHYTO-12-17-0424-R PMID: 29923802

Lalotra, S., Mukherjee, T., & Singh, A.. (2024). Unlocking sustainable agriculture: The crucial role of plant genetic engineering. In *Sustainable Agriculture and Biotechnology*. Springer. DOI: 10.1515/9783111369754-007

Lamichhane, S., & Thapa, S. (2022). Advances from conventional to modern plant breeding methodologies. *Plant Breeding and Biotechnology*, *10*(1), 1–4. DOI: 10.9787/PBB.2022.10.1.1

Lang, B. J., Guerrero, M. E., Prince, T. L., Okusha, Y., Bonorino, C., & Calderwood, S. K. (2021). The functions and regulation of heat shock proteins; key orchestrators of proteostasis and the heat shock response. *Archives of Toxicology*, *95*(6), 1943–1970. DOI: 10.1007/s00204-021-03070-8

LeBlanc, N., Salgado-Salazar, C., & Crouch, J. A. (2018). Boxwood blight: An ongoing threat to ornamental and native boxwood. *Applied Microbiology and Biotechnology*, *102*(10), 4371–4380. DOI: 10.1007/s00253-018-8936-2 PMID: 29610965

Ledford, H. (2022). Gene-edited tomatoes could provide new source of vitamin D. *Nature*. Advance online publication. DOI: 10.1038/d41586-022-01443-2

Lee, H. K., Oh, Y., Hong, J., Lee, S. H., & Hur, J. K. (2021). Development of CRISPR technology for precise single-base genome editing: A brief review. *BMB Reports*, *54*(2), 98–105. DOI: 10.5483/BMBRep.2021.54.2.217

Lee, J. K., Jeong, E., Lee, J., Jung, M., Shin, E., Kim, Y. H., Lee, K., Jung, I., Kim, D., Kim, S., & Kim, J. S. (2018). Directed evolution of CRISPR-Cas9 to increase its specificity. *Nature Communications*, *9*(1), 3048. DOI: 10.1038/s41467-018-05477-x PMID: 30082838

Lee, K.-R., Jeon, I., Yu, H., Kim, S.-G., Kim, H.-S., Ahn, S.-J., Lee, J., Lee, S.-K., & Kim, H. U. (2021). Increasing monounsaturated fatty acid contents in hexaploid Camelina sativa seed oil by FAD2 gene knockout using CRISPR-Cas9. *Frontiers in Plant Science*, *12*, 702930. DOI: 10.3389/fpls.2021.702930 PMID: 34267775

Lee, M. (2023). Deep learning in CRISPR-Cas systems: A review of recent studies. *Frontiers in Bioengineering and Biotechnology*, *11*, 1226182. DOI: 10.3389/fbioe.2023.1226182

Liang, Y., Zhang, Q., Wang, X., Gou, S., Lin, L., & Lai, L. (2022). AGBE: A dual deaminase-mediated base editor by fusing CGBE with ABE for creating a saturated mutant population with multiple editing patterns. *Nucleic Acids Research*, *50*(9), 5384–5399. DOI: 10.1093/nar/gkac353

Liao, Z., Chen, B., Boubakri, H., Farooq, M., Mur, L. A. J., Urano, D., & Aslam, M. M. (2025). The regulatory role of phytohormones in plant drought tolerance. *Planta*, *261*(5), 1–22. DOI: 10.1007/s00425-025-04671-8

Li, B., Sun, C., Li, J., & Gao, C. (2024). Targeted genome-modification tools and their advanced applications in crop breeding. *Nature Reviews. Genetics*, *25*(9), 603–622. DOI: 10.1038/s41576-024-00720-2

Li, C., Chen, M., Chao, S., Yu, J., & Bai, G. (2013). Identification of a novel gene, H34, in wheat using recombinant inbred lines and single nucleotide polymorphism markers. *Theoretical and Applied Genetics*, *126*(8), 2065–2071. DOI: 10.1007/s00122-013-2118-5 PMID: 23689741

Li, D., Liu, X., Wang, H., Yuan, S., & Jiang, L. (2024). The application and prospects of gene editing technology in enhancing crop yields: A review. *Advances in Resources Research*, *4*(4), 654–667.

Li, H., Wang, W., Yang, X., Kang, G., Zhang, Z., & Wu, K. (2024). Toxic effects of Bt-(Cry1Ab+ Vip3Aa) maize ("DBN3601T"event) on the Asian corn borer Ostrinia furnacalis (Guenée) in Southwestern China. *Agronomy (Basel)*, *14*(9), 1906. DOI: 10.3390/agronomy14091906

Li, H., Yang, Y., Hong, W., Huang, M., Wu, M., & Zhao, X. (2020). Applications of genome editing technology in the targeted therapy of human diseases: Mechanisms, advances and prospects. *Signal Transduction and Targeted Therapy*, *5*(1), 1. DOI: 10.1038/s41392-019-0089-y

Li, J., Scarano, A., Gonzalez, N. M., D'Orso, F., Yue, Y., Nemeth, K., Saalbach, G., Hill, L., de Oliveira Martins, C., Moran, R., Santino, A., & Martin, C. (2022). Biofortified tomatoes provide a new route to vitamin D sufficiency. *Nature Plants*, *8*(6), 611–616. DOI: 10.1038/s41477-022-01154-6

Limbalkar, O. M., Srivastava, P., Reddy, K. R., Lali, S., Lal, S. K., Bishi, S. K., Soren, K. R., Bhadana, V. P., Rakshit, S., & Tribhuvan, K. U. (2025). Genome editing and its impact on crop improvement: Current approaches and future prospects. *Discover Plants*, *2*(1), 358. DOI: 10.1007/s44372-025-00410-1

Limenie, A. D., & Alehegn, M. (2025). Genetic Engineering for Cereal Crop Yield Improvement and Disease Resistant Breeding. *TheScientificWorldJournal, 2025*(1), 6743917. DOI: 10.1155/tswj/6743917

Li, P., Li, X., & Jiang, M. (2021). CRISPR/Cas9-mediated mutagenesis of WRKY3 and WRKY4 function decreases salt and Me-JA stress tolerance in Arabidopsis thaliana. *Molecular Biology Reports, 48*(8), 5821–5832. DOI: 10.1007/s11033-021-06541-4

Li, S., Lin, D., Zhang, Y., Deng, M., Chen, Y., Lv, B., & Zhao, L. (2022). Genome-edited powdery mildew resistance in wheat without growth penalties. *Nature, 602*(7897), 455–460. DOI: 10.1038/s41586-022-04395-9

Li, T., Liu, B., Spalding, M. H., Weeks, D. P., & Yang, B. (2012). High-efficiency TALEN-based gene editing produces disease-resistant rice. *Nature Biotechnology, 30*(5), 390–392. DOI: 10.1038/nbt.2199 PMID: 22565958

Liu, Q. Cloning and characterizing candidate genes underlying major QTLs for wheat stem pith thickness and stem diameter (Doctoral dissertation, Murdoch University).

Liu, D. R., Komor, A. C., & Kim, Y. B. (2019). Development of a broadly applicable gene-editing tool by fusion of Cas9 with high-fidelity cytidine deaminase. *Nature.*

Liu, D., He, J., Li, Q., Zhang, X., Wang, Y., Sun, Q., & Wan, J. (2025). A WRKY transcription factor confers broad-spectrum resistance to biotic stresses and yield stability in rice. *Proceedings of the National Academy of Sciences of the United States of America, 122*(10), e2411164122. DOI: 10.1073/pnas.2411164122

Liu, H., Chen, W., Li, Y., Sun, L., Chai, Y., Chen, H., Nie, H., & Huang, C. (2022). CRISPR/Cas9 technology and its utility for crop improvement. *International Journal of Molecular Sciences, 23*(18), 10442. DOI: 10.3390/ijms231810442 PMID: 36142353

Liu, J., Liu, M., Wang, J., Zhang, J., Miao, H., Wang, Z., & Jin, Z. (2021). Transcription factor MaMADS36 plays a central role in regulating banana fruit ripening. *Journal of Experimental Botany, 72*(20), 7078–7091. DOI: 10.1093/jxb/erab341

Liu, J., Zhang, J., Kim, S. H., Lee, H.-S., Marinoia, E., & Song, W.-Y. (2021). Characterization of Brassica rapa metallothionein and phytochelatin synthase genes potentially involved in heavy metal detoxification. *PLoS One, 16*(6), e0252899. DOI: 10.1371/journal.pone.0252899

Li, Z., Huang, Q., Lin, B., Guo, B., Wang, J., Huang, C., Liao, J., & Zhuo, K. (2022). CRISPR/Cas9-targeted mutagenesis of a representative member of a novel PR10/ Bet v1-like protein subfamily significantly reduces rice plant height and defense against Meloidogyne graminicola. *Phytopathology Research*, *4*(1), 38. DOI: 10.1186/ s42483-022-00143-z

Lupi, A. C. D., Lira, B. S., Gramegna, G., Trench, B., Alves, F. R. R., Demarco, D., Peres, L. E. P., Purgatto, E., Freschi, L., & Rossi, M. (2019). Solanum lycopersicum GOLDEN 2-LIKE 2 transcription factor affects fruit quality in a light-and auxin-dependent manner. *PLoS One*, *14*(2), e0212224. DOI: 10.1371/journal.pone.0212224

Lu, Q., Zhao, H., Zhang, Z., Bai, Y., Zhao, H., Liu, G., & Song, W. (2024). Genomic variation in weedy and cultivated broomcorn millet accessions uncovers the genetic architecture of agronomic traits. *Nature Genetics*, *56*(5), 1006–1017. DOI: 10.1038/ s41588-024-01718-6

Lusser, M., Parisi, C., Plan, D., & Rodriguez-Cerezo, E. (2011). New plant breeding techniques. In *State-of-the-art and prospects for commercial development* (pp. 1–220). Joint Research Centre-Institute for Prospective Technological Studies, Publications Office of the European Union., DOI: 10.2791/54761

Ly, D. N. P., Iqbal, S., Fosu-Nyarko, J., Milroy, S., & Jones, M. G. K. (2023). Multiplex CRISPR-Cas9 gene-editing can deliver potato cultivars with reduced browning and acrylamide. *Plants*, *12*(2), 379. DOI: 10.3390/plants12020379 PMID: 36679094

Mafakheri, M., Kordrostami, M., & Al-Khayri, J. M. (2021). Abiotic stress in plants: socio-economic consequences and crops responses. *Nanobiotechnology: Mitigation of Abiotic Stress in Plants*, 1-28.

Magdy, M., Mostofa, M. G., Rahimi, M., & Abd El Moneim, D. (2023). Abiotic stress alleviation in plants: morpho-physiological and molecular aspects. In (Vol. 14, pp. 1295638): Frontiers Media SA.

Maharajan, T., Roch, G. V., & Ceasar, S. A. (2021). Recent advancements of molecular breeding and functional genomics for improving nitrogen-, phosphorus-and potassium-use efficiencies in wheat. In *Molecular breeding in wheat, maize and sorghum: Strategies for improving abiotic stress tolerance and yield* (pp. 170–196). CABI Wallingford UK. DOI: 10.1079/9781789245431.0009

Mahboob, W., Yang, G., & Irfan, M. (2023). Crop nitrogen (N) utilization mechanism and strategies to improve N use efficiency. *Acta Physiologiae Plantarum*, *45*(4), 52. DOI: 10.1007/s11738-023-03527-6

Maioli, A., Gianoglio, S., Moglia, A., Acquadro, A., Valentino, D., Milani, A. M., Prohens, J., Orzaez, D., Granell, A., Lanteri, S., & Comino, C. (2020). Simultaneous CRISPR/Cas9 editing of three PPO genes reduces fruit flesh browning in Solanum melongena L. *Frontiers in Plant Science*, *11*, 607161. DOI: 10.3389/fpls.2020.607161 PMID: 33343607

Malakar, P., & Chattopadhyay, D. (2021). Adaptation of plants to salt stress: The role of the ion transporters. *Journal of Plant Biochemistry and Biotechnology*, *30*(4), 668–683. DOI: 10.1007/s13562-021-00741-6

Malhi, G. S., Kaur, M., & Kaushik, P. (2021). Impact of climate change on agriculture and its mitigation strategies: A review. *Sustainability (Basel)*, *13*(3), 1318. DOI: 10.3390/su13031318

Mali, P., Yang, L., Esvelt, K. M., Aach, J., Guell, M., DiCarlo, J. E., Norville, J. E., & Church, G. M. (2013). RNA-guided human genome engineering via Cas9. *Science*, *339*(6121), 823–826. DOI: 10.1126/science.1232033

Mali, S., Dutta, M., & Zinta, G. (2023). Genome editing advancements in potato (Solanum tuberosum L.): Operational challenges and solutions. *Journal of Plant Biochemistry and Biotechnology*, *32*(4), 730–742. DOI: 10.1007/s13562-022-00812-2

Manghwar, H., Li, B., Ding, X., Hussain, A., Lindsey, K., Zhang, X., & Jin, S. (2020). CRISPR/Cas systems in genome editing: Methodologies and tools for sgRNA design, off-target evaluation, and strategies to mitigate off-target effects. *Advanced Science (Weinheim, Baden-Württemberg, Germany)*, *7*(6), 1902312. DOI: 10.1002/advs.201902312

Manghwar, H., Lindsey, K., Zhang, X. L., & Jin, S. X. (2019). CRISPR/Cas system: Recent advances and future prospects for genome editing. *Trends in Plant Science*, *24*(12), 1102–1125. DOI: 10.1016/j.tplants.2019.09.006 PMID: 31727474

Manna, M., Thakur, T., Chirom, O., Mandlik, R., Deshmukh, R., & Salvi, P. (2021). Transcription factors as key molecular target to strengthen the drought stress tolerance in plants. *Physiologia Plantarum*, *172*(2), 847–868. DOI: 10.1111/ppl.13268

Manzar, N., Kashyap, A. S., Maurya, A., Rajawat, M. V. S., Sharma, P. K., Srivastava, A. K., Roy, M., Saxena, A. K., & Singh, H. V. (2022). Multi-gene phylogenetic approach for identification and diversity analysis of *Bipolaris maydis* and *Curvularia lunata* isolates causing foliar blight of *Zea mays*. *Journal of Fungi (Basel, Switzerland)*, *8*(8), 802. DOI: 10.3390/jof8080802 PMID: 36012790

Manzoor, S., Nabi, S. U., Rather, T. R., Gani, G., Mir, Z. A., Wani, A. W., Ali, S., Tyagi, A., & Manzar, N. (2024). Advancing crop disease resistance through genome editing: A promising approach for enhancing agricultural production. *Frontiers in Genome Editing*, *6*, 1399051. DOI: 10.3389/fgeed.2024.1399051

Maqbool, M. A., Aslam, M., Beshir, A., & Khan, M. S. (2018). Breeding for provitamin A biofortification of maize (Zea mays L.). *Plant Breeding*, *137*(4), 451–469. DOI: 10.1111/pbr.12618

Marone, D., Mastrangelo, A. M., & Borrelli, G. M. (2023). From transgenesis to genome editing in crop improvement: Applications, marketing, and legal issues. *International Journal of Molecular Sciences*, *24*(8), 7122. DOI: 10.3390/ijms24087122

Marone, D., Russo, M. A., Mores, A., Ficco, D. B., Laidò, G., Mastrangelo, A. M., & Borrelli, G. M. (2021). Importance of landraces in cereal breeding for stress tolerance. *Plants*, *10*(7), 1267. DOI: 10.3390/plants10071267

Martínez-Fortún, J., Phillips, D. W., & Jones, H. D. (2017). Potential impact of genome editing in world agriculture. *Emerging Topics in Life Sciences*, *1*(2), 117–133. DOI: 10.1042/ETLS20170010 PMID: 33525764

Martínez-Fortún, J., Phillips, D. W., & Jones, H. D. (2022). Natural and artificial sources of genetic variation used in crop breeding: A baseline comparator for genome editing. *Frontiers in Genome Editing*, *4*, 937853. DOI: 10.3389/fgeed.2022.937853

Mattiello, L., Rütgers, M., Sua-Rojas, M. F., Tavares, R., Soares, J. S., Begcy, K., & Menossi, M. (2022). Molecular and computational strategies to increase the efficiency of CRISPR-based techniques. *Frontiers in Plant Science*, *13*, 868027. DOI: 10.3389/fpls.2022.868027 PMID: 35712599

Ma, X., Zhu, Q., Chen, Y., & Liu, Y.-G. (2016). CRISPR/Cas9 platforms for genome editing in plants: Developments and applications. *Molecular Plant*, *9*(7), 961–974. DOI: 10.1016/j.molp.2016.04.009

Ma, Y., Tang, M., Wang, M., Yu, Y., & Ruan, B. (2024). Advances in Understanding Drought Stress Responses in Rice: Molecular Mechanisms of ABA Signaling and Breeding Prospects. *Genes*, *15*(12), 1529. DOI: 10.3390/genes15121529

Mearaji, H. S., Ansari, A., Igdelou, N. K. M., Lajayer, B. A., & Pessarakli, M. (2021). Phytohormones and abiotic stresses: Roles of phytohormones in plants under abiotic stresses. In *Handbook of plant and crop physiology* (pp. 175–213). CRC Press. DOI: 10.1201/9781003093640-16

Medina-Puche, L., & Lozano-Durán, R. (2022). Plasma membrane-to-organelle communication in plant stress signaling. *Current Opinion in Plant Biology*, *69*, 102269. DOI: 10.1016/j.pbi.2022.102269

Mehta, S., Lal, S. K., Sahu, K. P., Venkatapuram, A. K., ... (2020). CRISPR/Cas9-edited rice: A new frontier for sustainable agriculture. In *CRISPR Crops for Durable Agriculture* (pp. 149–171). Springer.

Mei, F., Chen, B., Du, L., Li, S., Zhu, D., Chen, N., & Cheng, X. (2022). A gain-of-function allele of a DREB transcription factor gene ameliorates drought tolerance in wheat. *The Plant Cell*, *34*(11), 4472–4494. DOI: 10.1093/plcell/koac248

Mešić, A., Jurić, M., Donsì, F., Maslov Bandić, L., & Jurić, S. (2024). Advancing climate resilience: Technological innovations in plant-based, alternative and sustainable food production systems. *Discover Sustainability*, *5*(1), 423. DOI: 10.1007/s43621-024-00581-z

Mestanza, M., Hernández-Amasifuen, A. D., Pineda-Lázaro, A. J., Eriksson, D., & Guerrero-Abad, J. C. (2025). Genome editing for sustainable agriculture in Peru: Advances, potential applications and regulation. *Frontiers in Genome Editing*, *7*, 1574488. DOI: 10.3389/fgeed.2025.1611040 PMID: 40661867

Metje-Sprink, J., Sprink, T., & Hartung, F. (2020). Genome-edited plants in the field. *Current Opinion in Biotechnology*, *61*, 1–6. DOI: 10.1016/j.copbio.2019.08.007

Miedaner, T. (2016). Breeding strategies for improving plant resistance to diseases. In *Advances in plant breeding strategies: agronomic, abiotic and biotic stress traits* (pp. 561–599). Springer., DOI: 10.1007/978-3-319-22518-0_15

Mikhaylova, E., Kuluev, B., Gerashchenkov, G., Chemeris, D., Garafutdinov, R., Kuluev, A., Baymiev, A. K., Baymiev, A. K., & Chemeris, A. (2024). Prime-editing methods and pegRNA design programs. *Molecular Biology*, *58*(1), 17–32. DOI: 10.1134/S0026893324010084

Miller, J. C., Tan, S., Qiao, G., Barlow, K. A., Wang, J., Xia, D. F., & Hinkley, S. J. (2011). A TALE nuclease architecture for efficient genome editing. *Nature Biotechnology*, *29*(2), 143–148. DOI: 10.1038/nbt.1755

Mishra, R., & Zhao, K. (2018). Genome editing technologies and their applications in crop improvement. *Plant Biotechnology Reports*, *12*(2), 57–68. DOI: 10.1007/s11816-018-0472-0

Mohamed, H. I., Khan, A., & Basıt, A. (2024). CRISPR-Cas9 system mediated genome editing technology: An ultimate tool to enhance abiotic stress in crop plants. *Journal of Soil Science and Plant Nutrition*, *24*(2), 1799–1822. DOI: 10.1007/s42729-024-01778-x

Molla, K. A., Qi, Y., Karmakar, S., & Baig, M. J. (2020). Base editing landscape extends to perform transversion mutation. *Trends in Genetics : TIG*, *36*(12), 899–901. DOI: 10.1016/j.tig.2020.09.001

Molla, K. A., Sretenovic, S., Bansal, K. C., & Qi, Y. (2021). Precise plant genome editing using base editors and prime editors. *Nature Plants*, *7*(9), 1166–1187. DOI: 10.1038/s41477-021-00991-1 PMID: 34518669

Molla, K. A., & Yang, Y. (2020). Predicting CRISPR/Cas9-induced mutations for precise genome editing. *Trends in Biotechnology*, *38*(2), 136–141. DOI: 10.1016/j.tibtech.2019.08.002 PMID: 31526571

Mondal, S., Karmakar, S., Panda, D., Pramanik, K., Bose, B., & Singhal, R. K. (2023). Crucial plant processes under heat stress and tolerance through heat shock proteins. *Plant Stress*, *10*, 100227. DOI: 10.1016/j.stress.2023.100227

Moore, R., Chandrahas, A., & Bleris, L. (2014). Transcription activator-like effectors: A toolkit for synthetic biology. *ACS Synthetic Biology*, *3*(10), 708–716. DOI: 10.1021/sb400137b

Morales, F., Ancín, M., Fakhet, D., González-Torralba, J., Gámez, A. L., Seminario, A., Soba, D., Ben Mariem, S., Garriga, M., & Aranjuelo, I. (2020). Photosynthetic metabolism under stressful growth conditions as a bases for crop breeding and yield improvement. *Plants*, *9*(1), 88. DOI: 10.3390/plants9010088

Moscou, M. J., & Bogdanove, A. J. (2009). A simple cipher governs DNA recognition by TAL effectors. *Science*, *326*(5959), 1501. DOI: 10.1126/science.1178817

Movahedi, A., & Yang, L. (2025). CRISPR-Cas genome editing in plants: Revolutionizing precision agriculture and plant biotechnology. *Plant Innovation Journal*, *1*(1), 1–7. DOI: 10.62762/PIJ.2025.263896

Muhie, S. H. (2022). Novel approaches and practices to sustainable agriculture. *Journal of Agriculture and Food Research*, *10*, 100446. DOI: 10.1016/j.jafr.2022.100446

Munir, A., Salah, M. A., Ali, M., Ali, B., Saleem, M. H., Samarasinghe, K. G. B. A., De Silva, S. I. S., Ercisli, S., Iqbal, N., & Anas, M. (2024). Advancing Agriculture: Harnessing Smart Nanoparticles for Precision Fertilization. *BioNanoScience*, *14*(4), 3846–3863. DOI: 10.1007/s12668-024-01597-5

Mushtaq, M., Ahmad Dar, A., Skalicky, M., Tyagi, A., Bhagat, N., Basu, U., & Dar, T.-U.-H. (2021). CRISPR-based genome editing tools: Insights into technological breakthroughs and future challenges. *Genes*, *12*(6), 797. DOI: 10.3390/genes12060797

Nadakuduti, S. S., & Enciso-Rodríguez, F. (2021). Advances in genome editing with CRISPR systems and transformation technologies for plant DNA manipulation. *Frontiers in Plant Science*, *11*, 637159. DOI: 10.3389/fpls.2020.637159

Nadeem, M. A., Nawaz, M. A., Shahid, M. Q., Doğan, Y., Comertpay, G., Yıldız, M., Hatipoğlu, R., Ahmad, F., Alsaleh, A., Labhane, N., Özkan, H., Chung, G., & Baloch, F. S. (2018). DNA molecular markers in plant breeding: Current status and recent advancements in genomic selection and genome editing. *Biotechnology, Biotechnological Equipment*, *32*(2), 261–285. DOI: 10.1080/13102818.2017.1400401

Nagaraj, S., Rajasekaran, R., Palaniappan, J., Rangasamy, S., Narayanasamy, C., & Narayanan, M. B. (2024). Emerging technological developments to address pest resistance in Bt cotton. *Journal of Cotton Research*, *7*(1), 30. DOI: 10.1186/s42397-024-00192-z

Nambiar, T. S., Baudrier, L., Billon, P., & Ciccia, A. (2022). CRISPR-based genome editing through the lens of DNA repair. *Molecular Cell*, *82*(2), 348–388. DOI: 10.1016/j.molcel.2021.12.026

Nazari, M., Kordrostami, M., Ghasemi-Soloklui, A. A., Eaton-Rye, J. J., Pashkovskiy, P., Kuznetsov, V., & Allakhverdiev, S. I. (2024). Enhancing photosynthesis and plant productivity through genetic modification. *Cells*, *13*(16), 1319. DOI: 10.3390/cells13161319

Nazir, F., Peter, P., Gupta, R., Kumari, S., Nawaz, K., & Khan, M. I. R. (2024). Plant hormone ethylene: A leading edge in conferring drought stress tolerance. *Physiologia Plantarum*, *176*(1), e14151. DOI: 10.1111/ppl.14151

Ndudzo, A., Sibanda Makuvise, A., Moyo, S., & Bobo, E. D. (2024). CRISPR-Cas9 genome editing in crop breeding for climate change resilience: Implications for smallholder farmers in Africa. *Journal of Agriculture and Food Research*, *16*, 101132. DOI: 10.1016/j.jafr.2024.101132

Negi, C., Vasistha, N. K., Singh, D., Vyas, P., & Dhaliwal, H. S. (2022). Application of CRISPR-mediated gene editing for crop improvement. *Molecular Biotechnology*, *64*(11), 1198–1217. DOI: 10.1007/s12033-022-00507-y

Nekrasov, V., Wang, C., Win, J., Lanz, C., Weigel, D., & Kamoun, S. (2017). Rapid generation of a transgene-free powdery mildew resistant tomato by genome deletion. *Scientific Reports*, *7*(1), 482. DOI: 10.1038/s41598-017-00578-x PMID: 28352080

Nemudryi, A. A., Valetdinova, K. R., Medvedev, S. P., & Zakian, S. M. (2014). TALEN and CRISPR/Cas Genome Editing Systems: Tools of Discovery. *Acta Naturae*, *6*(3), 19–40. DOI: 10.32607/20758251-2014-6-3-19-40

Nerkar, G. A., Purankar, M. V., Sheelavantmath, S., & Devarumath, R. M. (2018). Biotechnological approach: a new dimension for sugarcane improvement. In K. Sengar (Ed.), *Biotechnology to enhance sugarcane productivity and stress tolerance* (pp. 1–24). CRC Press, Taylor & Francis Group. https://doi.org/DOI: 10.1201/9781315152776-1

Newman, A., Starrs, L., & Burgio, G. (2020). Cas9 Cuts and Consequences; Detecting, Predicting, and Mitigating CRISPR/Cas9 On-and Off-Target Damage: Techniques for Detecting, Predicting, and Mitigating the On-and off-target Effects of Cas9 Editing. *BioEssays : News and Reviews in Molecular, Cellular and Developmental Biology*, *42*(9), 2000047. DOI: 10.1002/bies.202000047

Nidhi, S., Anand, U., Oleksak, P., Tripathi, P., Lal, J. A., Thomas, G., Kuca, K., & Tripathi, V. (2021). Novel CRISPR–Cas systems: An updated review of the current achievements, applications, and future research perspectives. *International Journal of Molecular Sciences*, *22*(7), 3327. DOI: 10.3390/ijms22073327

Nie, H., Yang, X., Zheng, S., & Hou, L. (2024). Gene-Based Developments in Improving Quality of Tomato: Focus on Firmness, Shelf Life, and Pre-and Post-Harvest Stress Adaptations. *Horticulturae*, *10*(6), 641. DOI: 10.3390/horticulturae10060641

Nikolaou, G., Neocleous, D., Christou, A., Kitta, E., & Katsoulas, N. (2020). Implementing sustainable irrigation in water-scarce regions under the impact of climate change. *Agronomy (Basel)*, *10*(8), 1120. DOI: 10.3390/agronomy10081120

Nirmal, N. P., Anyimadu, C. F., Khanashyam, A. C., Bekhit, A. E. A., & Dhar, B. K. (2025). Alternative protein sources: Addressing global food security and environmental sustainability. *Sustainable Development (Bradford)*, *33*(3), 3958–3975. DOI: 10.1002/sd.3338

Nor A'azizam, N. M., Chopra, S., Guleria, P., Kumar, V., Abd Rahim, M. H., & Yaacob, J. S. (2024). Harnessing the potential of mutation breeding, CRISPR genome editing, and beyond for sustainable agriculture. *Functional & Integrative Genomics*, *24*(1), 15. DOI: 10.1007/s10142-024-01325-y PMID: 38421529

Ntsefong, G. N., Seyum, E. G., Kingsley, T. M., Kassie, F. C., Gabriel, M. S. T., Shariati, M. A., Karapetkovska-Hristova, V., Martin, B. J., & Youmbi, E. (2023). CRISPR-Cas-based genome editing for crop improvement: Progress, challenges and future prospects. *Global Journal of Botanical Science, 11*, 28–40. DOI: 10.12974/2311-858X.2023.11.3

Ochiai, H., & Yamamoto, T. (2014). Genome editing using zinc-finger nucleases (ZFNs) and transcription activator-like effector nucleases (TALENs). In *Targeted Genome Editing Using Site-Specific Nucleases: ZFNs, TALENs, and the CRISPR/Cas9 System* (pp. 3–24). Springer Japan., DOI: 10.1007/978-4-431-55200-4_1

Ochoa-Sanchez, A., Perez-Sanchez, G., Torres-Ledesma, A. M., Valdez, J. P. R., Rinaldi, G., Moguel, B. B., & Molina-Aguilar, C. (2021). Prime Editing, a novel genome-editing tool that may surpass conventional CRISPR-Cas9. *Re:GEN Open, 1*(1), 75–82. DOI: 10.1089/regen.2021.0016

Oh, J.-M., & Myung, K. (2022). Crosstalk between different DNA repair pathways for DNA double strand break repairs. *Mutation Research. Genetic Toxicology and Environmental Mutagenesis, 873*, 503438. DOI: 10.1016/j.mrgentox.2021.503438

Oliva, R., Ji, C., Atienza-Grande, G., Huguet-Tapia, J. C., Perez-Quintero, A., Li, T., Eom, J.-S., Li, C., Nguyen, H., Liu, B., Auguy, F., Sciallano, C., Luu, V. T., Dossa, G. S., Cunnac, S., Schmidt, S. M., Slamet-Loedin, I. H., Vera Cruz, C., Szurek, B., & Yang, B. (2019). Broad-spectrum resistance to bacterial blight in rice using genome editing. *Nature Biotechnology, 37*(11), 1344–1350. DOI: 10.1038/s41587-019-0267-z

Ortega-Salazar, D. C., Sbodio, A. O., Sugiyama, Y., Adaskaveg, A., Wang, D., Seymour, G. B., Li, X., Wang, S. C., & Blanco-Ulate, B. (2024). Double CRISPR knockout of pectin degrading enzymes improves tomato shelf-life while ensuring fruit quality. *Plants, People, Planet, 6*(2), 330–340. DOI: 10.1002/ppp3.10445

Ostmeyer, T., Parker, N., Jaenisch, B., Alkotami, L., Bustamante, C., & Jagadish, S. K. (2020). Impacts of heat, drought, and their interaction with nutrients on physiology, grain yield, and quality in field crops. *Plant Physiology Reports, 25*(4), 549–568. DOI: 10.1007/s40502-020-00538-0

Ousterout, D. G., & Gersbach, C. A. (2016). The Development of TALE Nucleases for Biotechnology. *Methods in Molecular Biology (Clifton, N.J.), 1338*, 27–42. DOI: 10.1007/978-1-4939-2932-0_3

Pacesa, M., Pelea, O., & Jinek, M. (2024). Past, present, and future of CRISPR genome editing technologies. *Cell, 187*(5), 1076–1100. DOI: 10.1016/j.cell.2024.01.042

Pallarès Masmitjà, M., Knödlseder, N., & Güell, M. (2019). CRISPR-gRNA design. In *CRISPR gene editing: methods and protocols* (pp. 3–11). Humana., DOI: 10.1007/978-1-4939-9170-9_1

Panda, D., Mishra, S. S., & Behera, P. K. (2021). Drought Tolerance in Rice: Focus on Recent Mechanisms and Approaches. *Rice Science*, *28*(2), 119–132. DOI: 10.1016/j.rsci.2021.01.002

Pandey, D. K., & Mishra, R. (2024). Towards sustainable agriculture: Harnessing AI for global food security. *Artificial Intelligence in Agriculture*, *12*, 72–84. DOI: 10.1016/j.aiia.2024.04.003

Pandey, S., Singh, A., Parida, S. K., & Prasad, M. (2022). Combining speed breeding with traditional and genomics-assisted breeding for crop improvement. *Plant Breeding*, *141*(3), 301–313. DOI: 10.1111/pbr.13012

Panigrahi, K. K., Mohanty, A., & Padhan, S. R.. (2023). Recent advances in genome editing towards sustainable agriculture. In *Genome Editing and Sustainable Crop Improvement*. CRC Press. DOI: 10.4324/9781003382102-7

Patel, A., Miles, A. J., Strackhouse, T., Cook, L., Leng, S., Patel, S., Klinger, K., Rudrabhatla, S., & Potlakayala, S. (2023). Methods of crop improvement and applications towards fortifying food security. *Frontiers in Genome Editing*, *5*, 1171969. DOI: 10.3389/fgeed.2023.1171969 PMID: 37484652

Pathi, K. (2021). Establishment of maize resistance to fungal diseases by host-induced gene silencing and site-directed mutagenesis.

Pattnaik, D., Avinash, S. P., Panda, S., & Bansal, K. C.. (2023). Accelerating crop domestication through genome editing for sustainable agriculture. *Journal of Plant Growth Regulation*, *42*(7), 4524–4545. DOI: 10.1007/s00344-022-10765-7

Pausch, P., Al-Shayeb, B., Bisom-Rapp, E., Tsuchida, C. A., Li, Z., Cress, B. F., Knott, G. J., Jacobsen, S. E., Banfield, J. F., & Doudna, J. A. (2020). CRISPR-casΦ from huge phages is a hypercompact genome editor. *Science*, *369*(6501), 333–337. DOI: 10.1126/science.abb1400 PMID: 32675376

Pavan, S., Jacobsen, E., Visser, R. G., & Bai, Y. (2010). Loss of susceptibility as a novel breeding strategy for durable and broad-spectrum resistance. *Molecular Breeding*, *25*(1), 1–12. DOI: 10.1007/s11032-009-9323-6 PMID: 20234841

Peng, A., Chen, S., Lei, T., Xu, L., He, Y., Wu, L., Yao, L., & Zou, X. (2017). Engineering canker-resistant plants through CRISPR/Cas9-targeted editing of the susceptibility gene CsLOB1 promoter in citrus. *Plant Biotechnology Journal*, *15*(12), 1509–1519. DOI: 10.1111/pbi.12733 PMID: 28371200

Penna, S., Shirani Bidabadi, S., & Jain, S. M. (2023). Mutation breeding to promote sustainable agriculture and food security in the era of climate change. In S. M. Jain & S. Penna (Eds.), *Mutation Breeding for Sustainable Agriculture and Food Security* (pp. 1–24). Springer. DOI: 10.1007/978-981-16-9720-3_1

Penna, S., & Jain, S. M. (2023). Fruit crop improvement with genome editing, in vitro and transgenic approaches. *Horticulturae*, *9*(1), 58. DOI: 10.3390/horticulturae9010058

Perochon, A., Kahla, A., Vranić, M., Jia, J., Malla, K. B., Craze, M., Wallington, E., & Doohan, F. M. (2019). A wheat NAC interacts with an orphan protein and enhances resistance to Fusarium head blight disease. *Plant Biotechnology Journal*, *17*(10), 1892–1904. DOI: 10.1111/pbi.13105 PMID: 30821405

Piergentili, R., Del Rio, A., Signore, F., Umani Ronchi, F., Marinelli, E., & Zaami, S. (2021). CRISPR-Cas and its wide-ranging applications: From human genome editing to environmental implications, technical limitations, hazards and bioethical issues. *Cells*, *10*(5), 969. DOI: 10.3390/cells10050969

Pixley, K. V., Cairns, J. E., Lopez-Ridaura, S., Ojiewo, C. O., Dawud, M. A., Drabo, I., Mindaye, T., Nebie, B., Asea, G., Das, B., Daudi, H., Desmae, H., Batieno, B. J., Boukar, O., Mukankusi, C. T. M., Nkalubo, S. T., Hearne, S. J., Dhugga, K. S., Gandhi, H., & Zepeda-Villarreal, E. A. (2023). Redesigning crop varieties to win the race between climate change and food security. *Molecular Plant*, *16*(10), 1590–1611. DOI: 10.1016/j.molp.2023.09.003

Poudel, B. (2024). CRISPR/Cas9 based gene editing in wheat and potato to enhance multiple disease resistance. [Unpublished manuscript].

Prado, J. R., Segers, G., Voelker, T., Carson, D., Dobert, R., Phillips, J., & Martino-Catt, S. (2014). Genetically engineered crops: From idea to product. *Annual Review of Plant Biology*, *65*(1), 769–790. DOI: 10.1146/annurev-arplant-050213-040039 PMID: 24579994

Prusty, A., Panchal, A., Singh, R. K., & Prasad, M. (2024). Major transcription factor families at the nexus of regulating abiotic stress response in millets: A comprehensive review. *Planta*, *259*(5), 118. DOI: 10.1007/s00425-024-04394-2

Puchta, H., Jiang, J., Wang, K., & Zhao, Y. (2022). Updates on gene editing and its applications. *Plant Physiology*, *188*(4), 1725–1730. DOI: 10.1093/plphys/kiac032

Pyott, D. E., Sheehan, E., & Molnar, A. (2016). Engineering of CRISPR/Cas9-mediated potyvirus resistance in transgene-free Arabidopsis plants. *Molecular Plant Pathology*, *17*(8), 1276–1288. DOI: 10.1111/mpp.12417 PMID: 27103354

Qaim, M. (2020). Role of new plant breeding technologies for food security and sustainable agricultural development. *Applied Economic Perspectives and Policy*, *42*(2), 129–150. DOI: 10.1002/aepp.13044

Qu, L., Huang, X., Su, X., Zhu, G., Zheng, L., Lin, J., & Xue, H. (2024). Potato: From functional genomics to genetic improvement. *Molecular Horticulture*, *4*(1), 34. DOI: 10.1186/s43897-024-00105-3

Rafiq, A. R. (2024). Genomic editing techniques for ensuring food security: CRISPR Cas, TALEN, ZFN, RNAi and mutagenesis. *Journal of Pure and Applied Agriculture*, *9*(2).

Rai, G. K., Khanday, D. M., Kumar, P., Magotra, I., Choudhary, S. M., Kosser, R., & Rouphael, Y. (2023). Enhancing crop resilience to drought stress through CRISPR-Cas9 genome editing. *Plants*, *12*(12), 2306. DOI: 10.3390/plants12122306

Rajput, M., Choudhary, K., Kumar, M., Vivekanand, V., Chawade, A., Ortiz, R., & Pareek, N. (2021). RNA interference and CRISPR/Cas gene editing for crop improvement: Paradigm shift towards sustainable agriculture. *Plants*, *10*(9), 1914. DOI: 10.3390/plants10091914 PMID: 34579446

Rallapalli, K. L., & Komor, A. C. (2023). The design and application of DNA-editing enzymes as base editors. *Annual Review of Biochemistry*, *92*(1), 43–79. DOI: 10.1146/annurev-biochem-052521-013938

Ran, F. A., Hsu, P. D., Lin, C. Y., Gootenberg, J. S., Konermann, S., Trevino, A. E., Scott, D. A., Inoue, A., Matoba, S., Zhang, Y., & Zhang, F. (2013). Double nicking by RNA-guided CRISPR Cas9 for enhanced genome editing specificity. *Cell*, *154*(6), 1380–1389. DOI: 10.1016/j.cell.2013.08.021 PMID: 23992846

Rasheed, A., Barqawi, A. A., Mahmood, A., Nawaz, M., Shah, A. N., Bay, D. H., Alahdal, M. A., Hassan, M. U., & Qari, S. H. (2022). CRISPR/Cas9 is a powerful tool for precise genome editing of legume crops: A review. *Molecular Biology Reports*, *49*(6), 5595–5609. DOI: 10.1007/s11033-022-07529-4

Rasheed, A., Gill, R. A., Hassan, M. U., Mahmood, A., Qari, S., Zaman, Q. U., & Li, H. (2021). A critical review: Recent advancements in the use of CRISPR/Cas9 technology to enhance crops and alleviate global food crises. *Current Issues in Molecular Biology*, *43*(3), 1950–1976. DOI: 10.3390/cimb43030135

Rasheed, A., Jie, H., Ali, B., He, P., Zhao, L., Ma, Y., & Hamid, M. R. (2023). Breeding drought-tolerant maize (Zea mays) using molecular breeding tools: Recent advancements and future prospective. *Agronomy (Basel)*, *13*(6), 1459. DOI: 10.3390/agronomy13061459

Rasheed, A., Raza, A., Jie, H., Mahmood, A., Ma, Y., Zhao, L., Xing, H., Li, L., Hassan, M. U., Qari, S. H., & Jie, Y. (2022). Molecular tools and their applications in developing salt-tolerant soybean (Glycine max L.) cultivars. *Bioengineering (Basel, Switzerland)*, *9*(10), 495. DOI: 10.3390/bioengineering9100495 PMID: 36290463

Rathod, A., & Verma, N. S. (2023). *Impact of abiotic stress on agronomical crops. Frontiers of Agronomy*. Elite Publishing House.

Rato, C., Carvalho, M. F., Azevedo, C., & Oblessuc, P. R. (2021). Genome editing for resistance against plant pests and pathogens. *Transgenic Research*, *30*(4), 427–459. DOI: 10.1007/s11248-021-00262-x

Ravikiran, K. T., Thribhuvan, R., Sheoran, S., Kumar, S., Kushwaha, A. K., Vineeth, T. V., & Saini, M. (2023). Tailoring crops with superior product quality through genome editing: An update. *Planta*, *257*(5), 86. DOI: 10.1007/s00425-023-04112-4

Ray, D. K., Mueller, N. D., West, P. C., & Foley, J. A. (2013). Yield trends are insufficient to double global crop production by 2050. *PLoS One*, *8*(6), e66428. DOI: 10.1371/journal.pone.0066428 PMID: 23840465

Ray, S., Sneha, K., & Jangid, C. (2023). CRISPR-Cas9 for sustainable food production: Impacts, recent advancements and future perspectives. *Food and Humanity*, *1*, 1458–1469. DOI: 10.1016/j.foohum.2023.10.014

Ray, U., & Raghavan, S. C. (2020). Modulation of DNA double-strand break repair as a strategy to improve precise genome editing. *Oncogene*, *39*(41), 6393–6405. DOI: 10.1038/s41388-020-01445-2

Raza, A., Anas, M., Bhardwaj, S., Mir, R. A., Charagh, S., Elahi, M., Zhang, X., Mir, R. R., Weckwerth, W., Fernie, A. R., Siddique, K. H. M., Hu, Z., & Varshney, R. K. (2025). Harnessing metabolomics for enhanced crop drought tolerance. *The Crop Journal*, *13*(2), 311–327. Advance online publication. DOI: 10.1016/j.cj.2025.01.001

Rehman, R. S., Zafar, S. A., Ali, M., Pasha, A. N., Naveed, M. S., Waseem, M., Ahmad, M., & Raza, A. (2022). CRISPR-Cas mediated genome editing: A paradigm shift towards sustainable agriculture and biotechnology. *Asian Plant Research Journal*, *9*(4), 32–48. DOI: 10.9734/aprj/2022/v9i130197

Ricroch, A., Clairand, P., & Harwood, W. (2017). Use of CRISPR systems in plant genome editing: Toward new opportunities in agriculture. *Emerging Topics in Life Sciences*, *1*(2), 169–182. DOI: 10.1042/ETLS20170085 PMID: 33525765

Rong, W., Qi, L., Wang, A., Ye, X., Du, L., Liang, H., Xin, Z., & Zhang, Z. (2014). The ERF transcription factor TaERF3 promotes tolerance to salt and drought stresses in wheat. *Plant Biotechnology Journal*, *12*(4), 468–479. DOI: 10.1111/pbi.12153

Roy, D., Deo, I., Kushwaha, U. K. S., & Singh, P. K. (2025). Speed Breeding: Space Inspired Plant Breeding for Crop Improvement. In *Smart Technologies in Sustainable Agriculture* (pp. 147-161): Apple Academic Press.

Roy, A., Moradkhani, H., Mekonnen, M., Moftakhari, H., & Magliocca, N. (2024). Towards strategic interventions for global food security in 2050. *The Science of the Total Environment*, *954*, 176811. DOI: 10.1016/j.scitotenv.2024.176811

Roy, S., Barman, M., & Das, A.. (2026). Application of genome editing for sustainable development of agriculture: A perspective view. In *Advances in Sustainable Crop Production*. CRC Press. DOI: 10.1201/9781779643155-5

Rozas, P., Kessi-Pérez, E. I., & Martínez, C. (2022). Genetically modified organisms: Adapting regulatory frameworks for evolving genome editing technologies. *Biological Research*, *55*(1), 55. DOI: 10.1186/s40659-022-00399-x

Sabat, M., & Tripathy, A. (2024). Genetically Modified and Gene-Edited Food Crops: Recent Status and Future Prospects. *Food Production, Diversity, and Safety Under Climate Change*, 211-222.

Saber Sichani, A., Ranjbar, M., Baneshi, M., Torabi Zadeh, F., & Fallahi, J. (2023). A review on advanced CRISPR-based genome-editing tools: Base editing and prime editing. *Molecular Biotechnology*, *65*(6), 849–860. DOI: 10.1007/s12033-022-00639-1

Sadikiel Mmbando, G., & Ngongolo, K. (2024). The recent genetic modification techniques for improve soil conservation, nutrient uptake and utilization. *GM Crops & Food*, *15*(1), 233–247. DOI: 10.1080/21645698.2024.2377408

Sahu, A., Verma, R., Gupta, U., Kashyap, S., & Sanyal, I. (2024). An overview of targeted genome editing strategies for reducing the biosynthesis of phytic acid: An anti-nutrient in crop plants. *Molecular Biotechnology*, *66*(1), 11–25. DOI: 10.1007/s12033-023-00722-1

Sander, J. D., & Joung, J. K. (2014). CRISPR-Cas systems for editing, regulating and targeting genomes. *Nature Biotechnology*, *32*(4), 347–355. DOI: 10.1038/nbt.2842

Sarfraz, S., Ali, F., Hameed, A., Ahmad, Z., & Riaz, K. (2023). Sustainable agriculture through technological innovations. In *Sustainable Agriculture in the Era of Climate Change*. Springer.

Sarma, B., Kashtoh, H., Lama Tamang, T., Bhattacharyya, P. N., Mohanta, Y. K., & Baek, K. H. (2023). Abiotic Stress in Rice: Visiting the Physiological Response and Its Tolerance Mechanisms. *Plants*, *12*(23), 3948. Advance online publication. DOI: 10.3390/plants12233948

Sarma, H. H., Borah, S. K., Dutta, N., Sultana, N., Nath, H., & Das, B. C. (2024). Innovative approaches for climate-resilient farming: Strategies against environmental shifts and climate change. *International Journal of Environment and Climate Change, 14*(9), 217–241. DOI: 10.9734/ijecc/2024/v14i94407

Sathee, L., Jagadhesan, B., Pandesha, P. H., Barman, D., & Adavi, B. (2022). Genome editing targets for improving nutrient use efficiency and nutrient stress adaptation. *Frontiers in Genetics, 13*, 900897. DOI: 10.3389/fgene.2022.900897

Sathiyabalan, A., Amir, M., Sandhu, R., & Bhattacharya, S. (2024). Evaluating CRISPR/Cas9's progressive role in crop enhancement and sustainable agriculture. *Journal of Advances in Biology & Biotechnology, 27*(11), 855–870. DOI: 10.9734/jabb/2024/v27i6949

Savary, S., Ficke, A., Aubertot, J. N., & Hollier, C. (2012). Crop losses due to diseases and their implications for global food production losses and food security. *Food Security, 4*(4), 519–537. DOI: 10.1007/s12571-012-0200-5

Savary, S., Willocquet, L., Pethybridge, S. J., Esker, P., McRoberts, N., & Nelson, A. (2019). The global burden of pathogens and pests on major food crops. *Nature Ecology & Evolution, 3*(3), 430–439. DOI: 10.1038/s41559-018-0793-y PMID: 30718852

Schaart, J. G., van de Wiel, C. C. M., & Smulders, M. J. M. (2021). Genome editing of polyploid crops: Prospects, achievements and bottlenecks. *Transgenic Research, 30*(4), 337–351. DOI: 10.1007/s11248-021-00251-0

Scheben, A., Wolter, F., Batley, J., Puchta, H., & Edwards, D. (2017). Towards CRISPR/Cas crops–bringing together genomics and genome editing. *The New Phytologist, 216*(3), 682–698. DOI: 10.1111/nph.14702

Schie, C. C., & Takken, F. L. (2014). Susceptibility genes 101: How to be a good host. *Annual Review of Phytopathology, 52*(1), 551–581. DOI: 10.1146/annurev-phyto-102313-045854 PMID: 25001453

Seleiman, M. F., Al-Suhaibani, N., Ali, N., Akmal, M., Alotaibi, M., Refay, Y., Dindaroglu, T., Abdul-Wajid, H. H., & Battaglia, M. L. (2021). Drought Stress Impacts on Plants and Different Approaches to Alleviate Its Adverse Effects. *Plants, 10*(2), 259. Advance online publication. DOI: 10.3390/plants10020259

Shah, T., Andleeb, T., Lateef, S., & Noor, M. A. (2018). Genome editing in plants: Advancing crop transformation and overview of tools. *Plant Physiology and Biochemistry : PPB, 131*, 12–21. DOI: 10.1016/j.plaphy.2018.05.009

Shah, Z., Iqbal, A., Khan, F. U., Khan, H. U., Durrani, F., & Ahmad, M. Z. (2020). Genetic manipulation of pea (Pisum sativum L.) with Arabidopsis's heat shock factor HsfA1d improves ROS scavenging system to confront thermal stress. *Genetic Resources and Crop Evolution, 67*(8), 2119–2127. DOI: 10.1007/s10722-020-00966-9

Shamshirgaran, Y., Liu, J., Sumer, H., Verma, P. J., & Taheri-Ghahfarokhi, A. (2022). Tools for efficient genome editing; ZFN, TALEN, and CRISPR. *Applications of genome modulation and editing*, 29-46.

Shan, Q., Wang, Y., Li, J., Zhang, Y., Chen, K., Liang, Z., & Gao, C. (2013). Targeted genome modification of crop plants using a CRISPR-Cas system. *Nature Biotechnology, 31*(8), 686–688. DOI: 10.1038/nbt.2650 PMID: 23929338

Sharma, A., Kumar, V., Shahzad, B., Ramakrishnan, M., Singh Sidhu, G. P., Bali, A. S., & Khanna, K. (2020). Photosynthetic response of plants under different abiotic stresses: A review. *Journal of Plant Growth Regulation, 39*(2), 509–531. DOI: 10.1007/s00344-019-10018-x

Sharma, U., Nisha, , & Ray, A. (2025). An insight into the journey of CRISPR-CAS9 and its application in crop improvement. *Discover Plants, 2*(1), 266. DOI: 10.1007/s44372-025-00343-9

Shelake, R. M., Kadam, U. S., Kumar, R., Pramanik, D., Singh, A. K., & Kim, J.-Y. (2022). Engineering drought and salinity tolerance traits in crops through CRISPR-mediated genome editing: Targets, tools, challenges, and perspectives. *Plant Communications, 3*(6), 100417. DOI: 10.1016/j.xplc.2022.100417

Shi, B., Li, K., Xu, R., Zhang, F., Yu, Z., Ding, Z., & Tian, H. (2025). Methionine-mediated trade-off between plant growth and salt tolerance. *Plant Physiology, 197*(3), kiaf074. DOI: 10.1093/plphys/kiaf074

Shi, J., Gao, H., Wang, H., Lafitte, H. R., Archibald, R. L., Yang, M., Hakimi, S. M., Mo, H., & Habben, J. E. (2017). ARGOS8 variants generated by CRISPR-Cas9 improve maize grain yield under field drought stress conditions. *Plant Biotechnology Journal, 15*(2), 207–216. DOI: 10.1111/pbi.12603 PMID: 27442592

Shi, J., Habben, J. E., Archibald, R. L., Drummond, B. J., Chamberlin, M. A., Williams, R. W., Lafitte, H. R., & Weers, B. P. (2015). Overexpression of ARGOS genes modifies plant sensitivity to ethylene, leading to improved drought tolerance in both Arabidopsis and maize. *Plant Physiology, 169*(1), 266–282. DOI: 10.1104/pp.15.00780

Shi, J., Zhao, B., Zheng, S., Zhang, X., Wang, X., Dong, W., Qi, J., Wang, E., & Wang, E. (2021). A phosphate starvation response-centered network regulates mycorrhizal symbiosis. *Cell, 184*(22), 5527–5540. DOI: 10.1016/j.ccll.2021.09.030

Shukla, V. K., Doyon, Y., Miller, J. C., DeKelver, R. C., Moehle, E. A., Worden, S. E., Mitchell, J. C., Arnold, N. L., Gopalan, S., Meng, X., Choi, V. M., Rock, J. M., Wu, Y.-Y., Katibah, G. E., Zhifang, G., McCaskill, D., Simpson, M. A., Blakeslee, B., Greenwalt, S. A., & Urnov, F. D. (2009). Precise genome modification in the crop plant maize using zinc-finger nucleases. *Nature*, *459*(7245), 437–441. DOI: 10.1038/nature07992

Singh, V. K., Pundir, S., Chaturvedi, D., Kaur, A., Pandey, A., Mandal, S., . . . Dhanda, P. S. (2023). Enhancing Maize (Zea mays L.) Crop through Advanced Techniques: A Comprehensive Approach. In *New Prospects of Maize*: IntechOpen.

Singh, A. K., Ramalingam, S., Rao, D. N., & Chandrasegaran, S. (2021). Genome editing revolution in life sciences. *Resonance*, *26*(7), 971–998. DOI: 10.1007/s12045-021-1195-z

Singh, A., Rajput, V. D., Rawat, S., & Sharma, R.. (2021). Emerging tools for sustainable agriculture and food security: An introduction. In *Emerging Tools for Sustainable Agriculture and Food Security* (pp. 1–15). Springer., DOI: 10.1007/978-981-16-0917-6_1

Singh, B., Lokya, V., Kaundal, P., & Tiwari, S. (2024). Genetic engineering for potato improvement: Current challenges and future opportunities. In *Genetic engineering of crop plants for food and health security* (Vol. 1, pp. 213–232). Springer Nature Singapore.

Singh, P., Markam, N. K., Shinde, S. B., Yadav, T., Mohanty, S., Prasad, M., Pagoch, S. S., Swain, S., Musmade, N. A., & Saikia, B. (2025). Applications of CRISPR/Cas gene editing in vegetable crop disease and stress resistance. *International Journal of Advanced Biochemistry Research*, *9*(1), 960–970. DOI: 10.33545/26174693.2025.v9.i8Sn.5290

Slaymaker, I. M., Gao, L. Y., Zetsche, B., Scott, D. A., Yan, W. X., & Zhang, F. (2016). Rationally engineered Cas9 nucleases with improved specificity. *Science*, *351*(6268), 84–88. DOI: 10.1126/science.aad5227 PMID: 26628643

Smirnikhina, S. A., Zaynitdinova, M. I., Sergeeva, V. A., & Lavrov, A. V. (2022). Improving homology-directed repair in genome editing experiments by influencing the cell cycle. *International Journal of Molecular Sciences*, *23*(11), 5992. DOI: 10.3390/ijms23115992

Smyth, S. J. (2022). Contributions of genome editing technologies towards improved nutrition, environmental sustainability and poverty reduction. *Frontiers in Genome Editing*, *4*, 863193. DOI: 10.3389/fgeed.2022.863193

Solis, C. A., Yong, M.-T., Zhou, M., Venkataraman, G., Shabala, L., Holford, P., Shabala, S., & Chen, Z.-H. (2022). Evolutionary significance of NHX family and NHX1 in salinity stress adaptation in the genus oryza. *International Journal of Molecular Sciences*, *23*(4), 2092. DOI: 10.3390/ijms23042092

Song, X., Dang, C., Wang, F., Yao, H., Song, Q., & Ye, G. (2026). *CRISPR/Cas gene editing for plant resistance to pests, diseases, and herbicide: application, risk assessment, and safety evaluation*. GM Crops Food. 2026 Mar 16;17(1):2641893. . eCollection 2026.DOI: 10.1080/21645698.2026.2641893

Song, H., Taylor, D. C., & Zhang, M. (2023). Bioengineering of Soybean Oil and Its Impact on Agronomic Traits. *International Journal of Molecular Sciences*, *24*(3), 2256. Advance online publication. DOI: 10.3390/ijms24032256

Song, W., Shao, H., Zheng, A., Zhao, L., & Xu, Y. (2023). Advances in roles of salicylic acid in plant tolerance responses to biotic and abiotic stresses. *Plants*, *12*(19), 3475. DOI: 10.3390/plants12193475

Song, Y., Hu, Z., Liu, S., Luo, S., He, R., Yang, X., Li, S., Yang, X., An, Y., & Lu, Y. (2025). Utilization of microalgae and duckweed as sustainable protein sources for food and feed: Nutritional potential and functional applications. *Journal of Agricultural and Food Chemistry*, *73*(8), 4466–4485. DOI: 10.1021/acs.jafc.4c11610 PMID: 39879156

Stover, E. D., Driggers, R., Richardson, M. L., Hall, D. G., Duan, Y., & Lee, R. F. (2014). Incidence and severity of Asiatic citrus canker on diverse citrus and citrus-related germplasm in a Florida field planting. *HortScience*, *49*(1), 4–9. DOI: 10.21273/HORTSCI.49.1.4

Subejano, M. S. E. P., Daquioag, J. E. L., & Penuliar, G. M. (2024). Genetically Modified Organisms in Research and Biotechnology. In *Biosafety and Biosecurity* (pp. 235–258). CRC Press. DOI: 10.1201/9781003426219-12

Suman, S., Bagal, D., Jain, D., Singh, R., Singh, I. K., & Singh, A. (2021). Biotic stresses on plants: reactive oxygen species generation and antioxidant mechanism. In *Frontiers in plant-soil interaction* (pp. 381–411). Elsevier. DOI: 10.1016/B978-0-323-90943-3.00014-6

Sun, N., & Zhao, H. (2013). Transcription activator-like effector nucleases (TALENs): A highly efficient and versatile tool for genome editing. *Biotechnology and Bioengineering, 110*(7), 1811–1821. DOI: 10.1002/bit.24890

Sun, S. R., Wang, Z. Q., Lian, M., Chen, J. L., Qin, Y. X., Chang, H. L., & Wang, Q. N. (2025). Systematic identification of sugarcane vacuolar H+-translocating pyrophosphatase (VPP) gene family and the role of ScVPP1 in salt resistance. *Plant Cell Reports, 44*(1), 11. DOI: 10.1007/s00299-024-03401-w

Sun, Z., Li, S., Chen, W., Zhang, J., Zhang, L., Sun, W., & Wang, Z. (2021). Plant dehydrins: Expression, regulatory networks, and protective roles in plants challenged by abiotic stress. *International Journal of Molecular Sciences, 22*(23), 12619. DOI: 10.3390/ijms222312619

Su, Q., Chen, L., Cai, Y., Wang, L., Chen, Y., Zhang, J., & Gao, Y. (2024). The FLOWERING LOCUS T 5b positively regulates photoperiodic flowering and improves the geographical adaptation of soybean. *Plant, Cell & Environment, 47*(1), 246–258. DOI: 10.1111/pce.14739

Su, T., Li, X., Yang, M., Shao, Q., Zhao, Y., Ma, C., & Wang, P. (2020). Autophagy: An intracellular degradation pathway regulating plant survival and stress response. *Frontiers in Plant Science, 11*, 164. DOI: 10.3389/fpls.2020.00164

Tabassum, J., Ahmad, S., Hussain, B., Mawia, A. M., Zeb, A., & Ju, L. (2021). Applications and potential of genome-editing systems in rice improvement: Current and future perspectives. *Agronomy (Basel), 11*(7), 1359. DOI: 10.3390/agronomy11071359

Tachikawa, M., & Matsuo, M. (2024). Global regulatory trends of genome editing technology in agriculture and food. *Breeding Science, 74*(1), 3–10. DOI: 10.1270/jsbbs.23046

Takagi, H., Abe, A., Yoshida, K., Kosugi, S., Natsume, S., Mitsuoka, C., Uemura, A., Utsushi, H., Tamiru, M., Takuno, S., Innan, H., Cano, L. M., Kamoun, S., & Terauchi, R. (2013). QTL-seq: Rapid mapping of quantitative trait loci in rice by whole genome resequencing of DNA from two bulked populations. *The Plant Journal : For Cell and Molecular Biology, 74*(1), 174–183. DOI: 10.1111/tpj.12105 PMID: 23289725

Takasaki, H., Maruyama, K., Kidokoro, S., Ito, Y., Fujita, Y., Shinozaki, K., Yamaguchi-Shinozaki, K., & Nakashima, K. (2010). The abiotic stress-responsive NAC-type transcription factor OsNAC5 regulates stress-inducible genes and stress tolerance in rice. *Molecular Genetics and Genomics, 284*(3), 173–183. DOI: 10.1007/s00438-010-0557-0

Tandzi, L., & Mutengwa, C. (2020). Factors Affecting Yield of Crops. In.

Tang, Z., Chen, S., Chen, A., He, B., Zhou, Y., Chai, G., Guo, F., & Huang, J. (2019). CasPDB: An integrated and annotated database for Cas proteins from bacteria and archaea. *Database : The Journal of Biological Databases and Curation, 2019*, baz093. DOI: 10.1093/database/baz093

Tan, X., Zeng, W., Yang, Y., Lin, Z., Li, F., Liu, J., Chen, S., Liu, Y.-G., Xie, W., & Xie, X. (2025). Genome-wide profiling of polymorphic short tandem repeats and their influence on gene expression and trait variation in diverse rice populations. *Journal of Genetics and Genomics*, *52*(6), 733–746. DOI: 10.1016/j.jgg.2025.03.005

Teper, D., White, F. F., & Wang, N. (2023). The dynamic transcription activator-like effector family of Xanthomonas. *Phytopathology*, *113*(4), 651–666. DOI: 10.1094/PHYTO-10-22-0365-KD

Tessema, N., Yadeta, D., Kebede, A., & Ayele, G. T. (2022). Soil and irrigation water salinity, and its consequences for agriculture in Ethiopia: A systematic review. *Agriculture*, *13*(1), 109. DOI: 10.3390/agriculture13010109

Thankappan, S., Binodh, A. K., & Kumar, P. R.. (2024). Genome editing of plant growth-promoting microbes (PGPM) towards developing smart bio-formulations for sustainable agriculture: Current trends and perspectives. In *Genome Editing in Microbial Biotechnology*. Bentham Science Publishers. DOI: 10.2174/9789815223798124010006

Thilakarathne, A. S., Liu, F., & Zou, Z. (2025). Plant Signaling Hormones and Transcription Factors: Key Regulators of Plant Responses to Growth, Development, and Stress. *Plants*, *14*(7), 1070. Advance online publication. DOI: 10.3390/plants14071070

Tian, L., Chen, X., Jia, X., Wang, S., Wang, X., Zhang, J., Zhang, Y., Wu, S., Chen, Y., & Wu, L. (2021). First report of antifungal activity conferred by non-conventional peptides. *Plant Biotechnology Journal*, *19*(11), 2147–2149. DOI: 10.1111/pbi.13691 PMID: 34455677

Tian, Y., Zhou, Y., Gao, G., Zhang, Q., Li, Y., Lou, G., & He, Y. (2023). Creation of two-line fragrant glutinous hybrid rice by editing the Wx and OsBADH2 genes via the CRISPR/Cas9 system. *International Journal of Molecular Sciences*, *24*(1), 849. DOI: 10.3390/ijms24010849

Tiwari, S., Patel, A., Singh, M., & Prasad, S. M. (2020). Regulation of temperature stress in plants. In *Plant life under changing environment* (pp. 25–45). Elsevier. DOI: 10.1016/B978-0-12-818204-8.00002-3

Tlak Gajger, I., & Dar, S. A. (2021). Plant Allelochemicals as Sources of Insecticides. *Insects*, *12*(3), 189. Advance online publication. DOI: 10.3390/insects12030189

Tong, H., Wang, H., Wang, X., Liu, N., Li, G., Wu, D., & Wei, Y. (2024). Development of deaminase-free T-to-S base editor and C-to-G base editor by engineered human uracil DNA glycosylase. *Nature Communications*, *15*(1), 4897. DOI: 10.1038/s41467-024-49343-5

Townsend, J. A., Wright, D. A., Winfrey, R. J., Fu, F., Maeder, M. L., Joung, J. K., & Voytas, D. F. (2009). High-frequency modification of plant genes using engineered zinc-finger nucleases. *Nature*, *459*(7245), 442–445. DOI: 10.1038/nature07845

Tripathi, L., Dhugga, K. S., Ntui, V. O., Runo, S., Syombua, E. D., Muiruri, S., Wen, Z., & Tripathi, J. N. (2022). Genome editing for sustainable agriculture in Africa. *Frontiers in Genome Editing*, *4*, 876674. Advance online publication. DOI: 10.3389/fgeed.2022.876697 PMID: 35647578

Tripathi, L., Ntui, V. O., & Tripathi, J. N. (2024). Application of CRISPR/Cas-based gene-editing for developing better banana. *Frontiers in Bioengineering and Biotechnology*, *12*, 1395772. DOI: 10.3389/fbioe.2024.1395772

Trump, B., Cummings, C., Klasa, K., Galaitsi, S., & Linkov, I. (2023). Governing biotechnology to provide safety and security and address ethical, legal, and social implications. *Frontiers in Genetics*, *13*, 1052371. DOI: 10.3389/fgene.2022.1052371

Tudi, M., Daniel Ruan, H., Wang, L., Lyu, J., Sadler, R., Connell, D., Chu, C., & Phung, D. T. (2021). Agriculture Development, Pesticide Application and Its Impact on the Environment. *International Journal of Environmental Research and Public Health*, *18*(3), 1112. Advance online publication. DOI: 10.3390/ijerph18031112

Tufail, U., Kanwal, R., Asghar, A., Ali, B., Shabbir, F., Rehman, K., & Raza, A. (2026). The Role of Calcium in Salinity Stress Alleviation: Mechanisms and Implications. *Plant Science Horizons*, *1*(1), 41–58. DOI: 10.66835/psh.2026.e0007

Tuncel, A., Pan, C., Clem, J. S., Liu, D., & Qi, Y. (2025). CRISPR–Cas applications in agriculture and plant research. *Nature Reviews. Molecular Cell Biology*, *26*(6), 419–441. DOI: 10.1038/s41580-025-00834-3 PMID: 40055491

Tuncel, A., Pan, C., Sprink, T., Wilhelm, R., Barrangou, R., Li, L., Shih, P. M., Varshney, R. K., Tripathi, L., Van Eck, J., Mandadi, K., & Qi, Y. (2023). Genome-edited foods. *Nature Reviews Bioengineering*, *1*(11), 799–816. DOI: 10.1038/s44222-023-00115-8

Turnbull, C., Lillemo, M., & Hvoslef-Eide, T. A. (2021). Global regulation of genetically modified crops amid the gene edited crop boom–a review. *Frontiers in Plant Science*, *12*, 630396. DOI: 10.3389/fpls.2021.630396

Tyagi, S., Kumar, R., Kumar, V., Won, S. Y., & Shukla, P. (2021). Engineering disease resistant plants through CRISPR-Cas9 technology. *GM Crops & Food*, *12*(1), 125–144. DOI: 10.1080/21645698.2020.1831729

Umar, O. B., Ranti, L. A., Abdulbaki, A. S., Bola, A. u. L., Abdulhamid, A. K., Biola, M. R., & Victor, K. O. (2021). Stresses in plants: Biotic and abiotic. In *Current trends in wheat research*: IntechOpen.

Umar, W., Ayub, M. A., Rehman, M. Z. u., Ahmad, H. R., Farooqi, Z. U. R., Shahzad, A., . . . Nadeem, M. (2020). Nitrogen and phosphorus use efficiency in agroecosystems. *Resources use efficiency in agriculture*, 213-257.

Uranga, M., Martín-Hernández, A. M., Storme, N. D., & Pasin, F. (2024). CRISPR–Cas systems and applications for crop bioengineering. *Frontiers in Bioengineering and Biotechnology*, *12*, 1483857. DOI: 10.3389/fbioe.2024.1483857 PMID: 39479297

Urnov, F. D., Miller, J. C., Lee, Y. L., Beausejour, C. M., Rock, J. M., Gregory, P. D., & Holmes, M. C. (2005). Highly efficient endogenous human gene correction using designed zinc-finger nucleases. *Nature*, *435*(7042), 646–651. DOI: 10.1038/nature03556

Urnov, F. D., Rebar, E. J., Holmes, M. C., Zhang, H. S., & Gregory, P. D. (2010). Genome editing with engineered zinc finger nucleases. *Nature Reviews. Genetics*, *11*(9), 636–646. DOI: 10.1038/nrg2842

Usman, B., Nawaz, G., Zhao, N., Liao, S., Liu, Y., & Li, R. (2020). Precise editing of the OsPYL9 gene by RNA-guided Cas9 nuclease confers enhanced drought tolerance and grain yield in rice (Oryza sativa L.) by regulating circadian rhythm and abiotic stress responsive proteins. *International Journal of Molecular Sciences*, *21*(21), 7854. DOI: 10.3390/ijms21217854

Vågsholm, I., Arzoomand, N. S., & Boqvist, S. (2020). Food security, safety, and sustainability—Getting the trade-offs right. *Frontiers in Sustainable Food Systems*, *4*, 487217. DOI: 10.3389/fsufs.2020.00016

van Dijk, M., Morley, T., Rau, M. L., & Saghai, Y. (2021). A meta-analysis of projected global food demand and population at risk of hunger for the period 2010–2050. *Nature Food*, *2*(7), 494–501. DOI: 10.1038/s43016-021-00322-9

Varotto, S. (2023). Current status and future prospective of genome editing application in maize. In *A roadmap for plant genome editing* (pp. 165–182). Springer Nature.

Venegas-Rioseco, J., Ginocchio, R., & Ortiz-Calderón, C. (2021). Increase in phytoextraction potential by genome editing and transformation: A review. *Plants*, *11*(1), 86. DOI: 10.3390/plants11010086

Vyroubalová, S., Václavíková, K. i., Turecková, V., Novák, O., Šmehilová, M., Hluska, T., Ohnoutková, L., Frébort, I., & Galuszka, P. (2009). Characterization of new maize genes putatively involved in cytokinin metabolism and their expression during osmotic stress in relation to cytokinin levels. *Plant Physiology*, *151*(1), 433–447. DOI: 10.1104/pp.109.142489

Waghmode, P., Pitha, C. C., & Banka, V. (2024). The role of genome editing technologies in sustainable agriculture. *Journal of Advances in Biology & Biotechnology*, *27*(4), 50–61. DOI: 10.9734/jabb/2024/v27i81168

Waites, J., Achary, V. M. M., Syombua, E. D., Hearne, S. J., & Bandyopadhyay, A. (2025). CRISPR-mediated genome editing of wheat for enhancing disease resistance. *Frontiers in Genome Editing*, *7*, 1542487. DOI: 10.3389/fgeed.2025.1542487

Wan, D.-Y., Guo, Y., Cheng, Y., Hu, Y., Xiao, S., Wang, Y., & Wen, Y.-Q. (2020). CRISPR/Cas9-mediated mutagenesis of VvMLO3 results in enhanced resistance to powdery mildew in grapevine (Vitis vinifera). *Horticulture Research*, *7*(1), 116. DOI: 10.1038/s41438-020-0339-8 PMID: 32821399

Wang, J., Song, L., Gong, X., Xu, J., & Li, M. (2020). Functions of jasmonic acid in plant regulation and response to abiotic stress. *International Journal of Molecular Sciences*, *21*(4), 1446. DOI: 10.3390/ijms21041446

Wang, J., Wu, B., Lu, K., Wei, Q., Qian, J., Chen, Y., & Fang, Z. (2019). The amino acid permease 5 (OsAAP5) regulates tiller number and grain yield in rice. *Plant Physiology*, *180*(2), 1031–1045. DOI: 10.1104/pp.19.00034

Wang, K., Gong, Q., & Ye, X. (2020). Recent developments and applications of genetic transformation and genome editing technologies in wheat. *Theoretical and Applied Genetics*, *133*(5), 1603–1622. DOI: 10.1007/s00122-019-03464-4

Wang, L., Chen, L., Li, R., Zhao, R., Yang, M., Sheng, J., & Shen, L. (2017). Reduced drought tolerance by CRISPR/Cas9-mediated SlMAPK3 mutagenesis in tomato plants. *Journal of Agricultural and Food Chemistry*, *65*(39), 8674–8682. DOI: 10.1021/acs.jafc.7b02745 PMID: 28873302

Wang, L., Yu, S., Tong, C., Zhao, Y., Liu, Y., Song, C., & Hua, W. (2014). Genome sequencing of the high oil crop sesame provides insight into oil biosynthesis. *Genome Biology*, *15*(2), R39. DOI: 10.1186/gb-2014-15-2-r39

Wang, M., Yuan, J., Qin, L., Shi, W., & Liu, S. (2020). TaCYP81D5, one member in a wheat cytochrome P450 gene cluster, confers salinity tolerance via reactive oxygen species scavenging. *Plant Biotechnology Journal, 18*(3), 791–804. DOI: 10.1111/pbi.13247

Wang, P., Si, H., Li, C., Xu, Z., Guo, H., Jin, S., & Cheng, H. (2025). Plant genetic transformation: Achievements, current status and future prospects. *Plant Biotechnology Journal, 23*(6), 2034–2058. DOI: 10.1111/pbi.70028

Wang, R., Yang, Q., Deng, Z., & Nian, W. (2025). The research on soil-plant-climate interactions: An integrated assessment of water management and drought resilience. *Advances in Resources Research, 5*(1), 456–476.

Wang, S.-W., Gao, C., Zheng, Y.-M., Yi, L., Lu, J.-C., Huang, X.-Y., Cai, J.-B., Zhang, P.-F., Cui, Y.-H., & Ke, A.-W. (2022). Current applications and future perspective of CRISPR/Cas9 gene editing in cancer. *Molecular Cancer, 21*(1), 57. DOI: 10.1186/s12943-022-01518-8

Wang, Y., Zafar, N., Ali, Q., Manghwar, H., Wang, G., Yu, L., Ding, X., Ding, F., Hong, N., Wang, G., & Jin, S. (2022). CRISPR/Cas genome editing technologies for plant improvement against biotic and abiotic stresses: Advances, limitations, and future perspectives. *Cells, 11*(23), 3928. DOI: 10.3390/cells11233928 PMID: 36497186

Wani, S. H., Choudhary, M., Barmukh, R., Bagaria, P. K., Samantara, K., Razzaq, A., Jaba, J., Ba, M. N., & Varshney, R. K. (2022). Molecular mechanisms, genetic mapping, and genome editing for insect pest resistance in field crops. *Theoretical and Applied Genetics, 135*(11), 3875–3895. DOI: 10.1007/s00122-022-04060-9 PMID: 35267056

Wan, L., Wang, Z., Tang, M., Hong, D., Sun, Y., Ren, J., Zhang, N., & Zeng, H. (2021). CRISPR-Cas9 gene editing for fruit and vegetable crops: Strategies and prospects. *Horticulturae, 7*(7), 193. DOI: 10.3390/horticulturae7070193

Wan, X., Hou, Q., & McConnell, L. L. (2022). Advances in genome editing for sustainable agriculture. *ACS Agricultural Science & Technology, 2*(12), 2390–2405. DOI: 10.1021/acsagscitech.2c00090

Wei, Z. (2024). Analysis Of Publicly Funded Intellectual Property Management To Support Agricultural Development And Food Security In Low-And Middle-income Countries.

Weih, M., Karley, A. J., Newton, A. C., Kiær, L. P., Scherber, C., Rubiales, D., & Pappagallo, S. (2021). Grain yield stability of cereal-legume intercrops is greater than sole crops in more productive conditions. *Agriculture, 11*(3), 255. DOI: 10.3390/agriculture11030255

Weinthal, D., Tovkach, A., Zeevi, V., & Tzfira, T. (2010). Genome editing in plant cells by zinc finger nucleases. *Trends in Plant Science, 15*(6), 308–321. DOI: 10.1016/j.tplants.2010.03.001

Wu, W., Miao, R., Li, Z., & Fang, Z. (2025). CRISPR/Cas9-mediated editing of BADH2 and Wx genes for the development of novel aromatic and soft-textured black and red rice. *Physiologia Plantarum, 177*(2), e70194. DOI: 10.1111/ppl.70194 PMID: 40171918

Xiao, L., Zhong, Y. P., Wang, B., & Wu, T. L. (2014). Mapping an aphid resistance gene in soybean [*Glycine max* (L.) Merr.] P746. *Genetics and Molecular Research, 13*(4), 9152–9160. DOI: 10.4238/2014.November.7.2 PMID: 25501137

Xia, X., Cheng, X., Li, R., Yao, J., Li, Z., & Cheng, Y. (2021). Advances in application of genome editing in tomato and recent development of genome editing technology. *Theoretical and Applied Genetics, 134*(9), 2727–2747. DOI: 10.1007/s00122-021-03874-3

Xiong, J. S., Ding, J., & Li, Y. (2015). Genome-editing technologies and their potential application in horticultural crop breeding. *Horticulture Research, 2*(1), 15019. DOI: 10.1038/hortres.2015.19

Yactayo-Chang, J. P., Tang, H. V., Mendoza, J., Christensen, S. A., & Block, A. K. (2020). Plant Defense Chemicals against Insect Pests. *Agronomy (Basel), 10*(8), 1156. Advance online publication. DOI: 10.3390/agronomy10081156

Yadav, M. R., Choudhary, M., Singh, J., Lal, M. K., Jha, P. K., Udawat, P., & Maheshwari, C. (2022). Impacts, tolerance, adaptation, and mitigation of heat stress on wheat under changing climates. *International Journal of Molecular Sciences, 23*(5), 2838. DOI: 10.3390/ijms23052838

Yang, Q., Tan, Y., Ye, Y., Zhao, D., & Liu, Q. (2023). Serotonin enrichment of rice endosperm by metabolic engineering. *The Crop Journal, 11*(6), 1943–1948. DOI: 10.1016/j.cj.2023.07.003

Yang, T., Ali, M., Lin, L., Li, P., He, H., Zhu, Q., & Deng, L. (2023). Recoloring tomato fruit by CRISPR/Cas9-mediated multiplex gene editing. *Horticulture Research, 10*(1), uhac214. DOI: 10.1093/hr/uhac214

Yang, Y., Zhu, K., Li, H., Han, S., Meng, Q., Khan, S. U., Fan, C., Xie, K., & Zhou, Y. (2018). Precise editing of genes in Brassica napus L. regulates multilocular silique development. *Plant Biotechnology Journal, 16*(7), 1322–1335. DOI: 10.1111/pbi.12872 PMID: 29250878

Yan, H., Jiao, X., Chen, Y., Liang, H., Liang, W., & Liu, C. (2024). Knockout of OsHMA3 in an indica rice increases cadmium sensitivity and inhibits plant growth. *Plant Growth Regulation, 103*(3), 635–646. DOI: 10.1007/s10725-024-01137-x

Yao, R., Liu, D., Jia, X., Zheng, Y., Liu, W., & Xiao, Y. (2018). CRISPR-Cas9/Cas12a biotechnology and application in bacteria. *Synthetic and Systems Biotechnology, 3*(3), 135–149. DOI: 10.1016/j.synbio.2018.09.004

Ye, G., McNell, D. L., & Hill, G. D. (2002). Breeding for resistance to lentil ascochyta blight. *Plant Breeding, 121*(3), 185–191. DOI: 10.1046/j.1439-0523.2002.00705.x

Yijun, G., Zhiming, X., Jianing, G., Qian, Z., Rasheed, A., Hussain, M. I., Ali, I., Shuheng, Z., Hassan, M. U., Hashem, M., Mostafa, Y. S., Wang, Y., Chen, L., Xiaoxue, W., & Jian, W. (2022). The intervention of classical and molecular breeding approaches to enhance flooding stress tolerance in soybean: A review. *Frontiers in Plant Science, 13*, 1085368. DOI: 10.3389/fpls.2022.1085368 PMID: 36643298

Yıldırım, K., Miladinović, D., Sweet, J., Akin, M., Galović, V., Kavas, M., Zlatković, M., & de Andrade, E. (2023). Genome editing for healthy crops: Traits, tools and impacts. *Frontiers in Plant Science, 14*, 1231013. DOI: 10.3389/fpls.2023.1231013

Yuan, L., Gai, W., Xuan, X., Ahiakpa, J. K., Li, F., Ge, P., & Zhang, Y. (2024). Advances in improving tomato fruit quality by gene editing. *Horticultural Plant Journal*. Advance online publication. DOI: 10.1016/j.hpj.2024.04.008

Yuan, P., Usman, M., Liu, W., Adhikari, A., Zhang, C., Njiti, V., & Xia, Y. (2024). Advancements in Plant Gene Editing Technology: From Construct Design to Enhanced Transformation Efficiency. *Biotechnology Journal, 19*(12), e202400457. DOI: 10.1002/biot.202400457

Yu, H., Lin, T., Meng, X., Du, H., Zhang, J., Liu, G., Chen, M., Jing, Y., Kou, L., Li, X., Gao, Q., Zhou, Y., Chen, Q., Liu, J., Huang, X., & Li, J. (2021). A route to de novo domestication of wild allotetraploid rice. *Cell, 184*(5), 1156–1170. DOI: 10.1016/j.cell.2021.01.013

Zafar, S. A., Zaidi, S. S.-A., Gaba, Y., Singla-Pareek, S. L., Dhankher, O. P., Li, X., Mansoor, S., & Pareek, A. (2020). Engineering abiotic stress tolerance via CRISPR/Cas-mediated genome editing. *Journal of Experimental Botany, 71*(2), 470–479. DOI: 10.1093/jxb/erz476

Zaidi, S. S. E. A., Mukhtar, M. S., & Mansoor, S. (2018). Genome editing: Targeting susceptibility genes for plant disease resistance. *Trends in Biotechnology*, *36*(9), 898–906. DOI: 10.1016/j.tibtech.2018.04.005 PMID: 29752192

Zaman, W., Khalil, A. A., & Amin, A. (2025). TALEN-Interceded Genome Editing in Plants: Unveiling New Frontiers in Secondary Metabolite Improvement and Genetic Diversity. *Plants*, *14*(19), 3024. DOI: 10.3390/plants14193024

Zeng, Y., Wen, J., Zhao, W., Wang, Q., & Huang, W. (2020). Rational improvement of rice yield and cold tolerance by editing the three genes OsPIN5b, GS3, and OsMYB30 with the CRISPR-Cas9 system. *Frontiers in Plant Science*, *10*, 1663. DOI: 10.3389/fpls.2019.01663 PMID: 31993066

Zetsche, B., Gootenberg, J. S., Abudayyeh, O. O., Slaymaker, I. M., Makarova, K. S., Essletzbichler, P., Volz, S. E., Joung, J., van der Oost, J., Regev, A., Koonin, E. V., & Zhang, F. (2015). Cpf1 is a single RNA-guided endonuclease of a class 2 CRISPR-Cas system. *Cell*, *163*(3), 759–771. DOI: 10.1016/j.cell.2015.09.038

Zhang, D., Hussain, A., Manghwar, H., Xie, K., Xie, S., Zhao, S., Larkin, R. M., Qing, P., Jin, S., & Ding, F. (2020). Genome editing with the CRISPR-Cas system: An art, ethics and global regulatory perspective. *Plant Biotechnology Journal*, *18*(8), 1651–1669. DOI: 10.1111/pbi.13383

Zhang, D., Zhang, Z., Unver, T., & Zhang, B. (2021). CRISPR/Cas: A powerful tool for gene function study and crop improvement. *Journal of Advanced Research*, *29*, 207–221. DOI: 10.1016/j.jare.2020.10.003

Zhang, H., Zhang, J., Lang, Z., Botella, J. R., & Zhu, J.-K. (2017). Genome editing—Principles and applications for functional genomics research and crop improvement. *Critical Reviews in Plant Sciences*, *36*(4), 291–309. DOI: 10.1080/07352689.2017.1402989

Zhang, H., Zhu, J., Gong, Z., & Zhu, J.-K. (2022). Abiotic stress responses in plants. *Nature Reviews. Genetics*, *23*(2), 104–119. DOI: 10.1038/s41576-021-00413-0

Zhang, L., Zhu, M., Ren, L., Li, A., Chen, G., & Hu, Z. (2018). The SlFSR gene controls fruit shelf-life in tomato. *Journal of Experimental Botany*, *69*(12), 2897–2909. Advance online publication. DOI: 10.1093/jxb/ery116

Zhang, Y., Berman, A., & Shani, E. (2023). Plant hormone transport and localization: Signaling molecules on the move. *Annual Review of Plant Biology*, *74*(1), 453–479. DOI: 10.1146/annurev-arplant-070722-015329

Zhang, Y., Chen, S., Yang, L., & Zhang, Q. (2023). Application progress of CRISPR/Cas9 genome-editing technology in edible fungi. *Frontiers in Microbiology*, *14*, 1169884. DOI: 10.3389/fmicb.2023.1169884 PMID: 37303782

Zhang, Y., Ren, Q., Tang, X., Liu, S., Malzahn, A. A., Zhou, J., Wang, J., Yin, D., Pan, C., Yuan, M., Huang, L., Yang, H., Zhao, Y., Fang, Q., Zheng, X., Tian, L., Cheng, Y., Le, Y., McCoy, B., & Qi, Y. (2021). Expanding the scope of plant genome engineering with Cas12a orthologs and highly multiplexable editing systems. *Nature Communications*, *12*(1), 1944. DOI: 10.1038/s41467-021-22330-w

Zhang, Y., & Showalter, A. M. (2020). CRISPR/Cas9 genome editing technology: A valuable tool for understanding plant cell wall biosynthesis and function. *Frontiers in Plant Science*, *11*, 589517. DOI: 10.3389/fpls.2020.589517 PMID: 33329650

Zhan, X., Lu, Y., Zhu, J.-K., & Botella, J. R. (2021). Genome editing for plant research and crop improvement. *Journal of Integrative Plant Biology*, *63*(1), 3–33. DOI: 10.1111/jipb.13063

Zhao, B., Liu, Q., Wang, B., & Yuan, F. (2021). Roles of phytohormones and their signaling pathways in leaf development and stress responses. *Journal of Agricultural and Food Chemistry*, *69*(12), 3566–3584. DOI: 10.1021/acs.jafc.0c07908

Zhao, B., Sun, C., Yang, D., Zhao, Y., Cao, J., & Wang, B. (2025). CRISPR/Cas9-mediated genomic editing of Brachytic2 creates semi-dwarf mutant alleles for tailored maize breeding. Plant Biotechnology Journal.Zhang, Q., Zhao, Z., & Li, P. (2024). Research and progress of gene editing technology in maize. *Zhiwu Xuebao*, *59*(6), 978.

Zhao, D., Li, J., Li, S., Xin, X., Hu, M., Price, M. A., Rosser, S. J., Bi, C., & Zhang, X. (2021). Glycosylase base editors enable C-to-A and C-to-G base changes. *Nature Biotechnology*, *39*(1), 35–40. DOI: 10.1038/s41587-020-0592-2

Zhao, P., Feng, Z., Cai, L., Phurbu, D., Duan, W., Xie, F., Li, X., & Liu, F. (2024). Development of an RPA-CRISPR/Cas12a Assay for Rapid and Sensitive Diagnosis of Plant Quarantine Fungus Setophoma terrestris. *Journal of Fungi (Basel, Switzerland)*, *10*(10), 716. DOI: 10.3390/jof10100716

Zheng, H., Wang, X., Qiu, J., & Zhou, S. (2025). The application of gene editing technology in reducing the use of pesticides and chemical fertilizers and mitigating the environmental impact of agriculture. *Advances in Resources Research*, *3*(1), 45–58.

Zhong, Q., Xu, Y., & Rao, Y. (2024). Mechanism of rice resistance to bacterial leaf blight via phytohormones. *Plants*, *13*(18), 2541. DOI: 10.3390/plants13182541

Zhou, J., Li, D., Wang, G., Wang, F., Kunjal, M., Joldersma, D., & Liu, Z. (2020). Application and future perspective of CRISPR/Cas9 genome editing in fruit crops. *Journal of Integrative Plant Biology*, *62*(3), 269–286. DOI: 10.1111/jipb.12793

Zhou, J., Peng, Z., Long, J., Sosso, D., Liu, B. O., Eom, J. S., & Yang, B. (2015). Gene targeting by the TAL effector PthXo2 reveals cryptic resistance gene for bacterial blight of rice. *The Plant Journal : For Cell and Molecular Biology*, *82*(4), 632–643. DOI: 10.1111/tpj.12838 PMID: 25824104

Zhu, H., Li, C., & Gao, C. (2020). Applications of CRISPR–Cas in agriculture and plant biotechnology. *Nature Reviews. Molecular Cell Biology*, *21*(11), 661–677. DOI: 10.1038/s41580-020-00288-9

Zhu, X. Y., Li, Y., Sun, Y., Zhu, Q., & Sun, J. (2018). Highly efficient targeted gene editing in upland cotton using the CRISPR/Cas9 system. *International Journal of Molecular Sciences*, *19*(10), 3000. DOI: 10.3390/ijms19103000 PMID: 30275376

Zou, S., Xu, Y., Li, Q., Wei, Y., Zhang, Y., & Tang, D. (2023). Wheat powdery mildew resistance: From gene identification to immunity deployment. *Frontiers in Plant Science*, *14*, 1269498. DOI: 10.3389/fpls.2023.1269498

Zsögön, A., Čermák, T., Naves, E. R., Notini, M. M., Edel, K. H., Weinl, S., Freschi, L., Voytas, D. F., Kudla, J., & Peres, L. E. P. (2018). De novo domestication of wild tomato using genome editing. *Nature Biotechnology*, *36*(12), 1211–1216. DOI: 10.1038/nbt.4272

About the Contributors

Sana Basharat is working as a postdoctoral researcher at college of tropical agriculture and forestry at hainan university, haikou, hainan, China. She got her doctoral degree from Department of botany, university of agriculture faislabad, Pakistan. Her research direction involved structural and functional Strategies in Cenchrus Species to Combat Environmental Extremities Imposed by Multiple Abiotic Stresses. She had published more than 15 research articles in renowned journals including Plants, Pakistan Journal botany, Acta Physiologiae Plantarum, so on.

Muhammad Waseem is Associate Professor/Researcher at College of Tropical Agriculture and Forestry, Hainan University, Haikou, Hainan, China. He received a doctoral degree (Ph.D., specialization in Plant Molecular Biology) from the School of Life Sciences, Chongqing University, Chongqing, PR China. He has five years of research experience including postdoctoral. His interests include transcriptional regulation of flowering time regulation, fruit ripening and development, and abiotic stresses such as salinity and drought. In addition, he has hands-on experience in next-generation sequencing data analysis, transcriptomics, and data visualization. Currently, Dr. Waseem is actively in investigating flowering time regulation in Brassica crops. He has been awarded with a National Science Foundation Project, RFIS-I in 2023. In brief, he has published more than 58 research articles, 5 review articles, and 9 book chapters in various high-impact factor publications including New Phytologist, International Journal of Macromolecules, Planta, GM Crops & Food, Environmental and Experimental Botany, Physiologia Plantarum, and Frontiers in Plant Science. He has also edited books with Springer publisher.

***

A. Mubarak Ali, PG and Research Department of Microbiology, Sacred Heart College (Autonomous), Tirupattur, Tamil Nadu, India

Banzeer Ahsan Abbasi received her graduate degree in Botany from Quaid-i-Azam University, Islamabad, Pakistan. Dr. Banzeer is currently working as an Assistant Professor at Department of Botany, Rawalpindi Women University, Rawalpindi, Pakistan. She has published more than 60 research articles in various international journals and 10 chapters in reputed publishers (ELSIEVER, RSC and SPRINGER). Currently, she is supervising 2 Ph.D, 6 MPhil and 6 BS students. She has expertise in the synthesis of novel nanomaterials from medicinal plants and their multiple biological and envirnemental applications. Her research is also focusing on phytochemistry, natural products of medicinal plants and Phylogenetics. She has a very decent research background with research published in well-reputed journals with H-index 28, i-10 index 44 and overall citations 3317.

Sumreen Dawood is an Assistant Professor in the Department of Botany, since 2023. Her research focuses on renewable energy generation, green nanotechnology, biofuel development, and plant taxonomy. She has made significant contributions to biofuel technology through the identification of promising biomass resources and the development of green nano-catalysts for sustainable biomass-to-biofuel conversion. Her work integrates plant sciences with nanotechnology to promote environmentally friendly and sustainable energy solutions.

Yonis Gulzar is an Associate Professor of Management Information Systems at King Faisal University, Saudi Arabia. His research lies at the intersection of artificial intelligence, computer vision, and intelligent information systems, with applications in healthcare analytics and smart agriculture. His recent work focuses on deep learning architectures for agricultural imaging, disease detection, seed classification, and AI-enabled decision support systems aimed at improving sustainability and productivity in precision agriculture. He also contributes to research on AI-driven knowledge systems and emerging technologies in digital transformation.

Amjid Khan is an interdisciplinary scientist and nanomaterials expert currently serving as a Postdoctoral Research Fellow under the prestigious UNESCO-UNISA Africa Chair in Nanoscience and Nanotechnology, College of Graduate Studies, at the University of South Africa, South Africa. Dr. Khan is a Senior Fellow under the prestigious UNESCO-UNISA-iTLABS/NRF Africa Chair in Nanoscience and Nanotechnology in Pretoria, Republic of South Africa. Dr. Khan's interdisciplinary research bridges advanced nanotechnology applications across biomedicine and sustainable agriculture. In biomedicine, his work focuses on the green and biomimetic synthesis of functional nanomaterials, including plasmonic nanoparticles, quantum dots, and metal-oxide nanocomposites, tailored for diagnostic nanoprobes, bio-imaging, targeted drug delivery, and precision therapeutics. In sustainable agriculture, he specializes in smart nanobiotechnology, abiotic stress mitigation, and the development of next-generation nano-biofertilizers to enhance crop resilience, soil health, and sustainable yield. A highly productive scholar, Dr. Khan has authored over 50 SCI-indexed peer-reviewed articles. Beyond his research appointments, he serves as an Advisory and Affiliate Member of the Paschim Medinipur Microbiological Innovation and Technological Advancement Society (PMMITAS). Demonstrating global editorial leadership, Dr. Khan serves as an Associate Editor for Frontiers in Systems Biology, a Reviewing Editor for the Springer Nature Reviewer Community, and a Lead Volume Editor for academic books with premier publishers, including IGI Global, John Wiley & Sons, and CRC Press/Taylor & Francis.

Waseem Ahmed Khattak is a plant scientist with research expertise spanning plant physiology, nanobiotechnology, and sustainable agriculture. His work focuses on the application of green-synthesized and biogenic nanoparticles to enhance crop productivity, strengthen antioxidant defense systems, and mitigate abiotic stresses such as heavy metal toxicity. He has conducted extensive field and pot experiments involving major crops, with particular emphasis on nutrient use efficiency, root morphology, biochemical profiling, and stress physiology. Dr. Khattak has authored and co-authored multiple research and review articles in reputable journals and has contributed chapters to academic books addressing sustainable agriculture, nanotechnology, and environmental remediation. His research integrates advanced biochemical, physiological, and nanotechnological approaches to develop eco-friendly strategies for resilient and sustainable crop production.

M. Manigandan, PG and Research Department of Microbiology,Sacred Heart College (Autonomous), Tirupattur, Tamil Nadu, India

P. Saranraj holds a Ph.D. in Microbiology from Annamalai University, Chidambaram, India, and is NET (ASRB–ICAR) qualified in Microbiology. He is currently serving as Assistant Professor and Head, Department of Microbiology, at Sacred Heart College (Autonomous), Tirupattur, Tamil Nadu, India. In addition, he is the Proprietor and Managing Director of JPS Scientific Publications, contributing actively to academic publishing and research dissemination. Dr. Saranraj has over 15 years of research experience and 10 years of teaching experience in the fields of Environmental Microbiology, Agricultural Microbiology, Industrial Microbiology, Bioremediation, Microbial Biotechnology, and Sustainable Ecosystem Management. His research work has focused on Microbial degradation of pollutants, Effluent treatment, Soil health improvement, Plant–microbe interactions, and Applied environmental sustainability. He has an extensive publication record, with more than 240 international research articles, over 90 books (international and national), and 70+ book chapters published with reputed academic publishers.

Qin Xiaopei, female, born in July 1988, is from Luoyang, Henan Province. She graduated from the School of Food and Biological Engineering of Jiangsu University with a master's degree. In September 2015, she joined the School of Food and Drug of Shanghai Zhiqiao Vocational University. Currently, she is the director of the Food Nutrition and Health major. She has published 5 core papers, obtained 3 patents, 1 computer software copyright, and co-authored 2 textbooks. She is an academic visitor of the Ministry of Education in domestic universities, a senior technician and assessor of agricultural product and food inspectors, an assessor of food safety managers, and a technician of public nutritionists.

Index

N

P

R

S

T

Y

Z

www.ingramcontent.com/pod-product-compliance
Lightning Source LLC
LaVergne TN
LVHW080859300926
849162LV00003B/480